THE MOLECULAR BIOLOGY OF BACTERIAL GROWTH

A Symposium held in honor of Ole Maaløe,
at the University of Alabama, Tuscaloosa

THE MOLECULAR BIOLOGY OF BACTERIAL GROWTH

A Symposium held in honor of Ole Maaløe,
at the University of Alabama, Tuscaloosa

Edited by

Moselio Schaechter
Tufts University School of Medicine

Frederick C. Neidhardt
University of Michigan Medical School

John L. Ingraham
University of California, Davis

Niels Ole Kjeldgaard
University of Aarhus, Denmark

With the Help of David Freifelder,
University of California, San Diego

Jones and Bartlett Publishers, Inc.
Boston   *Portola Valley*

Editorial offices: 30 Granada Court, Portola Valley, California 94025
Sales and customer service offices: 20 Park Plaza Boston, Mass. 02116

Library of Congress Cataloging in Publication Data

Molecular biology of bacterial growth.

Papers from a symposium honoring Ole Maaløe
held by the University of Alabama in April 1984.
1. Bacterial growth — Congresses. 2. Molecular
biology — Congresses. I. Schaechter, Moselio.
II. Maaløe, Ole. III. University of Alabama.
QR86.M65   1985   589.9′031   84-29978
ISBN 0-86720-049-9

56,908

Design/Production: Unicorn Production Services, Inc.
Printing/Binding: Alpine Press

Printed in the United States of America
Printing (last digit):   10   9   8   7   6   5   4   3   2

"Blot til lyst"

Ole Maaløe

INTRODUCTION

Several years ago, the University of Alabama, Tuscaloosa, Department of Microbiology and the Interdisciplinary Biochemistry Program initiated a new lecture series called Frontiers in Modern Biology. Well-known scientists were brought to Tuscaloosa nearly weekly for about a year, providing an exciting and educational experience for faculty, staff, and students. With unusual wisdom the administration of the University funded this program amply, recognizing the value that it would have for the scientific community of the university. Even though the program does not exist any more, the University continues to hold short courses on specific topics given by respected scientists. I have been given the responsibility to select the speakers for these special programs. In 1983 Professor Robert Haynes (York University, Toronto, Canada) gave a mini-course on biological repair processes.

Following the successful series by Dr. Haynes, I was asked to conceive of and organize a symposium, for the spring of 1984, that would honor someone whose is reknowned in both microbiology and biochemistry. It did not take more than a few minutes to call to mind Professor Ole Maaløe, University Institute of Microbiology, Copenhagen, Denmark. Ole was my mentor during a postdoctoral year in 1962 and has influenced my thinking about cell growth, and the thinking of many others, since that time. Ole Maaløe was the founder of what has become known as the "Copenhagen School" and was instrumental in making cell growth an analytical science. His influence on many young scientists and on the field of microbiology has been profound and felt worldwide. Furthermore, 1984 was Ole's 70th year, and for some time I had been thinking of some way to honor his accomplishments. I made the proposal to various faculty and administrators at the University of Alabama, and they were quickly infected with the notion of a "Maaløe Symposium". The plan was to bring together most of the people who had worked in Ole's lab since the 1950s, to summarize the work of the past thirty years, and to plan for future understanding of cell growth. Our model for the Symposium was to be the highly successful Cold Spring Harbor meetings. I contacted Ole, asking him if he would agreed to be so honored, if he would be available at a time convenient for the University of Alabama, and if he had suggestions for participants. Ole agreed and suggested that the Symposium be planned by four former colleagues, Niels Ole Kjeldgaard (with whom he had worked in Copenhagen for more than a decade), Moselio Schaechter (one of the earliest postdocs in the Copenhagen Institute), and Fred Neidhardt and John Ingraham, with whom he had just authored a book on bacterial growth. These four friends, feeling as I did that honoring Ole was long overdue, contacted Ole's former students and the Symposium began to materialize; it was held in April, 1984.

The University of Alabama was especially generous in providing airfare for many Europeans and Americans, for the general funding of the meeting, and for a fantastic banquet. Professor Harry Heath of the Department of Microbiology, his wife Lucie, and his graduate students, were responsible for local arrangements and cannot be thanked enough. It took very little talking to convince Donald Jones and Arthur Bartlett of Jones and Bartlett, Publishers, Inc., that the Symposium should be published, not only to honor Ole Maaløe, but because it is an up-do-date and valuable collection of information about cell growth. Authors, editors, and the members of the University of Alabama hope that the readers of the volume can appreciate the unique contribution of Ole Maaloe to this field and can sense some of the excitement that took place at the Symposium.

January, 1985   David Freifelder
San Diego, California

IN RETROSPECT

By October 1983 our institute in Copenhagen celebrated its 25th anniversary, and the occasion was marked by a great party. Sadly enough, not one of our many foreign friends and collaborators was in Denmark at that time. The symposium presented in this volume has made up for this in a splendid manner, and the silver anniversary in Copenhagen and the banquet in Tuscaloosa complement each other to give a true picture of the life and spirit in our laboratory, as I like to think of it.

In 1950 while I still worked in three small rooms at the State Serum Institute, fortune brought the first foreign scientists to the lab: Jim Watson, Günther Stent, Niels Jerne and I worked together in Copenhagen for a year at the end of which the group disintegrated. Jim went to Cambridge to meet Francis Crick, Günther went to André Lwoff's lab in Paris, and I joined the phage group in Max Delbrück's lab in Pasadena. Jerne stayed in Copenhagen, but not for long.

Back from Caltech it worried me that the exciting collaboration with scientists from other parts of the world might have come to an end. Nothing of the kind, our lab had been put on the map, and throughout the years of plenty, when fellowships to study abroad were easy to obtain, a large number of foreign scientists came to work with us (mostly post-docs and senior scientists on sabbatical leave from the US, but also quite a few from other countries in Europe.)

I like to think that our guests were attracted by the particular approach to the study of bacterial growth that grew out of work done in the small rooms in the Serum Institute by Gordon Lark, Victor Bruce, Elio Schaechter, Niels Ole Kjeldgaard and myself. Gordon and I first worked out a scheme involving shifts between 25 and 37 C that induced division synchrony in broth cultures of *S. typhimurium;* however, we went on to find that the temperature treatment distorted the normal pattern of DNA replication. At that point we abandoned the system, both being interested primarily in *normal growth* (it was a sad decision, for the synchrony curves looked quite nice).

Having seen how easy it is to introduce artifacts by submitting a growing culture to such "mild" treatments as shifts between two temperatures that both permit exponential growth, we went for the simplest possible experimental designs, banning all interference with the process of growth. Elio, Niels Ole and I began to analyze cultures in balanced growth in media supporting different rates of growth. The rigor with which we defined the state of balanced growth, and stressed the importance of the growth rate as a basic variable are spelled out at some length in the monograph Niels Ole and I wrote several years later (*Control of Macromolecular Synthesis,* Benjamin 1966). Our rule has therefore been, first, to measure at different growth rates the parameters that can be measured without disturbing the culture, and, second, only to interfere with growth if we know the *primary effect* of the agent or procedure used. Examples are starvation for a required amino acid, or addition of reagents that interfere with growth in a clearly defined way such as chloramphenicol, rifampicin or alpha-methyl-glucoside. Probably, our main achievement was to introduce experimental designs that emphasized simplicity and thereby facilitated the interpretation of the data obtained. This attitude in a sense defined the Copenhagen School of bacterial growth physiology, a recent product of which is the textbook by Ingraham, Maaløe and Neidhardt (*Growth of the Bacterial Cell*, 1983.).

These rules of conduct should, I think, be applied to studies of bacteria in general; we concentrated on *Escherichia coli* to permit us to use genetic analysis. I don't intend to present a chronology of people nor of their individual contributions over the last 25 years. However, looking back, a few key experiments stand out—most of them fruits of collaborative efforts. These experiments are referred to or implied in many of the contributions to this book.

Our work developed along two lines. One was based on the first data obtained with cultures in balanced growths at different rates (Schaechter, Maaløe and Kjeldgaard, 1958). The strong hint of a constant rate of protein synthesis per "nucleoprotein particle" (not yet known as a ribosome, let alone identified as the site of polypeptide synthesis) made us focus on the protein-synthesizing system (PSS) as a whole. Secondly, we were concerned with DNA replication *in vivo*. Shortly after the lab had moved to the present location in the Botanical Garden, a "run-out experiment" was done, which showed that in the absence of protein synthesis a round of DNA replication, once initiated, would run to completion, but without initiation of a new round (Maaløe and Hanawalt, 1961). The obvious is not always easy to spot, and only many years later was it realized by K. V. Rasmussen in our lab that the run-out technique offers direct estimates of the number of origins in a population of replicating genomes.

For at least 15 years after we moved we worked in the broad area of growth physiology. On the PSS project techniques were invented or adapted to permit quantitative measurements of parameters such as the elongation rates of RNA and polypeptide chains, the rate of synthesis of ribosomal proteins and other factors involved in protein synthesis, and the amounts and stability of mRNA. Extensive use was made of rifampicin to carry out run-out experiments on RNA synthesis by Pato, von Meyenburg and Molin, and 2-D gels to measure mRNA half-life and to estimate amounts of nonribosomal PSS proteins by Pedersen and Neidhardt. A project run exclusively by in-house people should be mentioned: The instantaneous down-shift caused by adding alpha-methyl-glucoside to a culture in glucose minimal medium was analyzed by Molin, von Meyenburg, Karlstrøm and a graduate student, Knud Johnsen (see Chapter 8 in Ingraham et al., 1983). Knud taught me a lesson; he declined an offer to continue a very promising scientific career. He was quite sure he wanted to teach biology in high school, and he was right, for he is a great teacher.

Together these measurements establish the high and nearly constant efficiency of ribosomes engaged in protein synthesis, and the more-or-less constant average protein yield per polysome (see my paper in this volume). Embedded in these studies was a thorough examination of ppGpp synthesis and levels at different growth rates in stringent as well as relaxed strains by Fiil, Friesen, and von

Meyenburg. They showed conclusively that ppGpp is of minor importance for the adjustment of the size of PSS to growth rate. This work was important in another way: it marked the beginning of genetic studies in our lab and involved isolation of mutants requiring elaborate selection techniques.

On the DNA project a new turn was marked by the visit to Copenhagen of Cooper and Helmstetter (1963-64), who taught us to use the "Baby-machine" to obtain division synchrony by selection. Their technique should introduce few, if any, artifacts, so this time we did not hesitate to use synchronized cultures. The experiments and thoughts of the following years were focussed on the initiation of replication, and the ideas of Pritchard and of Donachie played important roles. During this time the concept of autoregulation was introduced by Sompayrac and Maaløe (1973). It was suggested as a possible means of "timing" initiation.

During the last 10 years molecular genetics has been an increasingly important part of our work, but the link to growth physiology has been maintained. Most significantly, some of the autoregulations that seem to balance the synthesis of the different ribosomal components have been analyzed by Fiil, Pedersen, and Johnsen with strong inputs from a number of graduate students, and in collaboration with scientists in the U.S. (Nomura's group in Madison; Lindahl's lab in Rochester; Cathy and Craig Squires' lab in New York), in Canada (Friesen's group in Toronto; Dennis' lab. in Vancouver), and in the Wittmann group in West Berlin. Several papers in this volume illustrate these activities, the results of which have been so important for my own attempts to draw an integrated picture of a growing *E. coli* cell (see my paper in this volume).

On the DNA front, genetics came to be equally important. In the beginning there were a few mutants with altered DNA/mass ratios, some of them isolated the hard way by Knud Rasmussen and Flemming Hansen (they scanned agar plates with lots of colonies with only 50-100 cells, looking for abnormally large or small cells). The full switch to molecular genetics came with Kaspar von Meyenburg, who in 1975 turned his attention from ribosomes to DNA replication. As you know, we had already focussed on the act of initiation as the key to understanding the control of replication, and Kaspar's success in identifying and sequencing the site of initiation (*oriC*) was most encouraging. However, this turned out to be another case in which the sequence itself did not suggest a molecular mechanism. Scanning the region around *oriC*, the *dnaA* gene and its product were analyzed with two interesting results: (1) the DnaA protein was shown to be autoregulated, and (2) certain suppressors of *dnaA* mutants, isolated and studied by Tove Atlung, show that the DnaA protein interacts directly with the RNA polymerase, presumably in the act of synthesizing the RNA primer at the site of initiation of replication. This and more is presented by Kaspar, Tove, Flemming, Knud Rasmussen and others in this volume.

I have now brought the history of our activities up to the present, and I have included recent work in the two segregated groups: Kjeldgaard's group at the University of Aarhus (established in 1968), and von Meyenburg's group at the Technical University in Lyngby (1978). A third group has just been created by Søren Molin, next door to Kaspar's unit. It is time therefore to tell what I feel has been my own role during all these years. The two papers I wrote with Kjeldgaard and Schaechter and published in 1958 clearly shaped the course we have followed up to the present. Apart from this initial input, and a few later ones already mentioned, my main contribution has been to keep us on course. What this means is perhaps best seen by comparing my latest attempt to describe the properties of a growing *E. coli* cell (this volume) to a quote from the contribution I made to the 1960 Symposium of the Society for General Microbiology. The final paragraphs read:

The sketch of the growing bacterium presented here is based essentially on the idea of exchange of information between different molecular levels of organization in the cell. A flow of information is assumed to descend from a linear, genetic specification on a DNA strand, via RNA and protein and to give to a small molecule, such as an amino acid, its three-dimensional individuality. Equally specific information is believed to pass from the level of the small molecules back in the direction of the nucleus. This feedback of information, which produces the phenomenon of repression, is thought to be responsible for one of the remarkable properties of the cell: its ability to adjust the size and activity of different synthetic systems to the set of nutrients present in the medium; an adjustment that results in the establishment of a definite partitioning of energy and matter among the synthetic systems, to which corresponds a definite growth rate and cell composition.

In a paper like this, interpretations and generalizations certainly play an important role. It should therefore be made clear that we have adopted the view that RNA templates exist and are formed by direct contact with the DNA of the nucleus, and that repression involves the function but not the formation of the protein-synthesizing systems, because we find that alternative mechanisms, even if they cannot be excluded, appear less plausible."

These paragraphs at once show where we aimed and that we started out when molecular biology was in its infancy. They also give an idea of the great amount of work and thought contributed by my friends at home and abroad to put substance into the primitive 1960 sketch.

Sydney Brenner once offered a very brief description of our work. At a small symposium in Cambridge (England) he asked me what I wanted to talk about, and when I had explained the main idea, he said: "Oh, you are going to talk about the effects of banging on a network."

Ole Maaløe

CONTENTS

PART ONE
CONTROL OF THE PROTEIN-SYNTHESIS SYSTEM

PART TWO
CONTROL OF GENES
AND REGULONS

PART FOUR
GROWTH AND THE
HISTORY OF GROWTH

SYMPOSIUM PARTICIPANTS

Tove Atlung, *University Institute of Microbiology; Copenhagen, Denmark.*

Michael L. Berman, *Litton Institute of Applied Biotechnology; Rockville, Maryland.*

Patrick P. Dennis, *Department of Biochemistry, The University of British Columbia; Vancouver, B.C.*

William D. Donachie, *Department of Molecular Biology, Edinburgh University; Edinburgh, Scotland.*

Abraham Eisenstark, *University of Missouri; Columbia, Missouri.*

David Freifelder, *University of California, San Diego.*

James D. Friesen, *Department of Medical Genetics, University of Toronto; Toronto, Ontario.*

Jonathan Gallant, *Genetics Department, University of Washington; Seattle, Washington.*

Kirsten Gausing, *Department of Molecular Biology and Plant Physiology, University of Aarhus; Aarhus, Denmark.*

Larry Gold, *Department of Molecular, Cellular and Developmental Biology, University of Colorado; Boulder, Colorado.*

Flemming G. Hansen, *Department of Microbiology, The Technical University of Denmark; Lyngby-Copenhagen, Denmark.*

Harry E. Heath, *University of Alabama, Tuscaloosa.*

Lucie S. Heath, *University of Alabama, Tuscaloosa.*

Charles. E. Helmstetter, *Department of Experimental Biology, Roswell Park Memorial Institute; Buffalo, New York.*

John L. Ingraham, *University of California, Davis.*

Morten Johnsen, *Institute of Microbiology, University of Copenhagen Copenhagen, Denmark.*

Herman M. Kalckar, *Chemistry Department, Boston University; Boston, Massachusetts.*

Niels Ole Kjeldgaard, *Institute of Molecular Biology, University of Aarhus; Aarhus, Denmark.*

Tokio Kogoma, *Department of Biology, University of New Mexico; Albuquerque, New Mexico.*

Peter Kuempel, *Department of Molecular, Cellular and Developmental Biology, University of Colorado; Boulder, Colorado.*

Charles G. Kurland, *Department of Molecular Biology, The Biomedical Center; Uppsala, Sweden.*

Karl G. Lark, *University of Utah; Salt Lake City, Utah.*

Lasse Lindahl, *Department of Biology, The University of Rochester; Rochester, New York.*

Cyrus Levinthal, *Columbia University; New York, New York.*

Ole Maaløe, *University Institute of Microbiology; Copenhagen, Denmark.*

Millicent Masters, *Department of Molecular Biology, Edinburgh University; Edinburgh, Scotland.*

Agnete Munch-Petersen, *University Institute of Biological Chemistry B; Copenhagen, Denmark.*

Frederick C. Neidhardt, *Department of Microbiology and Immunology, University of Michigan Medical School; Ann Arbor, Michigan.*

Jan Neuhard, *Enzyme Division, University Institute of Biological Chemistry B; Copenhagen, Denmark.*

Donald P. Nierlich, *Department of Microbiology and Molecular Biology Institute, University of California, Los Angeles.*

Masayasu Nomura, *Institute for Enzyme Research and Departments of Genetics and Biochemistry, University of Wisconsin; Madison, Wisconsin.*

Martin Pato, *National Jewish Hospital; Denver, Colorado.*

Steen Pedersen, *Institute of Microbiology, University of Copenhagen, Copenhagen, Denmark.*

Robert H. Pritchard, *Department of Genetics, University of Leicester; Leicester, England.*

Moselio Schaechter, *Department of Molecular Biology and Microbiology, Tufts University; Boston, Massachusetts.*

Robert Schleif, *Biochemistry Department, Brandeis University; Waltham, Massachusetts.*

Catherine L. Squires, *Department of Biological Sciences, Columbia University; New York, New York.*

Gunther Stent, *Department of Molecular Biology, University of California, Berkeley.*

Annamaria Torriani, *Department of Biology, Massachusetts Institute of Technology; Cambridge, Massachusetts.*

Kaspar von Meyenburg, *Department of Microbiology, The Technical University of Denmark; Lyngby-Copenhagen, Denmark.*

Alvin L. Winters, *University of Alabama, Tuscaloosa*

Richard E. Wolf, Jr., *Department of Biological Sciences, University of Maryland, Baltimore County; Catonsville, Maryland.*

Andrew Wright, *Tufts University Medical School; Boston, Massachusetts.*

Charles Yanofsky, *Department of Biological Sciences, Stanford University; Stanford, California.*

Top Row (left to right): O. Maaløe; K.G. Lark.
Middle: R. Pritchard; R. Schleif.
Bottom: C. Squires; O. Maaløe, A. Wright, A. Torriani, P. Kuempel, W. Donachie.

Top Row: C. Levinthal; L. Lindahl,
Middle: F. Hansen, T. Atlung, R. Wolf, M. Johnsen; M. Nomura.
Bottom: T. Kogoma; J.D. Friesen, L. Gold.

Top Row: C. Yanofsky; J. Ingraham.
Middle: C.E. Helmstetter, P.P. Dennis; F.C. Neidhardt; D. Freifelder.
Bottom: M. Berman, J. Gallant, M. Pato; N.O. Kjeldgaard, J. Neuhard.

Top Row: K. von Meyenburg; M. Schaechter.
Bottom: K. Gausing; D. Nierlich; H. Kalckar.

Photos contributed by: Millicent Masters, Catherine L. Squires,
and Abraham Eisenstark.

THE MOLECULAR BIOLOGY OF BIOLOGY OF BACTERIAL GROWTH

A Symposium held in honor of Ole Maaløe,
at the University of Alabama, Tuscaloosa

PART ONE

CONTROL OF THE PROTEIN-SYNTHESIS SYSTEM

INTRODUCTION

A bacterial cell contains more than 1000 types of proteins that, in the aggregate, make up over 50 percent of its dry weight. Clearly, synthesizing protein is the major metabolic business of a growing bacterial culture. The assemblage of cellular components required to accomplish this task consists of ribosomes, tRNA, mRNA, and associated proteins, including RNA polymerase and the aminoacyl-tRNA synthetases. This collection comprises the cell's Protein-Synthesizing System, or as Ole Maaløe affectionately calls it, the PSS; and it reflects a major metabolic expenditure: 50 percent of the mass of rapidly growing culture is PSS. The realization and indeed the proof that this expenditure is made prudently came from the pair of papers that Schaechter, Maaløe, and Kjeldgaard published in 1958: in their experiments ribosomes were found to be made in quantitities that are optimal for the growth rate permitted by the external environment of the cell.

The fundamental nature of these observations was quickly appreciated, as was the importance of the primary question they posed: how are these controls mediated? With prescience, the authors stated "...no simple model can be suggested." In retrospect, they might well have added that answering the question would take a long time. The biochemistry, structure, and genetics of the components of PSS had to be understood first, and, indeed, in some of the papers that follow it can be seen that the essence of the control of synthesis of the PSS is becoming understood. As some of these papers show, an understanding of the complex and unique control mechanisms that modulate synthesis of protein components of the PSS came with remarkable rapidity; even some of the more subtle aspects of the regulation have already been revealed. In one of the papers the control of the protein component is found to depend on the synthesis of the RNA component, the control of which is understood only in broad outline.

Other aspects of the PSS are also considered here, namely: the accuracy with which it operates, the way it corrects translational mistakes, and its differences in organisms other than the well-studied enteric bacteria *Escherichia coli* and *Salmonella typhimurium*. Collectively, the associates of Ole Maaløe present a remarkably complete status report on PSS.

J. L. Ingraham

3

Understanding a 26-Year-Old Paper: Studies on the Relationship Between Autogenous Control and Growth-Rate Regulation of r-Protein Synthesis

Lasse Lindahl and Janice M. Zengel
The University of Rochester, New York

DISCOVERY OF REGULATION OF RIBOSOME ACCUMULATION

In 1958 Schaechter, Kjeldgaard, and Maaløe published measurements of the composition of *Salmonella typhimurium* grown in different media (Schaechter et al., 1958). One of their observations was that cells growing in rich growth medium devote a larger fraction of the total dry weight to ribosome formation than do cells growing in a poor medium. This observation, which suggests a very central role of ribosome formation in the regulation of cell growth rate, had a profound impact. Scores of scientists, ourselves amongst them, have attempted to unravel the molecular mechanisms behind that simple observation made in the "WHO Center of Biological Standardization" (of which Ole Maaløe was the chief) at Statens Seruminstitut in Copenhagen more than 25 years ago. Many of the important contributions have of course been made by workers in the University Institute of Microbiology in Copenhagen which Ole headed up a few years later.

SURVEYING THE REGULATION OF RIBOSOME ACCUMULATION

The first phase of the work following up on that important paper led to a very careful phenomenological description of ribosome accumulation. It became clear that under most conditions the synthesis of all the ribosomal components (53 proteins and 3 rRNA molecules) is regulated coordinately. This coordination in itself is remarkable because the genetic organization is complex: there are 7 rRNA operons, each coding for a copy of each of the three types of rRNA

4

molecules, and at least 16 protein operons, encoding from 1 to 11 ribosomal proteins (r-proteins). Another very important observation was that the rate of synthesis of ribosomal components changes almost instantaneously when the composition of the growth medium is changed. For example, after a shift from a relatively poor growth medium to a richer medium (a shift-up) the rate of r-protein synthesis is increased within a minute and then goes through several oscillations before assuming the new rate characteristic of the post-shift medium (for more detailed descriptions of the phenomenology of ribosome synthesis, see Maaløe and Kjeldgaard, 1966; Gausing, 1980).

REGULATORY PROCESSES AFFECTING RIBOSOME SYNTHESIS

Later work, particularly the work going on in the last 10 years or so, has been a more direct attack on the molecular mechanisms for regulating ribosome synthesis. Four major types of mechanisms have been suggested and documented to varying degrees: (1) *Stringent control,* by which transcription initiation at rRNA and r-protein promoters is inhibited by guanosine-5'-diphosphate-3'-diphosphate (ppGpp). (2) *Autogenous control of r-protein synthesis,* by which each of the r-protein operons is negatively controlled by the free form of an r-protein (i.e., protein not bound to rRNA) encoded by the operon. (3) *Autogenous control of rRNA synthesis,* by which rRNA or free ribosomes (i.e., ribosomes not bound to mRNA) inhibits the synthesis of rRNA. (4) *Passive control.* According to mainly theoretical arguments and calculations presented by Ole Maaløe over the last 10-15 years, much of the regulation of ribosome synthesis can be explained without assuming any regulatory substance that actively modulates the activity of the ribosomal genes. Instead Maaløe proposed that all types of promoters in some way compete for the total transcription activity and that the regulation of all the nonribosomal genes indirectly affects the transcription of the ribosomal genes. Figure 1 gives a schematic illustration of these mechanisms. Further details and discussion of the experimental evidence have been published by Gallant (1979); Lindahl and Zengel (1982); Maaløe and Bentzon (1984) and Nomura et al. (1984).

APPROACHES TO MODELING

It is possible to construct theoretical models explaining the phenomenological description of ribosome synthesis by various combinations of the four types of regulation listed in the previous paragraph. In fact, the problem is that too many models can be constructed on a purely theoretical basis, and it is difficult to know which ones to test first. Therefore, it seemed to us that in order to construct realistic models it was necessary to first determine experimentally how each of the mechanisms listed above (plus possibly other still undiscovered mechanisms) contribute to the overall rate of ribosome synthesis under various specific physiological conditions.

For some time now, we have done experiments to determine how autogenous control of r-protein synthesis actually contributes to the overall regulation of r-protein synthesis. The remainder of this paper summarizes what we have learned so far.

Most of our work has been concerned with the S10 operon. This operon encodes eleven r-proteins, with genes for 30S and 50S proteins

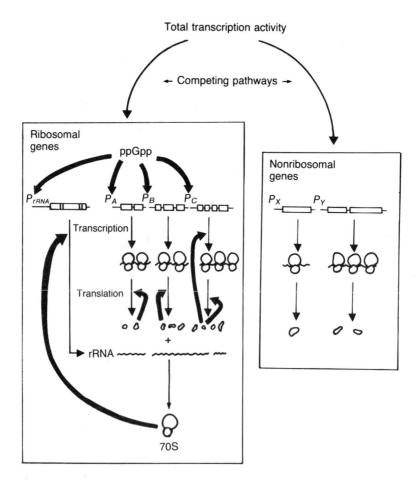

FIGURE 1 Schematic illustration of the regulatory processes that have been demonstrated or proposed to affect the synthesis of ribosomal components. Thin arrows indicate synthetic pathways; heavy arrows indicate negative regulatory circuits. Note that autogenous regulation of most r-protein operons is post-transcriptional. However, the S10 operon is regulated by L4 at both the transcriptional (Zengel et al., 1980; Lindahl et al., 1983) and post-transcriptional levels (Yates and Nomura, 1980).

intermingled with each other (Figure 2). The product of the third gene of the operon, protein L4, regulates the transcription of the entire S10 operon (Zengel et al., 1980). We have shown that this operon contains a site for premature termination of transcription (attenuation) about 30 bases upstream from the most proximal structural gene (Figure 2; Lindahl et al., 1983). In exponentially growing cells one of every two or three mRNA molecules initiated at the S10 promoter is elongated past the attenuator. When the regulatory protein L4 accumulates in excess, read-through at the attenuator decreases five- to tenfold, resulting in a dramatic decrease in the synthesis of r-proteins encoded by the S10 operon. Given the basal level of attenuation in exponentially growing cells, relief of attenuation could result in a two- to threefold increase in the expression of the S10 operon. Thus, the attenuation mechanism has the potential for regulating the S10 operon over a 20- to 30-fold range. We have determined how this capacity is used to regulate the operon under three different types of growth-medium-dependent regulation: nutritional shift-up, steady state growth in different media, and heat shock.

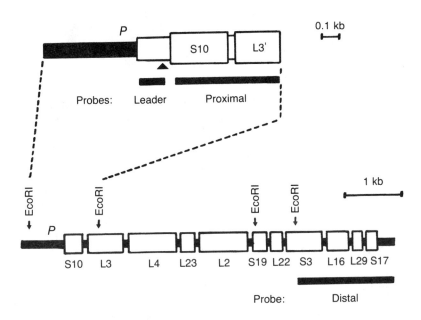

FIGURE 2 Map of the S10 operon of E. coli. The enlargement (top line) shows the proximal part of the operon with promoter P, leader, and proposed position of the attenuator (arrowhead).

Nutritional shift-up

We first followed the transcription of the S10 operon after a shift-up induced by adding glucose and amino acids to a glycerol minimal culture (Zengel et al., 1984). In the period immediately following the shift, the differential rates of transcription of both proximal and distal structural genes (i.e., the rate of incorporation into transcripts from the indicated genes relative to the total incorporation of radioactivity into RNA) fluctuate dramatically, going through oscillations (Figure 3; Zengel et al., 1984) in a pattern strikingly similar to the oscillations in the differential rate of r-protein synthesis (Gausing, 1980). Even though a small increase in the differential rate of leader transcription was consistently observed, this effect was much smaller than the fluctuations measured for the structural messenger transcription

(Figure 3). This suggests that regulation of read-through at the S10 attenuator is a major factor contributing to the oscillations of the structural messenger transcription. Since we know that the read-through is regulated by the concentration of free L4, we believe that the L4-mediated autogenous control of attenuation plays a crucial role in the regulation of protein synthesis from the S10 operon after a shift-up.

As pointed out above, there is a small but consistent increase in the differential transcription rate of the leader. This observation implies an additional regulatory mechanism at the level of initiation of transcription. Such a mechanism is also suggested by the fact that the rate of total RNA synthesis goes up within seconds of a shift-up (Gausing, 1980). Thus, the absolute rate of initiation of transcription at the S10 promoter goes up even more than is evident from the differential rate of leader transcription. We conclude that at least two mechanisms--one working at the level of initiation, the other at the level of attenuation--contribute to the regulation of the S10 operon after a shift-up.

Since the attenuation of the S10 operon is regulated by free L4, the most likely reason for a change in the increased read-through is a decrease in the concentration of free L4. Such a change can be explained if the synthesis of rRNA is increased after the shift, since free L4 is consumed by binding to 23S rRNA in the process of ribosomal assembly. We have investigated this hypothesis by measuring the rates of synthesis of two subsequences of rRNA after the shift-up. One of the rRNA subsequences is close to the rRNA promoter region, and the other is close to the L4 binding site on the 23S rRNA (J. M. Zengel and L. Lindahl, manuscript in preparation). The results of these experiments suggest that the differential rate of initiation of rRNA does increase immediately after the shift. The differential synthesis rate of the L4 binding site on 23S rRNA increases only after a delay of about 40 seconds, as would be expected because the DNA sequence encoding the L4 binding site is located about 2000 base pairs from the promoter region. A similar delay can be detected for the increase in read-through at the S10 attenuator, consistent with the idea that the increased read-through is induced by the change in the rRNA synthesis. Our detailed analysis of the transcription in the first several minutes after the shift also shows that the increase in initiation of transcription at the S10 promoter takes place within seconds of the shift and precedes the increase in the read-through. It therefore appears that the first thing occurring after a shift-up is stimulation of transcription initiation both at the S10 promoter and at the rRNA promoters. The increase in the rRNA synthesis in turn leads to a further boost of the transcription of the structural genes by relieving attenuation within less than a minute of the shift.

To conclude the discussion of the shift-up we should also comment on the observed oscillations in r-protein synthesis following shift-up. We do not know why the rate of r-protein synthesis fluctuates before achieving the steady state post-shift rate, but it seems to us that the most likely reason is that it takes several minutes to synthesize and assemble the ribosomal components into active ribosomes. This means that there is a delay between the time when a regulatory signal changes the rate of synthesis of ribosomal components and the time when this change is manifested as an increased accumulation of functional ribosomes. Such a delay could very well lead to an "overshoot," i.e., a stronger regulatory response than the post-shift growth medium can support in a steady state

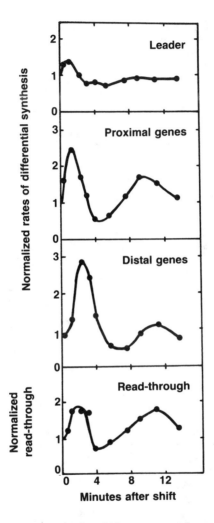

FIGURE 3 Transcription of the S10 operon after a nutritional shift-up (Zengel et al., 1984). The shift was induced by adding amino acids and glucose to a culture growing exponentially in glycerol minimal medium. The different panels indicate the differential transcription rates of the leader (top), the structural genes for S10 and L3 (second from top), and the structural genes for S3, L16, L29, and S17 (second from bottom). The bottom panel indicates the read-through at the S10 attenuator calculated from the transcription of the leader relative to the proximal structural genes (S10 and L3 genes). For details, see Zengel et al., 1984. This figure is modified from Zengel et al., (1984), copyright EMBO Journal.

situation. Several corrections in the regulatory parameters may be necessary before a final balance can be achieved.

Steady state of growth

We have also investigated how the attenuation of the transcription of the S10 operon might contribute to the regulation of protein synthesis from the operon during steady state growth. We found that there is little change in the read-through at the S10 attenuator for exponentially growing cultures with growth rates between 0.65 and 2.5 doublings per hour (37 C, Zengel et al., 1984), even though it has been shown by Dennis (1974) that the differential rate of synthesis of protein from the S10 operon increases almost twofold over the same range of growth rates. Therefore, it appears that the L4-mediated autogenous control of attenuation has a relatively insignificant role in the regulation of protein synthesis from the operon during steady state growth. The major contribution to the control of the r-protein synthesis under these circumstances seems to due to other regulatory processes that remain to be investigated.

Heat shock

Heat shock is another type of perturbation that activates the attenuation mechanism of the S10 operon. In analogy with the synthesis of EF-G, EF-Ts and r-protein S1 (Lemeaux et al., 1978), the differential synthesis of proteins from the S10 operon is temporarily depressed after a shift from 30 C to 42 C (J. M. Zengel, manuscript in preparation). The minimum differential rate, about 20-30% of the pre-shift rate, is observed about 2-5 minutes after the shift, whereupon the rate slowly recovers. Measurements of the transcription of the S10 leader and the proximal structural genes show that decreased read-through at the S10 attenuator significantly contributes to the depression of protein synthesis from the operon after the temperature shift (J. M. Zengel and L. Lindahl, manuscript submitted). Inhibition of initiation of transcription at the promoter appears to provide for additional regulation. Thus, heat shock is similar to the nutritional shift-up in the sense that the final response is accomplished by a combination of initiation control and regulation of attenuation.

   We do not know why attenuation is increased after a temperature shift. It does not appear to result from decreased rRNA synthesis, since the differential rate of rRNA synthesis continues at pre-shift levels immediately after the temperature shift (J.M. Zengel and L. Lindahl, manuscript submitted; Gallant, 1977). Thus, another interesting feature of the heat-shock response is the decoupling of autogenous control of the S10 operon from the synthesis of rRNA. This should not be possible if the attenuation control depends only on the rate of synthesis of L4-rRNA binding sites that can remove L4 from the pool of free protein. Other factors appear to be involved.

GROWTH MEDIUM REGULATION DEPENDS ON INTERACTIONS BETWEEN REGULATORY PROCESSES

The main conclusion from our experiments is that a unitary molecular mechanism for the regulation of r-protein synthesis does not exist.

The contribution of a particular mechanism depends on the physiological circumstances. For example, the attenuation control of the S10 operon is important during the transition phase following the perturbations of a nutritional or a temperature shift, but it is negligible in the regulation of the overall rate of synthesis of protein from the S10 operon during steady-state growth. We hypothesize that there is one (or perhaps several) mechanism responsible for the regulation in steady state of growth, i.e., when the synthetic rates of all cellular components are in a constant ratio. However, the nature of these processes remains obscure. When the cells are perturbed, additional regulatory mechanism(s) such as the L4-mediated autogenous control of S10 attenuation are activated by the imbalance caused by the perturbation. These additional regulatory process(es) may help the cells to approach the new steady-state rate faster than would be possible if the cells had to rely exclusively on the mechanism(s) responsible for steady-rate regulation.

It is important to note that even though our results suggest that autogenous control contributes little to the overall rate of r-protein synthesis during steady-state growth, this does not imply that autogenous control of the S10 (and other) r-protein operon is not functioning at all during steady-state growth. However, we believe that the role of autogenous control under these conditions is to compensate for imbalances between the synthesis of proteins from different r-protein operons. The necessary corrections probably vary from one cell to another: in some cells one r-protein operon might need a boost to keep up with other r-protein operons, whereas in other cells that operon may be temporarily overexpressed. Therefore, the net effect in the culture as a whole may be close to zero, and one would only be able to detect an effect by measuring expression in individual cells—of course, an impossible experiment.

CONCLUSION

The challenge for understanding at the molecular level the observations made 26 years ago by Maaløe and his coworkers still exists. We have uncovered several regulatory processes and understand these in limited molecular detail. However, not only are the mechanistic details of the known regulatory processes still incomplete, additional regulatory processes probably remain to be discovered. Finally, we still have to learn how the different regulatory processes interact to produce the overall regulation of ribosome accumulation.

ACKNOWLEDGEMENTS

We thank professor Ole Maaløe for his interest in our work and for stimulating discussions. The work in our lab was supported by a Public Health Service Research Grant and a Research Career Development Award (to L. L.) from the National Institute of Allergy and Infectious Diseases.

REFERENCES

Dennis, P. P. (1974). *J. Mol. Biol.*, 88, 25.

Gallant, J., Palmer, L., and Pao, C. C. (1978). *Cell,* 11, 181.

Gallant, J. (1979). *Annu. Rev. Genet.*, 13, 393.

Gausing, K. (1980). In *Ribosomes: Structure, Function, and Genetics,* G. Chambliss, G. R. Craven, J. Davies, K. Davis, L. Kahan, and M. Nomura (eds.), University Park Press, pp. 693-718.

Lemeaux, P. G., Herendeen, S. L., Bloch, P. L., and Neidhardt, F. C. (1978). *Cell,* 13, 427.

Lindahl, L. and Zengel, J. M. (1982). *Adv. in Genet.,* 21, 53.

Lindahl, L., Archer, R. H., and Zengel, J. M. (1983). *Cell,* 33, 241.

Maaløe, O. and Bentzon, M. W. (1985). This book.

Maaløe, O. and Kjeldgaard, N. O. (1966). *Control of Macromolecular Synthesis,* Benjamin.

Nomura, M., Gourse, R., and Baughman, G. (1984). *Annu. Rev. Biochem.,* 53, 75.

Schaechter, M., Maaløe, O., and Kjeldgaard, N. O. (1958). *J. Gen. Microbiol.,* 19, 592.

Yates, J. and Nomura, M. (1980). *Cell,* 21, 517.

Zengel, J. M., Archer, R. H., Freedman, L. P., and Lindahl, L. (1984). *EMBO J.,* 3, 1561.

Zengel, J. M., Mueckl, D., and Lindahl, L. (1980). *Cell,* 21, 523.

The Chain Growth Rate for Protein Synthesis is Variable in *Escherichia coli*

Steen Pedersen
University of Copenhagen, Denmark

INTRODUCTION

In 1958, Schaechter, Kjeldgaard, and Maaløe measured the content of macromolecules in bacteria growing in a variety of media and found the protein per genome-equivalent to be almost constant, whereas the amount of RNA varied proportionally to the growth rate. In cells growing at various rates, related parameters were then measured, e.g., the ribosomal protein content of the cells (Schleif, 1967; Gausing, 1977), the fraction of ribosomes in polysomes (Forchhammer and Lindahl, 1971), and the chain growth rate for protein synthesis (Engbaek et al., 1973; Jacobsen, 1974). These and a number of other measurements led Maaløe to develop his approach of looking at the cell (1966, 1969, 1979), which has also found its way into the textbooks (Stent and Calendar, 1978; Ingraham et al., 1983). Briefly, he argued that *Escherichia coli* obtain increased protein synthesis through a variation in the number of ribosomes and *not* by a variation of the chain growth rate for translation, and that bacteria are successful in regulating the protein-synthesizing machinery to operate with maximal and constant efficiency at all growth rates higher than that of acetate medium. I will not disturb this general picture but only point out that less emphasis should be put on "constant" and more on "maximal," in the sense of maximally obtainable, because I can show that the chain growth rate decreases by about 40% when the growth rate of the culture is decreased three- to fourfold when shifted from rich medium to acetate medium.

This result came from a series of experiments designed to test whether the ribosomal step-time differs for individual genes. In particular, I wanted to test the speculation by Grosjean and Fiers (1982) and by Gouy and Gautier (1982) that "rare" codons are translated more slowly. When the *lacI* gene and the first ribosomal genes were sequenced, it was discovered that for many different amino acids the frequency of usage of different codons varied considerably (Farabaugh, 1978; Post et al., 1979). This observation was verified by sequencing a number of genes, and the term "rare" codon was coined for those codons that are used very rarely in genes for abundant proteins, such as the ribosomal proteins (see review by Grosjean and Fiers, 1982). In the following discussion, I define a rare codon more specifically as one used with a frequency of less than 10% of the total number of codons for the same amino acid in the ribosomal protein genes. My experiments compare the translation times for the *lacI* gene and for four proteins having the codon usage of the ribosomal protein genes.

RESULTS

The method to measure the chain growth rate for translation was devised by Bremer and Yuan in 1968. A cultured is labeled with a pulse of radioactive amino acid (here [^{35}S]methionine), and then it is chased. The pulse must be much shorter than the time to synthesize the particular peptide. Samples are taken at various times, and the kinetics of appearance of radioactivity in the finished polypeptide is measured. This increases with time, until the first methionine in the molecule is located in the completed peptide, and then stays constant. If the length and the position of the first methionine in the protein are known, the step-time of the ribosome on this mRNA can be calculated. I separate the total cell extracts on two-dimensional gels and measure radioactivity in six proteins: four with codon usage typical for ribosomal protein genes, EF-Ts, EF-Tu, EF-G, and ribosomal protein S1, and two proteins with genes rich in rare codons, the *lac* repressor and beta-lactamase. The amount of *lac* repressor is normally too small to measure, and to increase it, the *lacI^{q1}* up-promoter and the high copy-number plasmid pBR322 were used. To have enough time to perform such an experiment, it was necessary to decrease the temperature to 25 C, which decreases the growth rate threefold. The detailed methods of the experiment, including the controls for the effectiveness of the chase and harvesting procedure, were published by Pedersen (1984). The result with cells grown in glucose minimal medium shows that EF-Ts, EF-Tu, EF-G, and ribosomal protein S1 are translated at rates of 0.18, 0.21, 0.22, and 0.23 sec per amino acid, respectively, whereas the *lacI* gene and the beta-lactamase gene, *bla,* were translated almost 50% slower (both about 0.29 sec per amino acid).

From working in Maaløe's lab, one knows that an experiment is not properly done until it has been carried out on cells growing in different media, so I decided to test the reproducibility of this finding in cells growing in acetate or in glucose medium, enriched with all amino acids except methionine. The results are shown in Figure 1. In both slow- and fast-growing cells the *lacI* and *bla* genes are translated about 50% slower than expected from the chain growth rate of the translation system proteins, measured in the same experiment. What is also evident, however, is that the chain growth rate differs for both classes of proteins in the three different media. The chain growth rates are summarized in Table 1.

The last experiment tries to address the problem that the effect of the rare codons might be exaggerated in my strain, because of a

TABLE 1   Step time for translation of individual genes
in different media at 25°C.*

Gene		Growth media		
		Acetate	Glucose	Rich
lacI		0.47	0.28	0.27
bla		0.39	0.29	0.31
	Average	0.43	0.29	0.29
tuf		0.30	0.22	0.20
rpsA		0.23	0.21	0.16
fus		0.25	0.23	0.17
	Average	0.26	0.22	0.18

*Units are seconds per amino acid.

high drain on the rare tRNAs, owing to the increased synthesis of
lac repressor and beta-lactamase. The experiment was done with a
strain having the *lacIq1* on an episome instead of the high copy
number-plasmid pBR322. In this strain the amount of *lac* repressor is
less than 0.1% of total protein, so it is difficult to get enough
radioactivity incorporated into the *lacI* spot. A parallel experiment
with a strain having a *lac* deletion was also carried out to make sure
that the right spot was measured. Both experiments were performed
with an MC1000 derivative (Casadaban and Cohen, 1980), an *E. coli*
K12 nonisogenic to NF929, used in the experiment from Figure 1. The
average of two independent experiments is shown in Figure 2. The
lac repressor is translated only about 10% slower than EF-G, which I
would not have considered significant had I not done the previous
experiments.

DISCUSSION

Translation rate of rare codons

When the *lacI* and *bla* genes are present on a high copy-number
plasmid, as shown in Figure 1, the slow translation of these genes
correlates with the high content of rare codons in those genes. If,
indeed, the codon usage was the cause of the slow translation, each
rare codon would be translated at a rate of 0.6 sec/amino acid
compared to 0.2 sec/amino acid for the common codons in this
experiment. If the effect was caused by the subset of the rare codons,
translated by rare tRNAs (Ikemura, 1981), each of such "rare tRNA
codons" would take about 2 sec to translate. Other mechanisms, such
as slow translation of a particularly difficult secondary structure of
the mRNA, could also explain this, but when the experiment is
performed in a strain in which the content of proteins with rare
codons is approximately the same as in a wildtype cell the *lacI* gene
is translated at about normal rate (Figure 2). Barring strain

differences, the slow translation of the *lacI* and *bla* genes shown in Figure 1 seems to be an artificial result, caused by an increased drain on the rare tRNAs, although the scattering of the points is

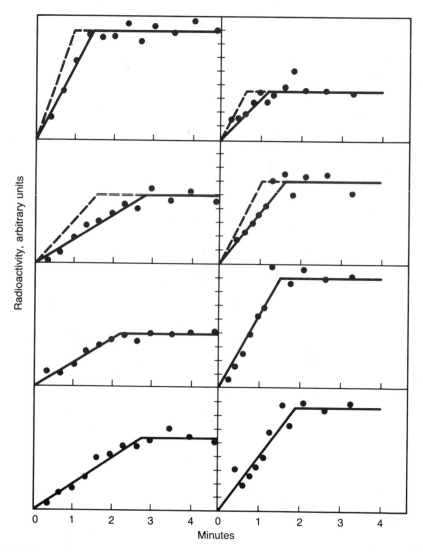

FIGURE 1 Synthesis of the bla, lacI, rpsA, and fus gene products (from top to bottom, respectively) in a pulse-chase experiment. Left column, acetate medium; right column, rich medium. The dotted lines in the bla and lacI panels indicate the calculated curves for these genes if synthesized with the average of the chain-growth rate, measured on the tuf, rpsA, and fus genes in the same experiment. Strain NF929/pR2172 was used (Pedersen, 1984).

larger in Figure 2 owing to the presence of a small amount of *lac* repressor. Implicit in the above speculations is the assumption that the cell is unable to regulate its content of rare tRNAs to suit the demand.

The data in Figure 2 show the *lacI* and *fus* genes to be translated at about the same rate in a haploid cell. However, the data do not exclude translation of the "rare tRNA codons" with one third of the rate of that of the common codons, and, therefore, the rare tRNA codons might still have the functions discussed previously (Pedersen, 1984). If so, drain of the rare codon tRNAs might create problems when high copy-number plasmids are used to obtain high expression of proteins with an unusual codon usage. Randall et al. (1980) also found evidence for translational pauses in the synthesis of the periplasmic maltose-binding protein, but the position of these pauses was not correlated with the presence of rare codons.

Variation of the step-time with growth rate

For proteins using the common codons the chain growth rate for translation also varies with the growth rate (Table 1), i.e., against the conviction in our laboratory. Examining more closely the previous data, e.g., those of Jacobsen (1974) and cited by Maaløe (1979), I found background problems to have influenced the conclusion. The data in the square-root plot (Schleif et al., 1973) clearly show a 35% increase in the chain growth rate when the chemostat culture is shifted to the normal glucose medium. Engbaek et al. (1973) showed the translation time for beta-galactosidase in both rich medium and in acetate medium to be longer than the translation time in glucose and concluded that chain growth rate was unaffected by the medium; however, in the case of the rich medium the conclusion is based on a limited set of samples. The results in Table 1 are very similar to those found indirectly for overall translation by Forchhammer and Lindahl (1971) and for translation of beta-galactosidase by Dalbow and Young (1975). This leads to the conclusion that the chain growth rate for translation varies continuously with the growth rate of the bacteria, increasing by about 40% when the bacterial growth rate changes from acetate to rich medium.

Is this now important, or should it be looked upon as a minor correction of our previous picture of the cell? It shows that the ribosome *in vivo* is not always working at saturated conditions, and that most likely the concentration of the ternary complex containing EF-Tu, GTP, and aminoacyl tRNA is limiting the translation rate in *E. coli*. A change in the ratio between charged and uncharged tRNA can explain more directly than previously the results of experiments involving shifts between media. The immediate increase in the rate of protein synthesis following a shift-up (Koch and Deppe, 1971) was attributed to an activation of idle ribosomes (Maaløe, 1979), but measurements showed most ribosomes in slowly growing cells to be present in polysomes (Forchhammer and Lindahl, 1971). At least a large fraction of the immediate increase in the rate of protein synthesis during the shift-up seems to be caused by speeding up the already engaged ribosomes. In a shift-up experiment, the basal level of ppGpp was found to drop transiently to zero (Friesen et al., 1975). This result is also hard to understand, unless a change in charging levels of tRNA between different media is allowed for. Yanofsky found such

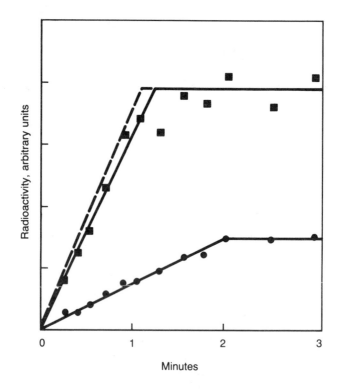

FIGURE 2 Syntheses of the lacI (squares) and the fus gene (circles) products in a pulse-chase experiment with NF1830: MC1000 recA1/F'lacIq1/lac::Tn5, grown in glucose medium. As in Figure 1, the dotted line represents the calculated curve if the lacI gene had been translated at the chain-growth rate of the fus gene.

a change by direct measurements of the charging of trp-tRNA in different media (this volume). Finally, the attenuation mechanism for regulating the pathways for amino acid biosynthesis becomes a more plausible and even ideal mechanism for regulation at varying growth rates, and not, as discussed by Maaløe (1979) and by Ingraham et al. (1983), a mechanism reserved only for severe starvation conditions. At the different growth rates a small variation in the charging level of tRNA could change the degree of attenuation, in particular, in those amino acid biosynthetic operons that have many codons in the leader peptide for the particular amino acid.

ACKNOWLEDGEMENTS

This work was supported by grants from the Carlsberg Foundation, the NOVO Foundation, and the Danish Natural Science Research Council (Nos. 11-3771 and 11-4305).

REFERENCES

Bremer, H. and Yuan, D. (1968). *J. Mol. Biol.*, 34, 527.

Casadaban, M. and Cohen, S. N. (1980). *J. Mol. Biol.*, 138, 179.

Dalbow, D. G. and Young, R. (1975). *Biochem. J.*, 150, 13.

Engbaek, F., Kjeldgaard, N. O. and Maaloe, O. (1973). *J. Mol. Biol.*, 75, 109.

Farabaugh, P. B. (1978). *Nature*, 274, 765.

Forchhammer, J. and Lindahl, L. (1971). *J. Mol. Biol.*, 55, 563.

Friesen, J. D., Fiil, N., and von Meyenburg, K. (1975). *J. Biol. Chem.*, 250, 304.

Gausing, K. (1977). *J. Mol. Biol.*, 115, 335.

Gouy, M. and Gautier, C. (1982). *Nucleic Acids Res.*, 10, 7055.

Grosjean, H. and Fiers, W. (1982). *Gene*, 18, 199.

Ikemura, T. (1981). *J. Mol. Biol.*, 146, 1.

Ingraham, J. L., Maaløe, O., and Neidhardt, F. C. (1983). *Growth of the Bacterial Cell*, Sinauer Associates.

Jacobsen, H. (1974). Thesis, University of Copenhagen.

Koch, A. L. and Deppe, C. S. (1971). *J. Mol. Biol.*, 55, 549.

Maaløe, O. and Kjeldgaard, N. O. (1966). *Control of Macromolecular Synthesis*, Benjamin.

Maaløe, O. (1969). *Dev. Biol.*, Suppl. 3, 33.

Maaløe, O. (1979). In *Biological Regulation and Development*, R. F. Goldberger (ed.), Plenum, Vol.1, p. 487.

Pedersen, S. (1984). In *Gene Expression*, Alfred Benzon Symp. 19, B. F. C. Clark and H. U. Petersen (eds.), Munksgaard, Copenhagen, p. 101.

Post, L. E., Strycharz, G. D., Nomura, M., Lewis, H., and Dennis, P. P. (1979). *Proc. Natl. Acad. Sci. USA*, 76, 1697.

Randall, L. L., Josefsson, L.-G., and Hardy, S. J. S. (1980). *Eur. J. Biochem.*, 107, 375.

Schaechter, M., Maaløe, O. and Kjeldgaard, N. O. (1958). *J. Gen. Microbiol.*, 19, 592.

Schleif, R. (1967). *J. Mol. Biol.*, 27, 41.

Schleif, R., Hess, W., Finkelstein, S., and Ellis, D. (1973). *J. Bacteriol.*, 115, 9.

Stent, G. S. and Calendar, R. (1978). *Molecular Genetics,* Freeman.

Control of Ribosome Biosynthesis in *Escherichia coli:* The Ribosome Feedback Regulation Model

Masayasu Nomura
University of Wisconsin, Madison

In his pioneering work, Ole Maaløe and his collaborators discovered that the amount of ribosomes in growing bacterial cells varies depending on culture conditions (see Maaløe and Kjeldgaard, 1966). They found a linear relationship between the growth rate and the cellular concentration of ribosomes (except in slow growth conditions), indicating that the efficiency of ribosomes is nearly constant under these conditions. Thus, bacterial cells adjust their ribosome content so that the maximum growth rate can be achieved under a given growth condition.

In more recent analyses, Maaløe has divided genes and gene expression in bacterial cells roughly into two parts: the protein-synthesizing system (PSS) and the nonprotein-synthesizing-system ("non-PSS"), which provides energy and substrates for protein synthesis and other polymerization reactions. Among polymerization reactions, protein synthesis is by far the most energy-consuming reaction (Ingraham et al., 1983). The question regarding the regulation of bacterial growth can now be rephrased: how do bacterial cells adjust the size of the PSS to match the growth rate under given growth conditions? Specifically, how do cells partition available materials and energy between the syntheses of the PSS and non-PSS to achieve the maximum growth rate? Maaløe suggested that this partition occurs by a "passive regulation model" (Maaløe, 1969; 1979; 1983). He suggested that there is a competition between the PSS and non-PSS in their gene expression, and that changes in the expression of the PSS genes take place passively as a result of the changes in the expression of non-PSS genes that take place by specific regulatory mechanisms in direct response to environmental changes. His thoughts on this problem strongly stimulated the field. In this article, I would like to discuss my own view on the growth rate-dependent regulation of ribosome biosynthesis.

THE TRANSLATIONAL FEEDBACK REGULATION MODEL OF RIBOSOMAL PROTEIN SYNTHESIS

It has now been established that the synthesis of most, if not all, ribosomal proteins (r-proteins) is feedback-regulated by certain key r-proteins at the level of translation, and that this feedback mechanism can explain the balanced and coordinated synthesis of ribosomal components (for reviews, see Nomura et al., 1982; 1984). Although there are almost certainly other regulatory mechanisms, such as regulation at the level of transcription (see below), it should be emphasized that regulation at the level of translation is very effective and able to prevent overproduction (or underproduction) over a wide range of transcriptional activities of r-protein genes. For example, many-fold increases in r-protein gene copy numbers result in corresponding increases in the rates of synthesis of r-protein mRNAs. Yet, no significant increase in the synthesis rate of r-protein is observed in most of the r-proteins studied, which is undoubtedly due to the operation of the translational feedback mechanism. The recent work on the expression of L11-L1 genes that are transcriptionally fused to the promoter for the *lac* operon clearly illustrates the nature of this regulation. Induction of transcription with a *lac* operon inducer, isopropylthiogalactoside (IPTG), increased the mRNA synthesis rate about tenfold without any significant increase in the synthesis rate of L11 or L1. A single-base alteration in the r-protein repressor-target site on the mRNA abolished this regulation and led to a large overproduction (up to tenfold) of these r-proteins, corresponding to the mRNA overproduction (Baughman and Nomura, 1983; 1984). These experiments convincingly demonstrate that *E. coli* cells are able to balance the synthesis rate of at least some, and probably most or all, r-proteins without regulating the synthesis rates of their mRNAs. The presence of *potentially productive* mRNA synthesis does not necessarily lead to the synthesis of r-proteins in corresponding amounts. The fact that the large increase in the synthesis of a potentially productive mRNA does not necessarily lead to an increase in the synthesis of r-proteins suggests that the partition of protein synthesis into PSS and non-PSS may not involve mechanisms operating at the level of transcription, such as competition between the transcription of PSS genes and that of non-PSS genes. Rather, the synthesis rates are determined by the (limiting) amounts of other ribosomal components through the translational feedback mechanism. In this sense, one can conclude that the synthesis rates of r-proteins are determined not by transcriptional mechanisms, but by the translational feedback mechanism and are balanced with the synthesis rate of ribosomes. Thus, it follows that the rates of synthesis of r-proteins are ultimately determined by the synthesis rate of a single component that is rate-limiting. We believe that this rate-limiting component is rRNA under normal growth conditions for the reasons described below.

The rate of synthesis of rRNA under normal growth conditions is balanced with the rate of accumulation of ribosomes. Of the r-protein mRNA molecules experimentally studied, most are synthesized in excess, and therefore the transcription of r-protein genes does not represent the rate-limiting reaction in ribosome biosynthesis. First, measurements for the *spc-* and alpha-operon mRNA indicate that the ratio of the *rate of synthesis* of r-protein mRNA to that of rRNA (per unit amount of protein) increases strongly with decreasing growth rate, while the ratio of the *steady-state amount* of r-protein mRNA to the rate of synthesis of rRNA (per unit amount of protein) changes only slightly (Dennis and Nomura, 1975; Gausing, 1977; Dennis, 1977; see also Nomura et al., 1984). These observations indicate that r-protein mRNA is always synthesized in excess (except possibly at the highest growth rate), and the efficiency of its utilization (and its half life) decreases with decreasing growth rate. Second, it is known that mutants with alterations in the repressor r-proteins (or their synthesis) overproduce r-proteins that are in the same translational regulatory unit as these repressor r-proteins. Again, the results show that synthesis of r-protein is not limited by the production of mRNA, but is usually kept at lower levels by translational repression. Of course, we cannot rigorously exclude the possibility that the synthesis of one or more (unidentified) r-protein is limiting the synthesis of ribosomes, and that rRNA synthesis takes place slightly in excess, and the excess rRNA breaks down. In fact, as will be discussed later in this article, this situation may be the case under conditions of slow growth, in which degradation of rRNA synthesis almost certainly takes place to a significant extent (Gausing, 1977; Norris and Koch, 1972).

THE RIBOSOMAL FEEDBACK REGULATION MODEL OF rRNA SYNTHESIS

As mentioned above, the regulation of the synthesis of rRNA is probably the most important factor in determining the synthesis rate of ribosomes under normal growth conditions (except slow growth conditions, see below). How is the synthesis regulated in response to growth conditions so that the amount of ribosomes synthesized is appropriate for the growth rate achieved in a given growth condition? One striking feature of the growth-rate-dependent regulation of ribosome synthesis is that many different sets of growth conditions (with respect to carbon and nitrogen sources, for example) producing the same growth rates also yield the same rate of ribosome synthesis. In order to explain this and other features of ribosome biosynthesis discussed later in this article, we have recently proposed a rather simple feedback model called "the ribosome feedback regulation model" (Jinks-Robertson et al., 1983; see Figure 1) and started a series of experiments to test the model. We suggest that bacterial cells regulate the ribosome synthesis rate (that is, the rRNA synthesis rate that is rate-limiting) as a result of a feedback mechanism that involves excess, nontranslating "free" ribosomes, rather than as a direct response to environmental conditions. We imagine that cells are inherently prone to make excess ribosomes relative to other non-PSS gene products that provide substrates and energy for macromolecular synthesis; overproduced nontranslating ribosomes monitor the outcome of ribosome biosynthesis and prevent further unnecessary synthesis of ribosomes by inhibiting rRNA synthesis (and/or the synthesis of some other rate-limiting ribosome component). In this way, cells are able to adjust the amount of ribosomes, so that the capacity for protein synthesis is just sufficient to maintain the growth rate, while the free nontranslating ribosome concentration is kept small.

Upon nutritional shift-up, preferential and immediate stimulation of rRNA (and r-proteins) is observed. According to the passive

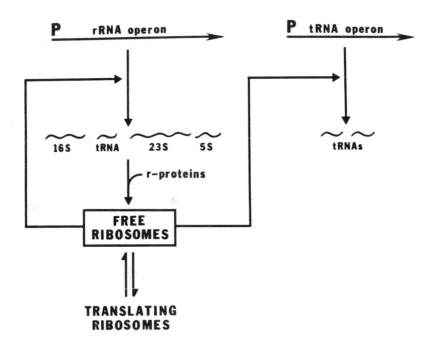

FIGURE 1 The ribosome feedback regulation model. Transcription of rRNA operons yields 16S RNA, 23S RNA, 5S RNA, and tRNA(s). The 16S, 23S, and 5S RNAs are assembled with the 52 r-proteins to form the 30S and 50S ribosomal subunits which together make up the 70S ribosome. The assembled ribosomes enter polysomes, ribosome-mRNA complexes that are actively synthesizing protein, but a portion of ribosomes exist in the pool without being engaged in protein synthesis. The concentration of these free ribosomes in the pool determines the degree of the expression of rRNA and tRNA operons; it is proposed that free ribosomes in the pool inhibit the transcription of these operons directly or indirectly.

regulation model, this stimulation results from repression of the biosynthetic operons used in poor medium. In contrast, we suggest that a stimulation of protein synthesis upon nutritional shift-up mobilizes the small amount of nontranslating ribosomes in the pool, thereby causing a burst of derepressed synthesis of rRNA and hence of ribosomes. Thus, the stimulation of rRNA synthesis (and r-protein synthesis) is very quick and transiently overshoots the new elevated steady-state rate in the rich media. In contrast, the total rate of protein synthesis increases gradually, utilizing newly synthesized ribosomes, and a new steady state is eventually reached at which ribosome synthesis is balanced with the total rate of protein synthesis, that is, with the growth rate under the new growth conditions. The situation after nutritional shift-down can also be explained in a similar way.

GENE DOSAGE EXPERIMENTS TO TEST THE RIBOSOME FEEDBACK REGULATION MODEL

A prediction of the ribosome feedback regulation model described above is that cells with an increased number of rRNA operons will not increase rRNA synthesis in proportion to the increase in gene copy numbers; rather, cells will maintain approximately the same (actually slightly higher, as discussed below) *total* rate of rRNA synthesis by reducing rRNA synthesis from *individual* rRNA operons. These predictions were essentially confirmed (Jinks-Robertson et al., 1983). In strains carrying extra rRNA operons on multicopy plasmids, the rate of total rRNA synthesis was about the same or only slightly higher compared to the control strains, and the reduction of transcription of individual rRNA operons was confirmed by monitoring the syntheses of tRNAs encoded by chromosomal rRNA operons (see Figure 2).

Although these experimental results are consistent with the ribosome feedback regulation model, they can also be explained by another model, which proposes that some factor essential for transcription of the rRNA operon (for example, RNA polymerase or some other hypothetical protein factor) is limiting, and the sum of the transcription rate of all rRNA operons is determined by the amount or availability of such a factor. Control experiments designed to distinguish these two alternatives were carried out using strains carrying extra rRNA operons that have large deletions within the rRNA coding region. If the total amount of rRNA transcription is controlled or limited by some positive factor, then the total amount of rRNA transcription (i.e., the synthesis of intact plus defective transcripts) should not change in cells containing such extra (defective) rRNA operons. On the other hand, according to the ribosome feedback regulation model, since the defective extra rRNA operons would not produce functional ribosomes, the extra defective rRNA synthesis would not be monitored as extra production of ribosomes. Hence, overproduction of rRNA synthesis would be observed by using suitable probes to detect both defective and intact rRNA synthesis. The results were clear. The intact rRNA genes were regulated normally to produce enough rRNA for the normal complement of ribosomes. In addition, the defective rRNA genes produced extra nonfunctional rRNA at a rate that simply reflected gene dosage, as predicted from the ribosome feedback regulation model (see Figure 2).

In the course of these gene-dosage experiments, we discovered that the synthesis of most, if not all, tRNA species is also subject to the same regulation by products of rRNA operons. In strains carrying extra rRNA operons on multicopy plasmids, the synthesis of tRNAs that are unrelated to rRNA operons was inhibited, as well as the synthesis of tRNAs encoded by chromosomal rRNA operons. When the plasmid-encoded rRNA operons contained deletions within the rRNA-coding region, no such inhibition was observed. Apparently, cells do not have regulatory systems that monitor production of tRNAs directly and regulate it to maintain the "normal" rate of synthesis appropriate for given growth conditions. In agreement with this conclusion, the regulation of the synthesis of tRNAs that are unrelated to rRNA operons is similar to that of rRNA, e.g., with respect to

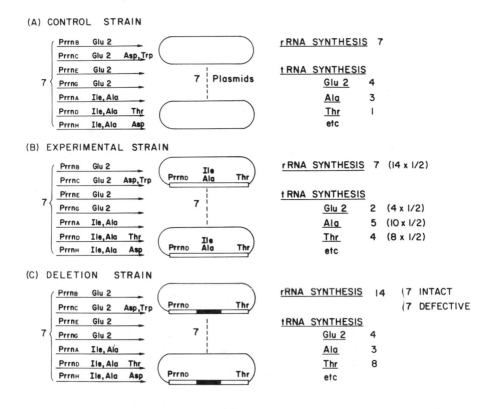

(A) CONTROL STRAIN

Prrnʙ	Glu 2	
Prrnc	Glu 2	Asp,Trp
Prrnᴇ	Glu 2	
Prrnɢ	Glu 2	
Prrnᴀ	Ile, Ala	
Prrnᴅ	Ile, Ala	Thr
Prrnʜ	Ile, Ala	Asp

7

7 Plasmids

rRNA SYNTHESIS 7

tRNA SYNTHESIS
Glu 2 4
Ala 3
Thr 1
etc

(B) EXPERIMENTAL STRAIN

Prrnʙ	Glu 2	
Prrnc	Glu 2	Asp,Trp
Prrnᴇ	Glu 2	
Prrnɢ	Glu 2	
Prrnᴀ	Ile, Ala	
Prrnᴅ	Ile, Ala	Thr
Prrnʜ	Ile, Ala	Asp

7

Prrnᴅ Ile Ala Thr

7

Prrnᴅ Ile Ala Thr

rRNA SYNTHESIS 7 (14 x 1/2)

tRNA SYNTHESIS
Glu 2 2 (4 x 1/2)
Ala 5 (10 x 1/2)
Thr 4 (8 x 1/2)
etc

(C) DELETION STRAIN

Prrnʙ	Glu 2	
Prrnc	Glu 2	Asp,Trp
Prrnᴇ	Glu 2	
Prrnɢ	Glu 2	
Prrnᴀ	Ile, Ala	
Prrnᴅ	Ile, Ala	Thr
Prrnʜ	Ile, Ala	Asp

7

Prrnᴅ Thr

7

Prrnᴅ Thr

rRNA SYNTHESIS 14 (7 INTACT
                    (7 DEFECTIVE

tRNA SYNTHESIS
Glu 2 4
Ala 3
Thr 8
etc

FIGURE 2 The rRNA gene dosage experiments. A control strain with the normal number of 7 rRNA operons per haploid chromosome was used (A). The experimental strain (B) carried an additional 7 copies of the rrnD operon encoded by a plasmid. Another strain ("deletion strain"; C) was identical to the experimental strain except that the plasmid-encoded rrnD operon had a deletion covering a part of 16S rRNA, spacer tRNA and a part of 23S rRNA genes. The relative synthesis rates of rRNA (measured using probes which do not overlap with the deleted region) and some pertinent tRNAs indicated in the figure are the values expected from the ribosome feedback regulation model. These values are calculated assuming that the copy number of the plasmid-encoded rrn operon was as indicated above and that all the rrn operon promoters, chromosomal as well as plasmid-encoded ones, function with an equal efficiency. For example, according to the model, the total rRNA synthesis rate in the experimental strain is kept about the same as that in the control strain (ignoring a small increase required for increasing the level of free ribosomes) by a feedback mechanism, reducing the efficiency of promoters for rRNA (and tRNA) operons by a factor of about 2. The relative synthesis rates are then calculated as the number of operons multiplied by the relative efficiency of the promoters. In deletion strains, the rrn operon on the plasmid does not contribute to the production of ribosomes and hence there is no reduction in the efficiency of promoters relative to the control strain. In the actual experiments (Jinks-Robertson et al.,

(FIGURE 2, continued)
1983), the results obtained were roughly consistent with those shown
in the figure and strongly support the ribosome feedback regulation
model.

---

dependency on growth rate, and yet, in contrast to the rRNA operons,
the rates of synthesis tRNAs depend on gene dosage; that is, there
is no feedback regulation by overproduced tRNAs. For example,
strains carrying extra genes for asparagine tRNA on a multicopy
plasmid overproduce this tRNA (per unit amount of cellular mass)
more than tenfold relative to control strains (Gourse and Nomura, 1984). Several papers that had been published earlier also
indicate synthesis of tRNAs that is dependent on gene dosage (Hill
et al., 1970; Ikemura and Ozeki, 1977; Comer, 1982). Thus, it appears
that the regulation of tRNA synthesis is carried out by the regulatory
system that has primarily evolved to regulate the synthesis of
ribosomes. In the above gene-dosage experiments, decreases in the
efficiency of individual rRNA promoters observed in cells carrying
extra intact rRNA operons take place, according to the ribosome
feedback regulation model, as a result of an increased level of free
nontranslating ribosomes in the cellular pool, and such an increase
should be detectable. The direct determination of the exact amounts
of free 70S ribosomes and their subunits has been technically difficult
mostly because of the presence of "70S ribosomes" produced from
polysomes *after* cell breakage and because of the existence of an
equilibrium between 70S ribosomes and 50S plus 30S subunits in
cellular extracts. Nevertheless, we have recently tested the prediction
of the ribosome feedback regulation model by comparing the
distribution of RNA in gently prepared cell extracts from cells carrying
extra rRNA operons with extracts from control cells. By using labeling
techniques with two isotopes, we have, in fact, found that the amounts
of polysomes and "70S ribosomes", which constitute the majority of
the ribosomes in the extracts, were approximately the same in the
two extracts, while the amount of both 30S and 50S ribosomal subunits
in the extract from cells with extra rRNA operons was almost twofold
higher than that in the extract from control cells. The amount of
RNA in the tRNA fraction in the extract from the plasmid carrying
strain was about half that found for the control cells, as was found
previously (Y. Takebe and M. Nomura, unpublished data). These
results are consistent with expectation from the ribosome feedback
regulation model and give additional support to the validity of this
model.

A NEW EXPERIMENTAL SYSTEM TO STUDY THE RIBOSOME FEEDBACK MODEL

In the gene-dosage experiments discussed in the previous section two
bacterial strains are compared, one strain containing extra rRNA
operons on a multicopy plasmid and the other with no such extra
rRNA operons or containing extra rRNA operons with deletions in
their coding region. In order to study further the mechanisms involved
in this feedback regulation, we have recently developed a system in
which overproduction of ribosomes can be achieved from a conditional

expression promoter. The tandem rRNA promoters were removed from the *rrnB* operon on a plasmid and replaced by the lambda *pL* promoter. This hybrid operon contains intact coding and adjacent regions required for rRNA processing and is heat inducible in strains carrying a temperature-sensitive lambda *cI* repressor. A control plasmid was also prepared, which is identical to this hybrid plasmid except for deletion of a large part of the rRNA coding region. In preliminary experiments, the strains carrying these plasmids were grown at 30 C, and then shifted to 42 C. Induction of the synthesis of extra rRNA and ribosomes was successfully achieved in this way, and the expected strong inhibition of the transcription of chromosomal rRNA operons (measured by following the synthesis of tRNAs encoded by these operons) and tRNA operons was shown in experimental, compared to control, strains. Also, it was shown that rRNA induction resulted in a twofold increase in 30S and 50S free ribosomal subunits, whereas the polysome fraction was unaffected (Gourse et al, 1985). We expect that the new experimental system will be useful in analyzing intermediate steps between induction (initiation of excess transcription of rRNA) and the final regulatory events (increased inhibition of transcription of chromosomal rRNA and tRNA operons.

OTHER OBSERVATIONS SUPPORTING THE RIBOSOME FEEDBACK MODEL

As already discussed, the ribosome feedback regulation model predicts that the conditions that lead to a decrease in free nontranslating ribosomes in the cellular pool would cause preferential stimulation of rRNA (and tRNA) synthesis, and alternatively, which conditions that lead to an increase in free ribosomes in the cellular pool would cause preferential inhibition of the synthesis of rRNA (and tRNA). One possible way to decrease free ribosomes is to inhibit ribosome assembly without inhibiting macromolecular synthesis. We have carried out studies of this type using two methods to block ribosome assembly. One method was to use a cold-sensitive mutant defective in ribosome assembly having a mutation in one of its r-protein genes, *rpsE*. It was shown by direct measurements that in contrast to the growth rate, the rRNA (and tRNA) rate of synthesis at 20 C relative to 37 C is much higher in the mutant than in control wildtype strains (A. Miura, Y. Takebe, S. Jinks-Robertson and M. Nomura, unpublished data). The results are in agreement with several earlier reports that have described overproduction of rRNA in other mutants defective in assembly of ribosomes (Tai et al., 1969; Buckel et al., 1972). The second method we used to inhibit ribosome assembly was to unbalance r-protein synthesis by overproducing translational repressor r-proteins from an inducible promoter, such as the promoter for the *lac* operon. It was previously shown that overproduction of S4 or S8 from the *lac* operon promoter by addition of the inducer IPTG leads to (1) strong inhibition of ribosome assembly without significant inhibition of total protein synthesis and to the accumulation of incomplete ribosomal particles (Dean and Nomura, 1980; A. Miura and M. Nomura, unpublished experiments). Under these conditions, cells continued linear growth and a significant stimulation of rRNA (and tRNA) synthesis rate was observed. Preferential stimulation of r-protein synthesis was also observed in the induced culture (Takebe et al, 1985). It should be noted that the primary event in this case is the overproduction of a repressor r-protein that leads to inhibition of ribosome assembly but that does not change the nutritional environment in any obvious way. These results are consistent with the ribosome feedback mechanism but are not easy to explain by other mechanisms suggested for rRNA regulation. Another condition in which one expects a decrease in free ribosome concentration is when the rate of elongation of protein chains is reduced without reducing the rate of chain initiation. In agreement with the model, stimulation of rRNA synthesis was observed in cells treated with inhibitors of elongation of protein chains, such as fusidic acid (Bennett and Maaløe, 1974) and chloramphenicol (Kurland and Maaløe, 1960; Shen and Bremer, 1977). (The original papers gave different possible explanations for this stimulation.) Similarly, the model predicts that specific inhibition of initiation of protein chains without inhibiting chain elongation would lead to an inhibition of rRNA synthesis. We have not succeeded in finding such conditions with sufficient specificity yet, and the prediction has not been critically tested. However, it is possible that upon nutritional shift-down, carbon starvation, or possibly amino acid starvation, cells might encounter such conditions (see e.g., discussion in O'Farrell, 1978), leading to an accumulation of free nontranslating ribosomes and the inhibition of rRNA synthesis.

Finally, I would like to comment on the situation with conditions of slow growth. According to the ribosome feedback regulation model, the pool of free ribosomes should increase as the degree of repression of rRNA synthesis increases. In fact, this expectation was confirmed in the gene-dosage experiments, as described above. Thus, under conditions of very slow growth, in which the degree of repression is very high, the amount of free ribosomes is expected to become a significant fraction of total ribosomes; thus, the breakdown of the linear relationship between the amount of ribosomes (or α_r) and the growth rate under slow growth conditions is not surprising. The published quantitative data on the relationship between cellular concentrations of ribosomes (measured as α_r) and growth rate (e.g., Gausing, 1977) are fully consistent with theoretical expectations from the feedback regulation model. The model also explains the paradoxical and seemingly harmful overproduction of ribosomes under conditions in which cellular growth is restricted by limiting the supply of phosphate (Alton and Koch, 1974). As pointed out by Koch (1971), the large overproduction of ribosomes at slow growth rates may be advantageous for cells that must be able to respond quickly to better growth conditions, but the overproduction might also be an unavoidable consequence of the feedback mechanism that has evolved in bacterial cells to regulate rRNA synthesis under other growth conditions. However, the ribosome feedback regulation model does not have a specific explanation for the reported large overproduction of rRNA in slow growth conditions, which is even higher than that required for the synthesis of the excess ribosomes discussed above (Norris and Koch, 1972; Gausing, 1977). It appears that rRNA synthesis is not the rate-limiting step in the synthesis of ribosomes under these conditions. One possible reason is that among the two rRNA promoters, only the P1 promoter is strongly feedback regulated and the P2 promoter that is more refractory to the feedback mechanism becomes dominant (cf., Sarmientos and Cashel, 1983). The synthesis of (certain) r-proteins might become the rate-limiting step under these slow growth conditions.

POSSIBLE MECHANISM FOR INHIBITION OF rRNA OPERON TRANSCRIPTION

As we have just shown, a variety of experimental observations are extremely suggestive that the ribosome feedback regulation model is essentially correct; that is, an increase in the concentration of free ribosomes leads to decreased transcription of rRNA genes. However, we have little information about possible molecular mechanisms involved in the inhibition of transcription. For example, it is not established whether the inhibition is due to a decrease in transcription initiation or increase in premature termination ("attenuation"). There is now experimental evidence for the presence of some mechanism ("antitermination") that would prevent premature termination in the process of transcription of the long untranslated rRNA operons (Morgan, 1980; Siehnel and Morgan, 1983; Li, Squires, and Squires, 1984). However, since such antitermination mechanisms are expected to be present in rRNA transcription, but not in tRNA transcription, and yet the feedback inhibition by free ribosomes apparently acts on tRNA synthesis as well as rRNA synthesis, the modulation of the degree of (hypothetical) antitermination by free ribosomes is not likely to be the mechanism for the feedback inhibition. In fact, recent deletion analyses using rrnB promoter fragments fused to lacZ indicate that the regulation of transcription of the rRNA operon that is dependent on growth rate probably does not involve the antitermination mechanism (R. Gourse and M. Nomura, unpublished data). The question whether the inhibition is caused directly by free ribosomes or indirectly by some other effectors is also not known. Efforts in our laboratory to show direct inhibitory effects of ribosomes on transcription from rRNA promoters in vitro have been negative so far. Thus, the real repressor could be a derivative of ribosomes that is always in equilibrium with free ribosomes. Alternatively, it is possible that the apparent feedback inhibition by free ribosomes is achieved indirectly; e.g., the increase in the concentration of free ribosomes somehow might increase the concentration of other effectors that act directly on rRNA gene transcription. For example, guanosine tetraphosphate, a long suspected and yet unproven inhibitor of rRNA synthesis, could be such an effector.

GROWTH-RATE-DEPENDENT REGULATION OF r-PROTEIN GENE TRANSCRIPTION

As discussed above, the synthesis of r-protein mRNA is usually in excess, and the rate of r-protein synthesis is adjusted at the level of translation to balance with the rate of rRNA synthesis. However, this does not mean that there is no transcriptional regulation of r-protein gene expression. In fact, the synthesis rate of r-protein mRNA (specifically the spc- and the alpha-operon mRNA) increases with increasing growth rate; the mRNA synthesis rate is approximately proportional to the growth rate rather than to the square of the growth rate (Gausing, 1977; Miura et al., 1981). Thus, one can ask about possible mechanisms for this growth-rate-dependent alteration of r-protein mRNA synthesis. Although detailed discussions (see Nomura et al., 1984) are not repeated here, we suggest the possibility that the feedback mechanism by free ribosomes acts not only on the transcription of rRNA and tRNA, but probably also on transcription from r-protein promoters, though to a weaker extent than that seen for rRNA and tRNA operons. Several observations support this suggestion. First, stimulation of transcription of r-protein genes was observed under conditions in which the assembly of ribosome is preferentially inhibited; namely, (1) in cold-sensitive strains defective in ribosome assembly at lower temperatures (A. Miura, Y. Takebe, S. Jinks-Robertson, and M. Nomura, unpublished), and (2) in cells in which the assembly of ribosomes is inhibited by overproduced repressor r-proteins (Takebe et al, 1985). Second, as for rRNA, stimulation was also observed in cells treated with chloramphenicol (Dennis, 1976). Analysis of the rates of synthesis of r-protein mRNA in the strain carrying extra intact rRNA operons showed a reduction in the synthesis rates of r-protein mRNA, but the degree of reduction was smaller than that observed in the rates of synthesis of rRNA (M. Tam and M. Nomura, unpublished). Of course, the degree of the negative feedback regulation, if it really exists as we suggest, should be weaker than that for rRNA transcription in order to allow the translational feedback regulation to operate. It seems to be reasonable to have such a dual mechanism: feedback regulation by free ribosomes at the transcription step for coarse regulation, and translational feedback regulation by repressor r-proteins as the final precise regulation. In this connection, it should be noted that for the regulation of the synthesis of r-proteins in the S10 operon, two different mechanisms have been discovered; one is an attenuation mechanism at the transcription step (Lindahl et al., 1983) and the other is the translational feedback mechanism by the repressor L4 (Yates and Nomura, 1980). In analogy to the above discussion, it is possible that the attenuation mechanism and the translational feedback mechanism could operate simultaneously, with the translational mechanism contributing the final adjustment at the last step of regulation.

CONCLUDING REMARKS

The ribosome feedback regulation model discussed in this article has now a considerable amount of experimental support. In addition, it can explain various known observations in the regulation of ribosome biosynthesis. Although the model is different from the passive regulation model proposed by Maaløe (1969, 1979, 1983), there is a similarity between the two; both consider the regulation of ribosome synthesis in the context of the expression of the entire cellular complement of genes. Control of ribosome biosynthesis cannot be considered separately from protein synthesis *in toto,* the major growth process of the cell. The passive regulation model has been very stimulating to the field in general and to me in particular. My own view on the control of ribosome synthesis described here is a result of continued efforts in my laboratory to answer the questions of control of ribosome biosynthesis posed by Maaløe and his coworkers in Denmark many years ago.

ACKNOWLEDGEMENTS

I thank my present, as well as my previous, coworkers, who have participated in the work described in this article. I also thank Drs. Richard L. Gourse, Carol Gross and Millard Susman for their useful comments on the manuscript. The work in this laboratory was supported by grants from the National Institutes of Health (GM-20427) and from the National Science Foundation (PCM 79-10616), and by the College of Agriculture and Life Sciences, University of Wisconsin, Madison. This article is paper number 2750 from the Laboratory of Genetics, University of Wisconsin, Madison, Wisconsin 53706.

REFERENCES

Alton, T. H. and Koch, A. L. (1974). *J. Mol. Biol.*, 86, 1.

Baughman, G. and Nomura, M. (1983). *Cell*, 34, 979.

Baughman, G. and Nomura, M. (1984). *Proc. Natl. Acad. Sci. USA*, 81, 5389.

Bennett, P. M. and Maaløe, O. (1974). *J. Mol. Biol.*, 90, 541.

Buckel, P., Ruffler, D., Piepersberg, W. and Böck, A. (1972. *Mol. Gen.Genet.*, 119, 323.

Comer, M. M. (1982). *Mol. Gen. Genet.*, 187, 132.

Dean, D. and Nomura, M. (1980). *Proc. Natl. Acad. Sci. USA*, 77, 3590.

Dennis, P. P. (1976). *J. Mol. Biol.*, 108, 535.

Dennis, P. P. (1977). *J. Mol. Biol.*, 115, 603.

Dennis, P. P. and Nomura, M. (1975). *J. Mol. Biol.*, 97, 61.

Gausing, K. (1977). *J. Mol. Biol.*, 115, 335.

Gourse, R.L. and Nomura, M. (1984). *J. Bact.*, 160, 1022.

Gourse, R.L., Takebe, Y., Sharrock, R.A., and Nomura, M. (1985). *Proc. Nat. Acad. Sci. USA*, 82, 1069.

Hills, C. W., Squires, C., and Carbon, J. (1970). *J. Mol. Biol.*, 52, 557.

Ikemura, T. and Ozeki, H. (1977). *J. Mol. Biol.*, 117, 419.

Ingraham, J. L., Maaløe, O., and Neidhardt, F. C. (1983). *Growth of the Bacterial Cell.* Sinauer Associates.

Jinks-Robertson, S., Gourse, R. L., and Nomura, M. (1983). *Cell*, 33, 865.

Koch, A. L. (1971). *Adv. Microb. Physiol.*, 6, 147.

Kurland, C. G. and Maaløe, O. (1962). *J. Mol. Biol.*, 4, 193.

Li, S.C., Squires, C.L., and Squires, C. (1984). *Cell*, 38, 851.

Lindahl, L., Archer, R., and Zengel, J. M. (1983). *Cell*, 33, 241.

Maaløe, O. and Kjeldgaard, N. O. (1966). *Control of Macromolecular Synthesis*, Benjamin.

Maaløe, O. (1969). *Dev. Biol.*, Suppl. 3, 33.

Maaløe, O. (1979). In *Biological Regulation and Development*. R. F. Goldberger (ed.), Plenum, p. 487.

Maaløe, O. (1983). In *Growth of the Bacterial Cell*. J. L. Ingraham, O. Maaløe, and F.C. Neidhardt, Chapter 2, p. 349. Sinauer Associates.

Miura, A., Krueger, J. H., Itoh, S., deBoer, H. A., and Nomura, M. (1981). *Cell*, 25 773.

Morgan, E. A. (1980). *Cell*, 21, 257.

Nomura, M., Jinks-Robertson, S., and Miura, A. (1982). In *Interaction of Translational and Transcriptional Controls in the Regulation of Gene Expression*. M. Grunberg-Manago and B. Safer (eds.), Elsevier, p. 91.

Nomura, M., Gourse, R., and Baughman, G. (1984). *Ann. Rev. Biochem.*, 53, 75.

Norris. T. E. and Koch, A. L. (1972). *J. Mol. Biol.*, 64, 633.

O'Farrell, P. H. (1978). *Cell*, 14, 545.

Sarmientos, P. and Cashel, M. (1983). *Proc. Natl. Acad. Sci. USA*, 80, 7010.

Shen, V. and Bremer, H. (1977). *J. Bacteriol.*, 130, 1098.

Siehnel, R. J. and Morgan, E. A. (1983). *J. Bacteriol.*, 153, 672.

Tai, P.-C., Kessler, D. P., and Ingraham, J. (1969). *J. Bacteriol.*, 97, 1298.

Takebe, Y., Miura, A., Bedwell, D.M., Tam, M., and Nomura, M. (1985). *J. Mol. Biol.*, 183, in press.

Yates, J. L. and Nomura, M. (1980). *Cell*, 21.,517.

The Role of Antitermination in the Control of *Escherichia coli* Ribosomal RNA Expression

Catherine L. Squires, Suzanne Li, Serap Aksoy, and Craig Squires
Columbia University, New York

It has been known for many years that the expression of ribosomal RNA (*rrn*) operons in *E. coli* is regulated as a function of the cellular growth rate and the stringent response (overproduction of the tetraphosphate nucleotide, ppGpp), and is tightly coordinated with ribosome biosynthesis (Gausing, 1980; Maaløe and Kjeldgaard, 1966). Recent studies have led to several important ideas about the molecular mechanisms involved in *rrn* expression. These include a model for feedback inhibition in which free ribosomes are responsible for the growth rate dependent regulation of ribosomal RNA (Jinks-Robertson, et al., 1983 and 1984), description of an antitermination system operating in *rrn* transcription (Brewster and Morgan, 1981; Aksoy, et al., 1984, and Li et al., 1984), and more detailed information about the interaction of ppGpp with the individual ribosomal promoters (Sarmientos, et al., 1983).

In formulating the model for feedback inhibition by free ribosomes, Nomura and coworkers examined the production of rRNA and tRNA genes in cells which harbor, on a plasmid, either an intact *rrn* operon or an operon in which deletions have been made. In these strains, the synthesis rate of rRNA is not significantly affected by an increase in gene dosage, thus excess rRNA is not produced. Cellular tRNAs (both *rrn* and non-*rrn* encoded species) are depressed 30–55 percent when an entire *rrn* operon is present on a multiple copy number plasmid. The depression is eliminated by deleting portions of the *rrn* operon on the plasmid. The conclusion drawn is that intact ribosomes are needed for the regulatory function, controlling levels of cellular tRNAs and autogenously controlling rRNA products (Jinks-Robertson et al., 1983 and 1984).

Evidence that an antitermination mechanism is involved in normal *rrn* transcription was first supplied by Morgan and coworkers who showed that several transposons and insertion elements are not completely polar when inserted into the *rrnC* operon (Brewster and Morgan, 1981; Siehnel and Morgan, 1983). Independent demonstration of antitermination by the *rrnG* control region was done using an *rrn-trp-lac* fusion system (Aksoy et al., 1984). The *rrn* antitermination system has been localized to a short DNA segment just after the P2 promoter (Li et al., 1984). This aspect of *rrn* expression, though initially unexpected, provides an explanation for the lack of polarity in *rrn* and perhaps other nontranslated operons in *E. coli.*

Several laboratories have made fusion plasmids carrying parts of *rrn* operons to study important aspects of *rrn* regulation. Cashel and coworkers found that of the two *rrn* promoters, only P1 is sensitive to ppGpp inhibition (Sarmientos et al., 1983). They also found a difference between P1 and P2 in binding of RNA polymerase and differences in response of the two promoters under fast growth and stationary phase conditions (Glaser et al., 1983; Sarmientos et al., 1983). Stark et al. (1983) have used *in vitro* mutagenesis of plasmids carrying the *rrnB* operon to study transcription and processing of mutant *rrn* operons in maxicell experiments. These experiments give some insights into the role that post-transcriptional events play in the overall rate of synthesis of mature ribosomal RNAs.

The mechanism of action of ppGpp on *rrn* expression is still not understood. It is clear that the rate of synthesis of stable RNA species is usually inversely related to the concentration of ppGpp in the cell (Ryals et al., 1982), but the details of how this occurs are not known. The main suggestions as to how ppGpp inhibits stable RNA synthesis have centered on inhibition of RNA polymerase initiation (Travers, 1980; Oostra et al., 1977, and Kingston et al., 1981). Kingston and Chamberlin (1981) have suggested that another effect of ppGpp is on elongation of transcripts, but this has only been shown *in vitro.* Other experiments by Kingston (1983) place the site of action of ppGpp just downstream of P2, and suggest that both promoters are affected by its presence. Sarimientos et al. (1983) find only P1 to be sensitive to ppGpp. The importance of ppGpp to *rrn* expression can be emphasized by the· fact that one group of researchers conclude that stable RNA genes are not controlled by factors other than the ppGpp-mediated system (Ryals et al., 1982). However, much more needs to be understood about the expression of stable RNA species and the role of ppGpp in order to support that conclusion, and it is most likely that ppGpp is just one of a number of regulatory features affecting the expression of stable RNAs.

DESCRIPTION OF ANTITERMINATION

Antitermination is a facet of transcription that has been studied extensively in the lambdoid bacteriophages and has been implicated in the regulation of several *E. coli* operons. In lambda, the transcription complex can be altered at specific sites on the genome such that it subsequently reads through otherwise efficient terminators. This read-through is essential for the expression of gene products during phage development, regulating the switch from early to late genes. After infection, two major promoters function to make short transcripts. These are the *pL* and *pR* promoters. Short messages, coding for one gene, are made on each strand and then terminated at the left (*tL*) and right (*tR*) terminators. This termination is dependent upon the host transcription-termination factor Rho. When the left message is translated into the N protein, this protein plus host factors called Nus proteins, interact at specific regions on the lambda DNA, called the *nut* loci, and alter the transcription complex such that it no longer terminates transcription at *tL* or *tR*. Thus, genes needed for further lambda development are now able to be transcribed and translated (Friedman and Gottesman, 1983).

The detailed molecular mechanism by which the transcription complex is altered is unknown. The lambda *nut* regions are composed of three features; box A, box B (a region of dyad symmetry) and box C. It has recently been shown that one of the Nus proteins, NusA, requires box A for its activity (Friedman and Olson, 1983). Box B has been shown to be required for N protein activity (Salstrom and Szybalski, 1978) and box C has been derived by comparing the sequences of five different *nut*-like loci (Friedman and Gottesman, 1983). Another important fact about the lambda antitermination system is that translation affects antitermination. Olson et al. (1983) sequenced three lambda *nutR* mutants that were deficient in antitermination. All three were deletions in the *cro* gene just preceding *nutR*. The mutations caused a frameshift at the end of the *cro* gene such that translation terminated very close to *nutR*. They concluded from these studies that *nut* sites can be rendered inactive by translation in their vicinity. To summarize antitermination in lambda: the N protein, host Nus factors and a *nut* region are required to convert a "regular" transcription complex into a "antitermination" transcription complex. The lambda system provides the background for the paradigm upon which thinking about antitermination systems in *E. coli* is based.

ANTITERMINATION STUDIES WITH THE *rrn* OPERONS

The stable RNA operons of *E. coli* do not exhibit polarity even though they make an RNA product that is not translated (Morgan and Nomura, 1979). By contrast, most *E. coli* operons that specify proteins exhibit polarity if their translation is interrupted. Polarity is defined as the decreased expression of a distal gene (or sequences) of an operon resulting from specific genetic signals in proximal sequences. There are both translational (Oppenheim and Yanofsky, 1980; Aksoy et al., 1984) and transcriptional (Adhya and Gottesman, 1978) components of polarity. The transcriptional component requires Rho protein and exposed messenger RNA. In addition, it is thought that specific Rho-protein binding sites and RNA polymerase pause sites on the exposed mRNA segment are required for premature transcription termination (Adhya and Gottesman, 1978). The absence of transcriptional polarity in stable RNA genes might be caused either by factors intrinsic to the structure of the stable RNA (e. g., absence of Rho-specific termination sites) or by a system similar to bacteriophage lambda *N*-gene antitermination.

To determine whether transcription termination signals are missing from the *rrn* structural genes, or, if something unique about the *rrn* transcription complexes renders them able to ignore termination signals, transcription from the *ara* (arabinose operon) and the *rrnG* promoters through a segment of 16S gene DNA was compared (Aksoy et al., 1984). In each case, the promoter-16S DNA sequences were joined to a *trp-lac* fusion, and *lacZ* mRNA was examined in wildtype (rho⁺) and Rho-protein mutant (*rho115*) backgrounds. This system showed significant Rho-dependent termination when the *16S-trp-lac* sequences are transcribed from the *ara* promoter. However, Rho protein has little or no effect when the same *16S-trp-lac* sequences are transcribed from the *rrnG* promoter. These results provided evidence that segments of rRNA are not intrinsically resistant to Rho-mediated termination and suggested that an antitermination system is involved in transcription of the *rrnG* operon (Aksoy et al., 1984).

Next, an experimental system that examined the ability of specific areas of the *rrnG* control region to convert an ordinary transcription complex into an antitermination transcription complex was employed. The plasmid pKK232-8, made by Jurgen Brosius (1984), was used as a modular system to insert different promoter fragments, putative antiterminator fragments and a terminator. Plasmid pKK232-8 contains the chloramphenicol acetyltransferase (*cat*) and beta-lactamase (*bla*) genes and a multilinker site at the proximal end of the *cat* gene. Combinations of promoters and putative antiterminator sequences, inserted into proximal sequences, were tested for their ability to transcribe through a terminator inserted into a distal site. *In vivo cat* expression was used to measure the amount of transcription that reads through the terminator in these plasmids. The assay system measures chloramphenicol acetyltransferase (Cat) and beta-lactamase (Bla) activities, which are then expressed as the ratio of Cat to Bla units in the cell extracts. The beta-lactamase serves as an internal control for variations in plasmid copy number and other factors unrelated to antitermination.

Three promoters were used: the *rrnG* P2 promoter, the *lac* operon promoter and the *tac* (a hybrid *trp-lac*) promoter. As antitermination fragments, various pieces of the *rrnG* control region were used. As a terminator, the same segment of the 16S gene (used previously with the arabinose promoter, (Aksoy et al., 1984)), except in the reverse orientation, was employed. It terminates transcription more efficiently in the reverse orientation than in the forward direction.

The expected result with this system is for transcription from the promoters inserted to be efficiently stopped by the terminator unless a fragment conferring antitermination ability upon the transcription complex has been inserted. In that case, *cat* mRNA and protein should be produced, and the Cat-to-Bla ratio will rise.

In choosing potential antitermination fragments, we concentrated on the region between the P2 promoter and the start of the 16S gene. This region contains sequences similar to box A and box C of the lambdoid phages (all *rrn* operons sequenced have identical box A and box C regions), and before the box A sequence, there is a region of dyad symmetry resembling a box B structure. Also, Kingston and Chamberlin (1981) found that when they transcribed the *rrnB* promoter region *in vitro*, termination of transcription after box C dependent on the NusA protein was observed. They refer to this as "turnstile attenuation". The presence of the *nut*-like locus and the involvement of NusA in this area made it the most likely region in the *rrn* control sequence to be required for antitermination.

Using the fusion plasmids, the antitermination system involved in ribosomal RNA transcription was localized and compared with antitermination in the lambdoid bacteriophages (Li et al., 1984). A 67-base-pair restriction fragment immediately following the *rrnG* ribosomal RNA operon P2 promoter, containing sequences similar to the box B, box A and box C found in the lambda *nut* loci, decreases transcription termination by about 50 percent. Fragments following this region are not necessary for antitermination. In addition to conferring antitermination properties on transcripts initiated at the *rrnG* P2 promoter, the 67-bp region also causes transcripts from *lac* and hybrid *trp-lac* promoters to read through a transcription terminator. Translation of the box-A box-C-like sequences results in the loss of antitermination activity. This suggests that ribosomes interfere with antitermination if they encroach on the *nut*-like sequences in the *rrn* transcripts and thus that the transcript is involved in the antiterminator mechanism. Translation of half of the box-B-like sequence does not alter antitermination. Inversion of the 67-bp segment with respect to the *rrnG* P2 promoter also results in loss of antitermination. We have concluded that the *E. coli* ribosomal RNA operons possess an antitermination system that is both structurally and functionally similar in many of its features to the antitermination system used by the bacteriophage lambda (Li et al., 1984).

It is interesting to consider the significance of *rrn* antitermination to the regulation of expression of these operons. It may be necessary to have an antitermination system in *rrn* operons simply to avoid Rho-dependent termination. Alternatively, antitermination may provide another means of control by regulating the number of polymerase molecules that transcribe the entire operon. According to this idea, transcription complexes that are not converted to the antitermination mode would have a greater likelihood of stopping early at terminators that are present in the 16S genes. We have only examined antitermination under rich-medium growth conditions. Different growth conditions may affect the level of antitermination and play a role in the modulation of *rrn* expression.

ACKNOWLEDGEMENTS

We would like to thank Olaf Nielsen for reading the manuscript.

REFERENCES

Adhya, S., and Gottesman, M. (1978). *Ann. Rev. Biochem.*, 47, 967.

Aksoy, S., Squires, C. L., and Squires, C. (1984). *J. Bacteriol.*, 157, 363.

Aksoy, S., Squires, C. L., and Squires, C. (1984). *J. Bacteriol.*, 159, 260.

Brewster, J., and Morgan, E. (1981). *J. Bacteriol.*, 148, 897.

Brosius, J. (1984). *Gene,* 27, 151.

Friedman, D., and Gottesman, M. (1983). In *Lambda II*. R. Hendrix,

J. Roberts, F. Stahl, and R. Weisberg (eds.), Cold Spring Harbor Laboratory. pp. 21–52.

Friedman, D., and Olson, E. (1983). *Cell*, 34, 143.

Gausing, K. (1980). In *Ribosomes: Structure, Function and Genetics*. G. Chambliss, G. Craven, J. Davies, K. Davis, L. Kahn and M. Nomura (eds.). University Park Press. pp. 693–718.

Jinks-Robertson, S., Grouse, R., and Nomura, M. (1983). *Cell*, 33, 865.

Jinks-Robertson, S., Baughman, G., and Nomura, M. (1984). In *Gene Expression*. Alfred Benzon Symposium 19, B. Clark and H. Peterson (eds.). Munksgaard, Copenhagen.

Kingston, R., and Chamberlin, M. (1981) *Cell*, 27, 523.

Kingston, R. (1983). *Biochemistry*, 22, 5249.

Li, S., Squires, C. L., and Squires, C. (1984). *Cell*, 38, 851

Morgan, E., and Nomura, M. (1979). *J. Bacteriol.*, 137, 507.

Olson, E., Flamm, E., and Friedman, D. (1983). *Cell*, 31, 61.

Oostra, B., van Ooyen, A., and Gruber, M. (1977). *Mol. Gen. Genetics*, 152, 1.

Oppenheim, D., and Yanofsky, C. (1980). *Genetics*, 95, 785.

Ryals, J., Little, R., and Bremer, H. (1982). *J. Bacteriol.*, 151, 1261.

Salstrom, J., and Szybalski, W. (1978). *J. Mol. Biol.*, 124, 195.

Sarmientos, P., Sylvester, J., Contente, S., and Cashel, M. (1983). *Cell*, 32, 1337.

Sarmientos, P., and Cashel, M. (1983). *Proc. Natl. Acad. Sci. USA*, 80, 7010.

Siehnel, R., and Morgan, E. (1983). *J. Bacteriol.*, 153, 672.

Stark, M. Gourse, R., and Dahlberg, A. (1982). *J. Mol. Biol.*, 159, 417.

Travers, A. (1980). *J. Bacteriol.*, 141, 973.

Alternate mRNA Secondary Structures as a Mechanism of Translation-Inhibition

Linda Mamelak*, Tove Christensen**, Niels Fiil**, James D. Friesen*, and Morten Johnsen[†]

 * *University of Toronto, Ontario*
 ** *NOVO Research Institute; Bagsvaerd, Denmark*
 † *University of Copenhagen, Denmark*

Autoregulation is one of the biosynthetic mechanisms involved in the balanced production of ribosomal RNA (rRNA) and of ribosomal proteins (r-proteins) in *Escherichia coli*. In growth conditions in which there is a surplus of r-protein mRNA compared to the amount of rRNA present, certain r-proteins act as inhibitors for further synthesis of themselves as well as of other r-proteins (see review by Nomura et al., 1982). The molecular mechanism that drives this regulation is thought to be a competition between binding of the regulatory r-protein to the rRNA and to its own polycistronic mRNA.

This model lends itself to experimental verification. It should be possible to demonstrate *in vitro* the specific binding of a regulatory r-protein to its own mRNA and this binding should be inhibited in the presence of rRNA. Moreover, it should be possible to isolate mutants in which either the binding site on the mRNA or the regulatory protein is altered, so that the regulation is rendered nonfunctional. These mutations would be comparable to the *lacO*c and the *lacI*- mutations, respectively.

We have studied the regulation of ribosomal proteins L10 and L12, whose genes, *rplJ* and *rplL,* are organized in a transcriptional unit that also contains the *rpoB* and *rpoC* genes for the subunits of RNA polymerase. The regulatory protein in this operon is L10, possibly in a complex with L7/12. This complex (sometimes referred to as L8) is stable and forms part of the structure of the *E. coli* ribosome (Liljas and Petterson, 1979). We have demonstrated both binding of the regulatory r-protein (Johnsen et al., 1982) and the existence of classes of predicted mutants (Friesen et al., 1983). Here we intend to review this evidence very briefly and restate the model of translational regulation of L10 synthesis that we published recently (Christensen et al., 1984). We then present additional evidence in its favor and discuss some further implications of it.

The model was devised to account for the following observations: (1) L10-L7/12 binds stably to the *rplJ* leader of mRNA in a specific location, some 180 nucleotides upstream from the translation initiation site of *rplJ* (Johnsen et al., 1982). (2) There are two point mutations in the leader mRNA that fail to register feedback (Friesen et al., 1983); one of these mutations completely abolishes formation of a stable protein-mRNA complex, although a second retains significant binding activity of the mRNA (Christensen et al., 1984). (3) There are six point mutations, lying 80–200 nucleotides upstream from the translation initiation site of *rplJ*, that reduce the translation efficiency of *rplJ* (Fiil et al., 1980). (4) A 96-nucleotide deletion whose rightward endpoint lies upstream from the translation initiation site of *rplJ* completely abolishes *rplJ* expression (Johnsen et al., 1982, Friesen et al., 1983).

In order to account for these observations, we propose the following hypothesis. The portion of *rplJ* leader mRNA that lies between nucleotides 1505 and 1721 (the start codon of *rplJ*) exists normally in a configuration that we call Form I (Figure 1, left). In this form the region of the *rplJ* ribosome binding site (Shine and Dalgarno, 1974) is not base-paired. and hence *rplJ* is open for translation. Form I is stabilized by stems A, B and C (Figure 1, left), which we suggest are recognized by L10-L7/12 and to which the protein complex might bind (see below). The alternate configuration we call Form II (Figure 1, right). Stems A and B of Form I, or stems A and B' of Form II comprise the stable binding site of L10-L7/12, since they are protected *in vitro* by L10-L7/12 (Johnsen et al., 1982). The transition between forms I and II is brought about by a shift in base-pairing (Figure 1, lower) such that nucleotides 1584 to 1589, which formerly were in stem C of Form I, now base-pair with nucleotides 1537 to 1541 in stem B' of Form II. Nucleotides 1627 to 1638 in stem C of Form I base-pair with nucleotides 1709 to 1718 in the mRNA region near the ribosome-binding site of *rplJ* to sequester it and to reduce *rplJ* translation. This is the mechanism whereby binding of L10-L7/12 in the central region of the *rplJ* leader signals the inhibition of *rplJ* translation. Below we discuss possible mechanisms for the transition between Forms I and II.

The phenotype of the 96-nucleotide deletion that completely removes stems A and B (or B') is particularly interesting. This deletion entirely abolishes *rplJ* expression *in vivo* (Friesen et al., 1983), as might be expected, since the structure of the remaining leader RNA is strongly in favor of the nonexpressing configuration, which has base-pairing in the region of the Shine-Dalgarno sequence (Figure 2D). If our idea regarding the importance of the base-pairing with the Shine-Dalgarno sequence is correct, then we predict that one ought to be able to isolate phenotypic revertants of the 96-nucleotide deletion that have regained expression of *rplJ* because of weakening of this base-pairing. Such revertants are quite easy to obtain since we have the deletion associated with a *rplJ-lacZ* fusion on a plasmid (Friesen et al., 1983); cells with this plasmid are unable to grow on lactose as sole carbon source. Thus, it is straightforward to isolate variants capable of growth on lactose. (We found that this

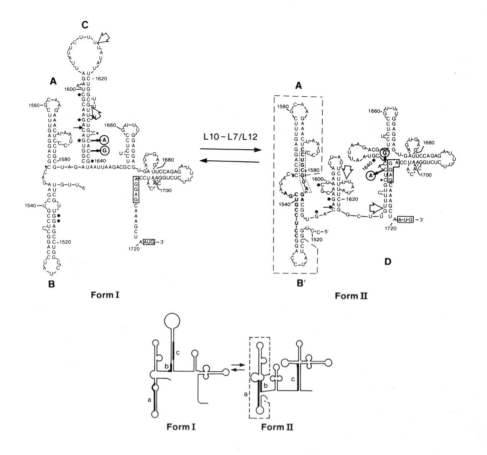

FIGURE 1 Upper panel. Possible secondary structures of the rplJ leader mRNA. Base numbers are as in Post et al., (1979). Mutations that inhibit translation of rplJ (asterisks) or escape the translational inhibition (solid circles) are indicated. The rightward endpoint of a 96-nucleotide deletion (the leftward endpoint does not appear in the diagram) is shown by an arrow between nucleotides 1592 and 1593. In Form II, a 7-base homology to 23S rRNA is shown in a dashed box, and a direct repeat (8 of 10 nucleotides) is shown in boldface. Stem D of Form II blocks the rplJ ribosome-binding site; the Shine-Dalgarno sequence and the initiator codon are enclosed in a box. Lower panel. A diagram showing the main features of Forms I and II. The thick lines labelled a, b, and c indicate the regions in which base-pairing is shifted by binding of L10-L7/L12. For both the upper and lower panels the large dashed box in Form II encloses the area that is protected in vitro by L10-L7/L12. The encircled nucleotides in Stem D of Form II (or Stem C of Form I) are mutations that restore the expression of rplJ that had been abolished by the 96-nucleotide deletion (see text). A short deletion between nucleotides 1613 and 1627 that similarly restores expression of rplJ is indicated by open arrows.

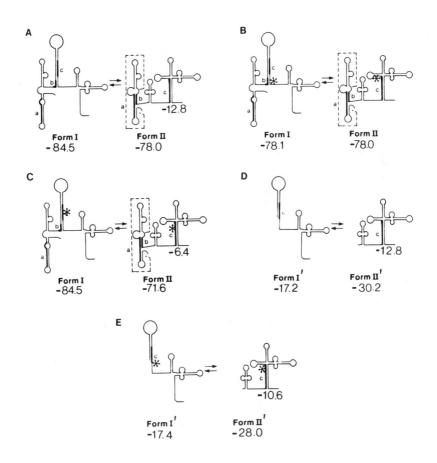

FIGURE 2 Calculated mean free energies (according to Tinoco et al., 1973) of Form I and Form II for the wildtype <u>rplJ</u> leader and selected mutants. (A) Wildtype configurations. (B) A single base change at nucleotide 1640 that reduces <u>rplJ</u> expression. (C) A single base change at nucleotide 1634 that abolishes <u>rplJ</u> feedback. (D) A 96-nucleotide deletion that abolishes <u>rplJ</u> expression. (E) A single base change at nucleotide 1636 that partially reverses the effect of the 96-nucleotide deletion.

must be carried out in a *recA* strain; otherwise a frequent event occurred in which apparently the deleted *rplJ* leader region carried on the plasmid was replaced by the normal region from the chromosome by a "gene conversion."). Eleven mutants were isolated, of which three have been characterized. Their phenotypes are given in Table I. The activity of these mutants has been restored to between 50 and 75% of wildtype, and they respond very poorly to feedback. Analysis of the nucleotide sequence indicated that all three mutations lie in the region that is capable of forming a base pair with the Shine-

TABLE 1   Revertants of $\Delta 96$.

Lac fusion plasmid	Mutations	β-Galactosidase Activity	
		pBR322	pNF1344
pGA189	Wildtype	123	13
pJF3359	$\Delta 96$	3	3
pJF4149	$\Delta 96$; 1614–1627	72	65
pJF4150	$\Delta 96$; C→A 1636	82	75
pJF4154	$\Delta 96$: C→G 1638	76	58

Dalgarno sequence in Form II (Figure 1); in all three cases the base-pairing is weakened, either through point mutations that destroy CG base pairs at the top of the stem, or by a deletion that weakens base pairing at its base. It is to be expected that feedback inhibition in these mutants would be reduced since: (a) they cannot base-pair efficiently with the Shine–Dalgarno region and, (b) they do not have the binding site of the L10–L7/12 complex. The characteristics of these revertants strongly support the idea, proposed above, that base-pairing with the Shine-Dalgarno region is important in the determination of *rplJ* activity; they also are consistent with the observation that the binding site for the regulatory protein lies in a region well-separated from the site for initiation of translation.

Mean free energies of the base-paired regions of the *rplJ* leader, including approximately 200 nucleotides upstream of the *rplJ* translation start site, can be calculated from the empirical values of Tinoco et al.(1973). Although these calculations will not result in the true absolute values of the free energies, they will presumably reflect differences in stabilities of the structures. The calculated values are shown in Figure 2 for the wildtype configuration and those of several mutants. We stress that the alternative configurations (Form I and Form II) shown in Figures 1 and 2 are only two of many that can empirically be devised from the sequence. However, it is important to note that other configurations that satisfy the dual criteria of providing: (a) base-pairing and nonbase-pairing in the region of the Shine–Dalgarno sequence and (b) highly stable base-paired structures in other regions of the mRNA leader differ in detail rather than substance from the scheme we have proposed. While it is quite likely that details of the structure that we have suggested are subject to modification, we feel confident in the general picture. We note that the calculated difference in free energies between wildtype Forms I and II is about -6.5 kcal/mole, yielding a dissociation constant of about 10^{-6}. The strength of the base-pairing that includes the Shine–Dalgarno region (-13.2 kcal/mole) is equal to the binding energy (-13.2 kcal/mole) of the initiation sequence (CCUG) on the 3' end of 16S ribosomal RNA. Thus, it is quite plausible that the formation of stem D of Form II could compete with base-pairing of 16S rRNA to reduce translation initiation frequency of *rplJ*.

We can analyze mutations in the leader mRNA in light of the calculated values of the mean free energy. The mutations that reduce translation of *rplJ* (indicated by asterisks in Figure 1) generally weaken Form I with little effect on Form II; this results in nearly equal free energies in the two configurations and thus will shift the equilibrium towards Form II, that is, no translation of *rplJ*.

One of the mutations that abolishes feedback inhibition (indicated by a dot at nucleotide 1634 in Figure 1) but retains some ability to bind the L10–L7/12 complex (Christensen et al., 1984) reduces the free energy of base-pairing with the Shine–Dalgarno sequence nearly two-fold, thus rendering translation permanently on, whether L10–L7/12 is bound or not.

As noted above, a 96–nucleotide deletion removes stable portions of the mRNA structure; the remaining structure is overall much less stable than the wildtype and is strongly tilted in favor of the closed configuration (Figure 2D). As expected, second-site revertants of this deletion have reduced stabilities of the Shine–Dalgarno base-pairing (Figure 2E), resulting in 50–80% restoration of *rplJ-lacZ* expression. The calculated reduction in mean free energy owing to the second-site revertants is relatively small (about –2.2 kcal/mole).

How can the L10–L7/12 complex influence the RNA configuration of the *rplJ* leader region? We suggest three possibilities (Figure 3). First, Form I and Form II might be in spontaneous equilibrium (Figure 3A). A calculation of the equilibrium constant for this indicates a value of approximately 10^{-6}, meaning that under undisturbed conditions, 1 in 10^6 mRNA molecules is in the closed state. In this model, we suppose that L10–L7/12 binds only to Form II and freezes these molecules in the closed state. For this model to merit serious consideration, the time taken for the molecules to convert between Forms I and II must not exceed a few minutes in order that the regulatory mechanism react reasonably swiftly to changes in growth conditions. At the present time we do not know if this is a reasonable proposition.

The second possibility is that L10–L7/12 binds to Form I, reducing its free energy so that Form II becomes favored (Figure 3B). The binding energy of bacteriophage R17 coat protein to the replicase inhibitor site (–11 kcal/mole; Carey and Uhlenbeck, 1983) might afford a reasonable comparison to the present system. If a similar energy for binding of L10–L7/12 to stems A and B of Form I were available, then the base-pairs in stem C could be melted to reanneal as Form II. When the concentration of L10–L7/12 falls below the dissociation constant for protein–RNA binding, the protein dissociates from the mRNA leader and spontaneous reversion to the energetically favored Form I occurs. Note that in this model, the effect of L10–L7/12 is not catalytic: (i) it acts stoichiometrically, and (ii) it drives the reaction away from the equilibrium state. An evaluation of the likelihood of this model must await a measurement of the protein–RNA binding energy of this system. However, we note that the minimum binding energy that would suffice to promote conversion of the leader mRNA to Form II is approximately –6.5 kcal/mole; this corresponds to a K_d of about 10^{-6} M; preliminary determinations of the binding constant indicate a value in this range (Tove Christensen, unpublished).

The third possible model is that only newly synthesized mRNA is capable of assuming Form II (Figure 3C). Since the RNA chain grows from the 5' end, the first structures formed are stems A and B. This event could be followed immediately by the binding of L10–L7/12 to it, which could sterically hinder formation of stem C, but

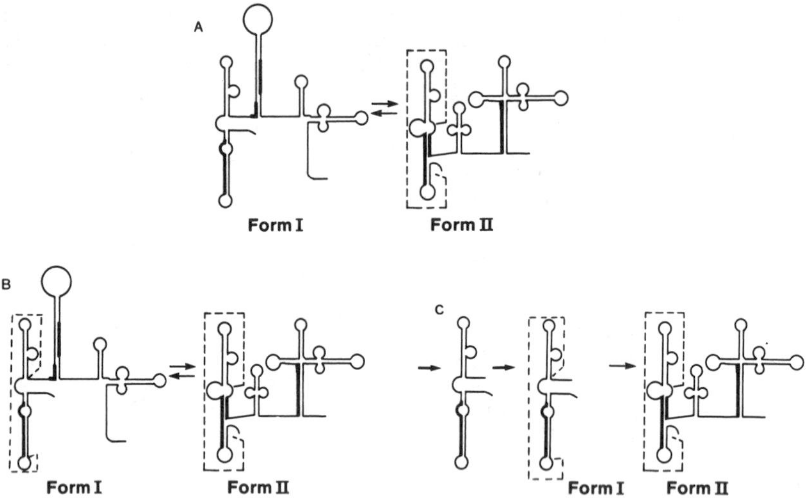

FIGURE 3 Three possible models for the influences of L10-L7/L12 on the transition between Forms I and II. The dashed boxes represent the site of L10-L7/L12 binding. See text for further explanation.

not of stem D and the remainder of Form II. A possible problem with this model is that *in vitro* studies show clearly that mRNA whose synthesis is complete can be inhibited by added L10 (Brot et al., 1980).

We do not yet have sufficient data to choose among these three possibilities.

We note that the mechanism that we are proposing for the regulation of *rplJ* is much more complex than has been proposed for several other r-protein operons (Nomura et al., 1982). In those cases it is suggested that the regulator protein binds to its target structure on the mRNA leader and sterically hinders loading of ribosomes of the translation initiation site. One significant difference between the *rplJ-rplL* operon and all other r-protein operons is that four copies of L7/12 are produced for every one of L10. It is possible that the elaborate structure of the *rplJ* mRNA leader is involved in setting this ratio of gene activities, though it is not obvious how the secondary structure would influence this.

Finally, we have not taken into account possible tertiary interactions, nor, at our current state of understanding, can we do so. We cannot confidently rule out the possibility that the phenotype of some of our mutants can be accounted for by higher-order nucleotide interactions.

ACKNOWLEDGEMENTS

This work was supported by the National Cancer Institute of Canada, the Danish Natural Science Research Council (grants 11-3771 and 11-4305), and the NOVO Foundation.

REFERENCES

Brot, N., Caldwell, P. and Weissbach, H. (1980). *Proc. Natl. Acad. Sci. USA,* 77, 2592.

Carey, J. and Uhlenback, O. C. (1983). *Biochemistry,* 22, 2610.

Christensen, T., Johnsen, M., Fiil, N. P., and Friesen, J. D. (1984). *EMBO J.,* 3, 1609.

Fiil, N. P., Friesen, J. D., Downing, W. L., and Dennis P. P. (1980). *Cell,* 19, 837.

Friesen, J. D., Tropak, M., and An, G. (1983). *Cell,* 32, 361.

Johnsen, M., Christensen, T., Dennis, P. P., and Fiil, N. P. (1982). *EMBO J.,* 1, 999.

Nomura, M., Jinks-Robertson, S., and Miura, A. (1982). In *Interaction of Translational and Transcriptional Controls in the Regulation of Gene Expression,.* M. Grunberg-Manago and B. Safer (eds.), Elsevier Biomedical, pp. 91-104.

Petterson, I. and Liljas, A. (1979). *FEBS Lett.,* 98, 139.

Post, L. G., Strychartz, G. D., Nomura, M., Lewis, H., and Dennis, P. P. (1979). *Proc. Natl. Acad. Sci. USA,* 76, 1697.

Shine, J. and Dalgarno, L. (1974). *Proc. Natl. Acad. Sci. USA,* 71, 1342.

Tinoco, I., Borer, P. N., Dengler, B., Levine, M. R., Uhlenbeck, O. C., Crothers, D. M., and Gralla, J. (1973). *Nature New Biol.,* 246, 40.

Promoter Collectives in Bacteria

Ole Maaløe
The University Institute of Microbiology;
Copenhagen, Denmark

M. Weis Bentzon
State Serum Institute, Copenhagen

INTRODUCTION

Bacteria grow mainly by making protein of approximately 1000 different kinds. The protein-synthesizing system (PSS) is large and complex, but the same processes of transcription and translation account for the great diversity among proteins in terms of size, structure, and function.

In order to make a specific polypeptide chain a transcript of the corresponding gene(s) must associate with a ribosome, where it is translated by aminoacylated tRNA molecules. The overall process requires the participation of the DNA-dependent RNA polymerase (RNA-P), a complete set of aminoacyl-tRNA synthetases, and a host of initiation, elongation, and termination factors. *In toto,* the PSS consists of about 100 proteins (of which about 50 are in the ribosome), the rRNA molecules, and the numerous tRNA species.

When *Escherichia coli* cells grow at the maximum rate ($\mu = 2.8$ doublings per hour, at 37 C) slightly more than half the total protein produced is PSS protein. At lower growth rates, which means lower rates of protein synthesis, the size of PSS is reduced (Figure 1). This is an old observation, and we know now that bacterial ribosomes function at near-maximum efficiency, except at the lowest growth rates.

Apparently, bacteria have developed means by which they adjust the size and capacity of the PSS to match the potential of the growth medium. Intuitively such an adjustment makes sense, and several attempts have been made to explain how it is achieved.

The ideas invoked to solve this problem naturally reflect the changing fashion in science during the last 30 years. The fortuitous advent of "relaxed" mutants of *E. coli* (1955) focussed attention on control of rRNA synthesis as a way of regulating ribosome synthesis. Since shortage of any of the natural amino acids reduces RNA synthesis in cells with stringent control, a common effector, (called the "catholic inducer" by G. Stent), was thought to operate on a multiple induction/repression system. When the guanosine tetra- and pentaphosphates were discovered, one or both of them were thought to be the missing effector(s). This idea had to be abandoned when elaborate quantitative measurements showed that the concentrations of ppGpp and pppGpp in exponentially growing cells did not change nearly enough to account for the differences in the rate of rRNA synthesis between growth rates. At present, it seems most likely that ppGpp elicits an SOS-type response when protein synthesis is compromised, for example, during amino acid starvation (Ingraham et al., 1983).

Other schemes have been suggested that involve modifying RNA polymerase in a way that at high growth rates a greater fraction of these molecules increase their affinity for rRNA and tRNA promoters (e.g., Travers, 1980). As in the case first described, this model presupposes the existence of an unidentified effector, the concentration of which would have to change appropriately with growth rate. In addition, all models that assume primary regulation of rRNA synthesis leave open the problem of balancing the syntheses of rRNA against ribosomal protein (r-protein). Only small amounts of free rRNA and r-protein are present in growing cells, but it is not known how this balance is achieved. A temporary excess of rRNA might bind free r-proteins and thus reduce repression of r-protein operons, but it is equally possible that excess r-proteins, of one or more types, interact with RNA-P in a way that stimulates rRNA synthesis.

We shall leave this unsolved problem behind and focus instead on protein synthesis in the hopes that we will be able to learn something about how the size of PSS is adjusted to the growth rate. Research on the protein aspects of the problem began with the introduction of a technique for measuring rates of synthesis of r-protein *in vivo* (Schleif, 1967), and gradually the methods of molecular genetics were introduced to complement those of growth physiology. This combined approach is described in Chapters 6, 7 and 8 in "Growth of the Bacterial Cell" by Ingraham et al. (1983). The present paper extends Chapter 8 in which we introduced an essentially statistical analysis based on a great variety of measurements that describe protein synthesis in relation to growth rate

Experiments in growth physiology have shown that once the formation of a polysome is initiated, protein synthesis from whatever transcript is involved proceeds at nearly the same rate and with the same average yield (same number of protein molecules produced); both parameters are more or less independent of growth rate. Thus, individual proteins are synthesized in packages with the same average number of copies, one package per polysome. Acts of transcription *not* followed up by translation may occur, but they are irrelevant because they do not contribute to protein synthesis. During balanced growth total cell mass (M) and total protein (P) increase at the same relative rate, and μ (mass doublings per hr) can therefore be expressed in terms of the frequency ($F\mu$) with which new polysomes are created: that is, the number of unit packages of protein synthesized per sec. This is described in the legend to Figure 1 and shown on the graph.

From molecular genetics we have learned how the PSS genes; and operons map on the *E. coli* chromosome, and how some of the; control mechanisms are constructed which serve to balance the

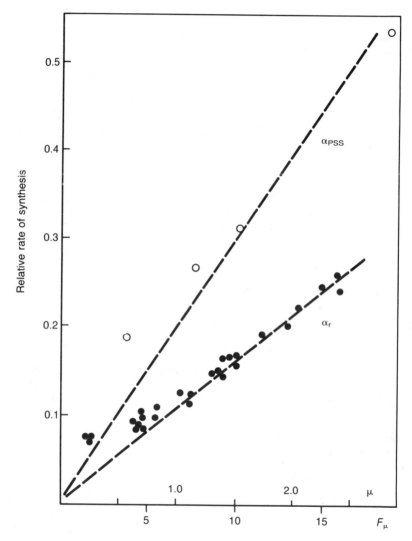

FIGURE 1 Relative rates of synthesis of ribosomal proteins (α_r) and all PSS proteins (α_{PSS}). Reproduced from Chapter 6 (p. 277) in Ingraham et al. (1983), with addition of a scale showing the equivalence of μ (doublings per hour) and $\underline{F}_\mu$ (frequency of initiation of new polysomes per sec).

synthesis of proteins whose genes map in different operons. The roughly 50 r-protein genes are transcribed from about 20 positions scattered throughout the *E. coli* genome. At some but not all positions more than one promoter can be seen to serve a single transcriptional unit. However, we need not be concerned with particulars of this kind. Our data are estimates of the relative expression of different genes, and we do not distinguish between genes, or operons, preceded by one or by several promoters. What matters is the overall efficiency with which transcription *and* translation can be initiated.

Ribosomal protein genes have been located close to the origin (*oriC*) and also near the terminus of replication. These locations imply that the ratio between the numbers of early and late replicated r-protein genes in slow- and in fast-growing cells varies almost threefold. A single effector can therefore hardly be imagined to balance the production of all the r-proteins at all growth rates. Indeed, the r-protein genes and operons so far examined in detail have been found to be regulated individually by autorepression at the transcriptional and/or at the translational level (cf. Chapter 7 in Ingraham et al., 1983).

In the most general terms, PSS serves to express the phenotype of the cell. Anatomically PSS is not an organelle (such as a mitochondrion), but functionally it forms a system held together by a network of regulatory bonds. We assume that this network maintains constant ratios between the components of PSS whether the size of the system as a whole is large or small relative to total cell mass.

THE TWO DOMAINS OF THE *E. coli* CELL

In growing *E. coli* cells all monomers and the energy for polymerization of amino acids and nucleotides are supplied by the biosynthetic and fueling enzymes that make up the non-PSS domain. The controls operating within this domain are different in purpose and design from the autoregulations that maintain *fixed* ratios between the PSS proteins. In the non-PSS domain the controls are designed to *vary* the ratios between enzyme systems in response to effectors that signal the need for changing the relative amounts of individual sets of enzymes. This difference in design reflects the roles as suppliers and consumers, respectively, of the non-PSS and of the PSS proteins.

When both energy and the flow of monomers are adequate, PSS will operate at near-maximum efficiency, as long as its components are correctly balanced. In the domain of biosynthesis and fueling the rules are different: if the capacity of one of the supply lines either exceeds or falls below the demand for its product, synthesis of the enzymes involved will either decrease or increase in response to changes in the concentration of the cogent effector. A case in point are the enzymes of the *trp* operon; during growth in a minimal medium they are synthesized at a rate that evidently satisfies the demand for tryptophan for growth, but that is repressed to about 20% of the rate measured in constitutive mutants.

The properties of the PSS and the non-PSS domains are complementary and suggest that during balanced growth they are matched in the sense that neither is developed beyond the capacity of the other for consuming and supplying, respectively. This seems indeed to be the case, as indicated by the fact that polypeptide and nucleotide chains grow at matching and near-maximum rates, independently of the growth rate of the cell. In other words, whether the ratio of PSS protein to total protein is high or low, the flow of monomers and energy is sufficient to maintain nearly the same high efficiency of protein synthesis. At the same time, significant overproduction of the biosynthetic and fueling enzymes is prevented by induction and repression mechanisms.

ADJUSTMENT OF THE SIZE OF PSS

We now return to the observation that bacteria tend to match the size and capacity of PSS to the potential of the medium in which they grow. In conformity with the classic operon model, this adjustment has usually been ascribed to *active* control, implying the existence of an effector to gauge the level of PSS activity. Actually, the autoregulations observed within PSS fit such a model: if the synthesis of one element of PSS were controlled directly, then the rest might follow suit because all belong to an interconnected network. However, effective adjustment of the size of PSS can, and probably does, occur without being mediated by an effector. This so-called *passive* mode of regulation was described in Chapter 8 of Ingraham et al. (1983), and it will now be analyzed further.

Note first that active and passive control are not mutually exclusive. It is well known that high levels of ppGpp inhibit the synthesis not only of RNA but of r-protein as well. This is sufficient to show that some active control is exerted when ppGpp is produced at abnormally high rates due to interference in one way or another with protein synthesis. This phenomenon may be important under conditions of stress, whether imposed from outside in a "shift experiment" or created by random fluctuations in cell metabolism. However, we shall first examine balanced growth at various rates.

With this important restriction we begin by considering the equilibrium that must exist between initiation of new polysomes and loss of old ones. For the purpose of estimating the size of PSS (that is, the ratio of PSS proteins to total protein) at different growth rates we note that the number of PSS genes and operons is small: 20 for r-proteins, RNA-P subunits, and the elongation factors (Tu, Ts and G), plus 20 or 30 to account for other abundant PSS proteins. This total of 40-50 is not more than approximately 1/10 the total number of active genes and operons in the *E. coli* genome. Nevertheless, we know that at maximum growth rate at least half the total protein produced is PSS protein. In this extreme case more than half of the polysomes in the cell represent a small subset of all the genes being expressed; at lower growth rates the relative abundance of PSS polysomes decreases. It is this relationship between growth rate and expression of the PSS genes that we want to derive from equilibrium considerations.

Let the number of promoters in the PSS and in the non-PSS domains be k_1, and k_2, respectively, and let S_1 and S_2 denote the average promoter strengths in the same domains. An S-value is here defined as the probability (sec^{-1}) of initiating a new polysome by starting transcription from a gene whose promoter(s) is "free," i.e., not "closed," either for intrinsic, structural reasons, or because a regulatory switch blocks access to transcription. Free promoters in the two domains are designated f_1, and f_2, respectively, and we assume that initiations take place at random among free promoters. Thus, α PSS equals $f_1/(f_1+f_2)$. The equilibria between free and closed promoters in the PSS and the non-PSS domains can now be written:

$$S_1(k_1 - f_1) = F_\mu \left[f_1/(f_1 + f_2)\right] \tag{1a}$$

and

$$S_2(k_2 - f_2) = F_\mu \left[1 - (f_1/f_1 + f_2)\right] \tag{1b}$$

The products on the left side represent closed promoters $(k-f)$ that make the transition to the free state during the next second. These products are equated to the total number that makes the reverse transition by becoming occupied, i.e., F_μ times the fractions of all initiations that occur in the PSS (α_{PSS}) and in the non-PSS domain $(1 - \alpha_{PSS})$, respectively. Equations (1a) and (1b) can be solved for F_μ to give,

$$F_\mu = \left[(k_2/(1 - \alpha_{PSS})) - (k_1/\alpha_{PSS})\right] \left[1/S_2 - 1/S_1\right]^{-1} \tag{2}$$

We first note that F_μ is proportional to the growth rate μ (see Figure 1), and that α_r ($\sim 0.5\ \alpha_{PSS}$) has been measured for many μ values. What remains is to assign numbers to the values of k and S. In the standard cell with two genome equivalents of DNA $k_1 \sim 80$ and $k_2 \sim 800$ (the latter is judged from the fact that the cell produces about 10^3 different proteins from operons with an average of 2-3 genes). A minimum value of 1/4 can be assigned to S_1, because, at maximum growth rate ($\mu = 2.8$ at 37 C), initiation at PSS promoters must take place at average intervals of about 4 sec to account for the rate of r-protein synthesis. Independent estimates of S_2 are not available, but an S_2 value representative of all the non-PSS promoters can be calculated for the average cell growing under standard conditions, $\alpha_r \sim 0.15$ and $F_\mu \sim 10$ (Figure 1), and a value of 1/80 for S_2 satisfies equation (2).

With the numerical values assigned to k_1, k_2, S_1 and S_2, equation (2) generates the theoretical relation between F_μ and α_r ($\sim 0.5\ \alpha_{PSS}$). Figure 2 shows the result (the hatched zone) and the measured α_r values taken from Figure 1. The fit is remarkable and shows that equation (2) is adequate to describe the ratio of PSS protein to total-protein as a function of F_μ, and that a free parameter such as an effector is not required. Recently, Steen Pedersen (this volume) has shown that cgr_p instead of being independent of μ (as we used to believe) decreases by 30-40% between growth rates differing by a factor of 3. When this is taken into account F_μ is no longer strictly proportional to μ, and the effect is to displace the points representing the low F_μ values on Figure 2 a little to the right. This correction is small, but it does in fact improve the fit to the theoretical curve.

It should be noted that Equations 1a and 1b imply that α_{PSS} increases with F_μ because high initiation frequencies favor strong promoters. Beyond F_μ values of about 30 no equilibrium can be defined, because F_μ then exceeds the maximum number of promoters that can enter the "free" state in one sec ($S_1k_1 + S_2k_2 \sim 30$).

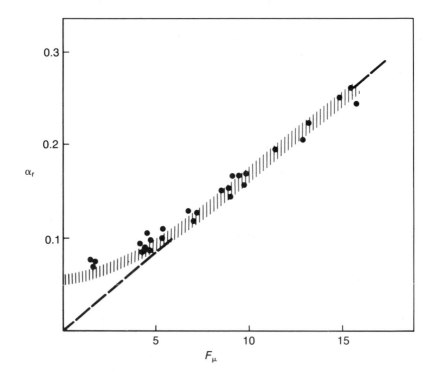

FIGURE 2 The α_r measurements from Figure 1, and a hatched zone centered on the curve calculated from Equation 2 using a single (average) $\underline{S_2}$ value of 1/80.

PROMOTERS WITHIN AND WITHOUT PSS

The promoters serving the genes and operons within PSS form a collective of strong promoters to which the same strength (S_1) has been assigned. Here we must recall that strength of a promoter is a function, on the one hand, of the fixed base sequence of that promoter (which we often know without knowing how to "read" it), and, on the other hand, of the effect of repression or attenuation. An S_1-value thus combines the *intrinsic,* or structural property of a PSS promoter and the *extrinsic* effect of autoregulation. With this definition we confidently assign the same S_1-value to the PSS promoters, knowing that the output of the genes they serve is balanced.

With equal confidence we can argue that the S_2-values (promoter strengths in the non-PSS domain) must vary greatly. Consider first constitutive synthesis. In such cases the extrinsic effect(s) on transcription is eliminated (e.g., by deletion of a repressor gene) leaving the base sequence of the promoter to define its strength. Calculations show that a fully induced *lac* operon (with a promoter that is insensitive to catabolites) is transcribed at average intervals of 30–40 sec at $\mu = 0.5$ (at 37 C); the *lacI* gene, which is also constitutive, is transcribed at much longer intervals. Thus, the promoters outside PSS comprise some with low and some with very low intrinsic strength. In addition, we note that the extrinsic effects on transcription from non-PSS promoters vary between operons and depends to a large extent on the growth conditions. An extreme case is the *lac* operon with and without induction. Fully induced, S_2 at μ = 0.5 is 1/30 to 1/40; without induction this value is reduced 100-1000 times.

It is clear that an S_2 *distribution* must take the place of the single S_2 *value* fitted to Equation 2. The width and shape of this distribution might be estimated from intensities and positions of the individual spots on an 2-D gel displaying most of the proteins synthesized by *E. coli* (O'Farrell, 1975). The published data are not sufficiently detailed for this purpose, and we have therefore calculated α_r (and α_{PSS}) for distributions of S_2 composed such as to yield the same quantity of protein as would the identical number of operons with the S_2 value (1/80) used to construct the curve of α_r vs. F_μ shown in Figure 2. The results illustrated in Figure 3 show that a balanced distribution of S_2 values does not significantly change the picture obtained using the unrealistic "average" S_2 of 1/80.

COLLATERAL EVIDENCE

A strong point of the analysis is that the parameters of Equation 2 are derived from a variety of measurements of protein synthesis and from related genetic evidence. None of these experiments was specifically designed to test the hypothesis of passive regulation of the synthesis of the elements of PSS. What we now have is therefore an overall description of protein synthesis in *E. coli* which predicts the relationship between F_μ and α_{PSS}, and Figure 2 has shown that this prediction agrees with actual measurements of α_r (and α_{PSS}).

Considering that our analysis combines a large number of independent measurements, all of which are subject to an error of about ±10%, the agreement might be more or less fortuitous. Therefore, we have looked for collateral evidence. Figure 4 illustrates how the relative rate of synthesis of a given protein (α_i) is expected to change as a function of F_μ. Panel a shows the obvious complementarity of the α_{PSS} and α (nonPSS) curves; in panel b the latter is resolved into curves relating α_i to F_μ for different S_2 values. The curves for intermediate and low S_2 values show the gradual fall in α_i with increasing F_μ (or μ) first observed by Rose and Yanofsky (1972) for constitutive synthesis of the tryptophan enzymes. The phenomenon was called "metabolic regulation," and it has been shown to apply to constitutive synthesis of a handful of enzymes (cf. Ingraham et al., 1983, Chapter 8). Thus, the fall of α_i with increasing growth rate can be interpreted to mean that the operon in question is transcribed from a promoter of intermediate or low intrinsic strength.

A second case in support of the general usefulness of Equation 2 is presented in Chapter 6 of Ingraham et al. (1983). In brief, the quantities of DNA, RNA and protein in *E. coli* strain B/r were measured at several growth rates realized either by using different carbon sources *or* in a glucose-limited chemostat. Identical cell compositions were observed whether the biosynthetic and fueling reactions were based on, say, succinate or on glucose in limited supply, *provided* the cultures grew at the same rate, i.e., had the same μ and F_μ values. In other words, Equation 2 reliably predicts

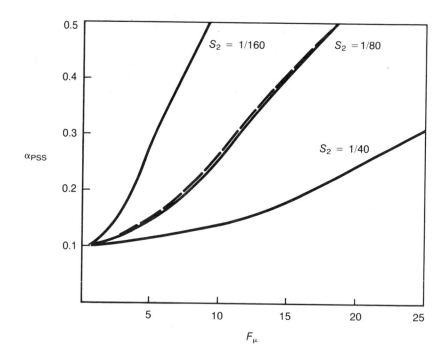

FIGURE 3   The fully drawn curves show α_{PSS} as function of $\underline{F}_\mu$ for $\underline{K}_1 = 800$ and average $\underline{S}_2$ values of 1/160, 1/80, and 1/40, respectively. The broken curve represents a simple case: 200 promoters with $\underline{S}_2 = 1/160$, 500 with $\underline{S}_2 = 1/80$, and 100 with $\underline{S}_2 = 1/40$ ($\underline{K}_1 = 800$). Wider distributions have been analyzed as follows:

S_2^{-1}	20	40	80	160	320
1	0	0	800	0	0
2	16	40	600	80	64
3	32	80	400	160	128
4	48	120	200	240	192

With increasing variance, the curves are displaced upwards. The α_{PSS} corresponding to $\underline{F}_\mu = 10$ increases by 25% between distribution 1 and 4.

composition, including α_r, whether the biochemical activities in the non-PSS domain are those required for growth on succinate or on glucose.

TRANSIENTS DURING GROWTH

As emphasized above, only states of balanced growth were initially considered. This was indeed a prerequisite for setting up the equilibrium equations (1a and 1b) from which the analysis proceeded. However, shifts between growth media have been studied extensively:

(a)

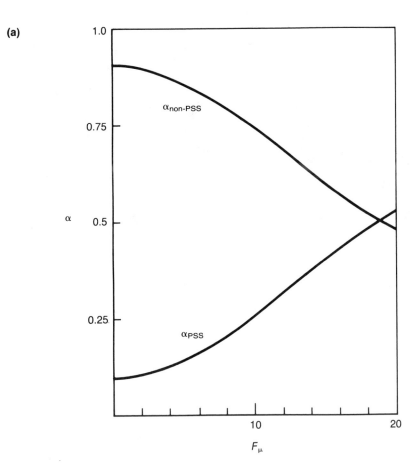

FIGURE 4 (a) Complementarity between the α_{PSS} and the $\alpha_{non-PSS}$ curves. (FIGURE 4 continues)

in particular, shifts from minimal to rich medium. The reason is that some immediate effects of this shift are obvious: within seconds many biosynthetic operons are more or less fully closed down by repression or attenuation. In the terms used in the present study, this is equivalent to drastic reductions of the S_2 values pertaining to the operons in question, and hence to added emphasis on transcription from the strong PSS promoters.

The *S10* operon, which comprises 11 r-protein genes, has been analyzed in detail by Lindahl and coworkers and shown to be autoregulated by attenuation. Their most recent paper (Zengel et al., 1984) reveals the striking difference between the setting of the attenuator, on the one hand during balanced growth at various rates, and on the other hand following a shift to rich medium. Figure 5 (also their Figure 5) shows that the setting ("read-through") is almost independent of growth rate, implying that the autoregulation

(b)

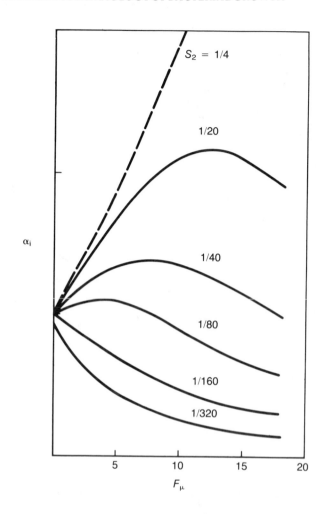

(FIGURE 4, continued) (b) The relation between F_μ and α_i (relative rate of synthesis of protein i) depends on arbitrarily chosen S_2 values for promoter i.

contributes little, if anything, to the regulation of α_r. This agrees with the assumption that in a steady state of growth the autoregulations serve to balance the output from the 20-40 operons of the PSS domain.

In contrast, the shifts to rich medium we now discuss perturb the balance in favor of transcription from the PSS promoters. Figure 6 (Figure 3 in Zengel et al., 1984) illustrates the oscillations that are elicited by the shift. This pattern was first demonstrated by Gausing (1980) for the entire class of r-proteins. The data in Figure 6 offer new insight: the top panel shows an immediate but modest increase of transcription of the leader segment of the $S10$ mRNA

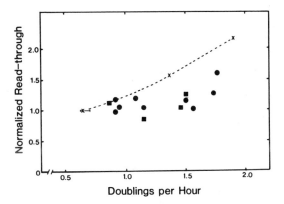

FIGURE 5   Read-through at the S10 attenuator during steady-state growth in different media. Strain LL308 containing plasmid pLL36 (circles) or pLFl (triangles) was grown exponentially at 37°C in minimal medium supplemented with (1) glycerol, (2) glycerol and 19 amino acids (minus methionine), (3) glucose, (4) glucose and 19 amino acids, or (5) glucose and casamino acids (0.2% or 1%). The read-through was calculated from the relative transcription rates of the leader and, of the structural genes (S10 and L3, in the case of pLL36; lacZ' in the case of pLFl. The cultures were labeled for 30 sec with [³H]uridine, and the radioactivity incorporated into the pertinent transcripts was determined by hybridization. Each experiment utilized a glycerol-minimal culture and one or more cultures growing in other media. The read-through values determined for each culture were normalized to the value obtained for the glycerol culture in the same experiment. Thus, the relative read-through in glycerol-minimal medium is defined as 1. Most of the points represent the average of two or three labelings of a given culture. The dashed line illustrates the differential rate of r-protein synthesis (α_r) at different growth rates; the three X's on the dashed line represent actual measurements of the differential rate of synthesis of protein S10 at the indicated growth rates.

followed at 2–4 min after the shift by a fall to a nearly constant level. This early jump in the frequency with which transcription is initiated at the S10 promoter may be a direct consequence of turning down many biosynthetic operons. The bottom panel shows that with a delay of 1–1.5 min read-through at the attenuator increases nearly two-fold and by 4 min it has again dropped to slightly below the preshift value. The two center panels illustrate the sharp increase and subsequent fall in the relative rates of synthesis of early and late segments of the S10 mRNA. These rates of synthesis obviously reflect changes in the product of the frequency of initiation at the S10 promoter and the frequency of read-through. The immediate increase in transcription frequency is magnified by the release of attenuation setting in shortly after. The result is an overshoot in the synthesis of the r-proteins in the S10 operon (and probably the rest

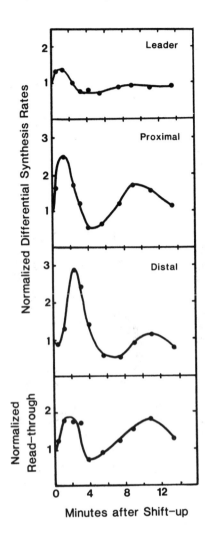

FIGURE 6 Transcription of the S10 operon after a nutritional shift-up of a haploid strain. RNA was labeled with [³H]uridine in pulses of 45 sec and hybridized to appropriate DNA probes. The read-through at the S10 attenuator after the shift-up was calculated by dividing the radioactivity in the structural gene messages by the radioactivity in the leader transcript. The resulting quotients were then normalized to the preshift value.

of the r-proteins), relative to the modest increase in transcription frequency. This combined effect probably causes the concentration of the effector (r-protein L4) to increase over the first few minutes with the result that attenuation is again tightened.

To understand the early post-shift events we shall consider how the release of attenuation comes about. The answer is at hand: the concentration in the cells of r-protein L4 must initially have gone down. This may be explained (tentatively) by taking into account the synthesis of rRNA. At maximum growth rate transcription at the promoters of the seven *rrn* operons is initiated at intervals of 1-2 sec; and these very strong promoters would be expected to respond at least as effectively as the r-protein promoters to a sudden closing down of operons in the non-PSS domain. The net result seems to be that, for a short time, the rate at which free L4 molecules are withdrawn from the pool by being bound to 23S RNA exceeds the rate of L4 synthesis, leading to depletion of the pool and hence to release of attenuation.

The drop in the differential synthesis rates following the initial rise probably reflect (1) lowered read-through owing to overproduction of r-protein L4, and (2) the observed small drop in initiation frequency at the *S10* promoter. The latter may be due to readjustments in the repression pattern in the non-PSS domain in response to the early overproduction of ribosomes with concomitant increases in demands on biosynthetic and fueling reactions, as suggested in Chapter 8 of Ingraham et al. (1983). We can take the reasoning one step further; the drop in read-through, which must be attributed to swelling of the L4 pool, suggests that, at this time during the transient, the synthesis of rRNA does not keep up with that of r-protein. Thus, the unsolved problem we deliberately left out in the beginning of this paper comes up again, begging for an answer; we need to know the mechanism(s) by which the syntheses of r-protein and rRNA are balanced in the steady state of growth.

REFERENCES

Gausing, K. (1980). In *Ribosomes: Structure, Function and Genetics,* G. Chambliss, G. R. Craven, J. Davies, K. Davis, L. Kahan and M. Nomura (eds.), University Park Press, p. 693.

Ingraham, J. L., Maaløe, O., and Neidhardt, F. C. (1983). *Growth of the Bacterial Cell,* Sinauer Associates, Inc., Sunderland, Massachusetts.

O'Farrell, P. H. (1975). *J. Biol. Chem.,* 250, 4007.

Rose, J. K. and Yanofsky, C. (1972). *J. Mol. Biol.,* 69, 103.

Schleif, R. (1967). *J. Mol. Biol.* 27, 41.

Travers, A. A., Buckland, R., and Debenham, P. B. (1980). *Biochemistry* 19, 656.

Zengel, J. M., Archer, R. H., Freedman, L. P., and Lindahl, L. (1984). *EMBO J.,* 3, 1561.

Regulation of IF3 Expression in *E. coli*

Larry Gold, Gary Stormo, and Roger Saunders
University of Colorado, Boulder

INTRODUCTION

For a very long time we have been interested in the mechanisms by which ribosomes select mRNAs for translation (Singer and Gold, 1976; Gold et al., 1981). We have tried to study the elements that control translation rates for different messages and have tried to rationalize the "recognized" with the "recognizer" (Schneider et al., in preparation). A portion of the story is explained by the Shine and Dalgarno hypothesis (1974), which seems to be largely correct (Gold et al., 1981). Shine and Dalgarno chose to sequence 16S RNA, because available biochemical data implicated the small ribosomal subunit in translational initiation. One would never have predicted that the important elements for mRNA recognition would lie within the first 12 nucleotides sequenced!

For several reasons, the Shine and Dalgarno hypothesis seems too simple, given the complexity of the ribosome. Why should only a few of the 1542 16S-RNA nucleotides be involved in base pairing with mRNAs, since surely there are other 16S-RNA domains that lie in appropriate locations for mRNA interactions? Why should a ribosome that covers about 22 nucleotides 5' to the initiation codon and another 12 or so 3' to the initiation codon use so few of those messenger nucleotides to achieve site selection? How can one achieve the observed 1000-fold differences in rates of translation with as little information as appears to reside in the sequences complementary to the 3' end of 16S RNA (Gold et al., 1981)? Questions such as these were on our minds when we first inspected a set of *E. coli* ribosome binding sites to ask about new locations of information; we found that ribosome binding sites are in fact nonrandom over the entire region covered by ribosomes (Figure 1), as though mRNAs might often have determinants that interact with other (unknown) non-Shine-Dalgarno domains on the ribosome (Gold et al., 1981; Stormo et al., 1982a,b).

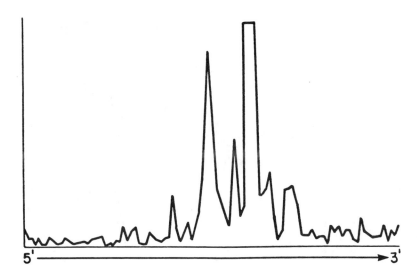

5'————————————————————————————————▶3'

FIGURE 1 Information within translational initiation domains. The non-randomness of nucleotides 5' and 3' to E. coli initiation codons is displayed (see Gold et al., 1981; Stormo et al., 1982a,b; Schneider et al., in preparation). The graph contains a large peak coinciding with the initiation codon, and another large peak corresponding to the Shine and Dalgarno region.

One should note that the 5' polypurine tracts within mRNAs were not appreciated (but see Steitz, 1969; 1973) until Shine and Dalgarno published their idea; differences in spacing between the initiation codon and the polypurine tracts hid the now-obvious "consensus". We once attempted to find other 16S RNA domains that would also anneal, on paper, to a statistically significant set of ribosome binding sites; computers allow such "experiments," but the great length of 16S RNA yielded more domains than are likely to be used. Nevertheless, we have always believed that 16S RNA contains other domains that potentiate mRNA binding, at least for some mRNAs. The sequence of the mRNA encoding the initiation factor IF3 was recently published (Sacerdot et al., 1982); the translational initiation domain breaks many "rules" followed by most other E. coli mRNAs (Gold et al., 1981; Stormo et al., 1982a,b). Nevertheless, the IF3 protein is abundant (Howe and Hershey, 1983). Therefore, we asked if another mode of translational initiation, utilizing novel 16S RNA interactions, might account for IF3 expression.

INFERENCES

RNA sequences available on the 30S ribosomal particle

We wish to identify 16S RNA sequences that lie on an available surface of a 30S particle; we will demand that these RNA sequences sit close to regions known to interact with mRNA. The secondary structure of 16S RNA (Figure 2) is now well established (Woese et al., 1983). The

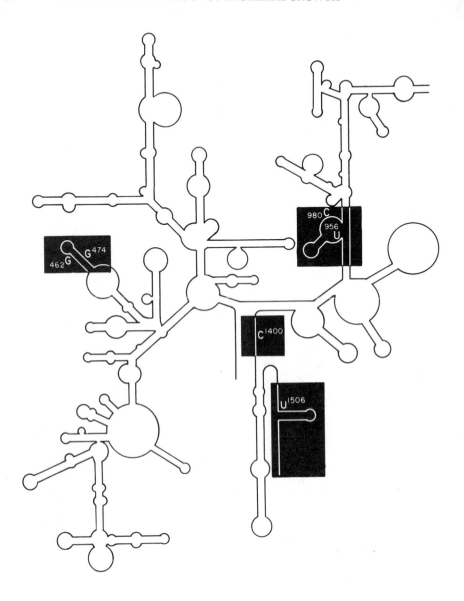

FIGURE 2 Secondary structure of <u>E. coli</u> 16S RNA. The proposed secondary structure of 16S RNA is given, according to Woese et al. (1983). Boxes are placed around regions of the molecule that lie close to the mRNA binding domain and which, we will argue, are involved in IF3 expression.

picture highlights the regions that we will argue, lie close to the groove that captures mRNA during translational initiation.

We start with a conventional representation of a 30S particle (with the 3' end of 16S RNA, plus the P and A tRNA binding sites).

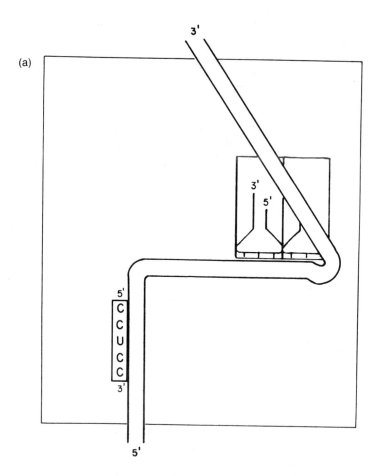

FIGURE 3 Plausible locations for some 16S RNA domains on 30S particles. (a) A 30S particle is shown, with the positions of the 3' end of 16S RNA and the P and A sites for tRNA binding. A generic mRNA is shown winding through the 30S particle. The spacing between the polypurine domain of the mRNA and the initiation codon reflects the average spacing for <u>E. coli</u> mRNAs (Gold et al., 1981; Stormo et al., 1982a,b). All subsequence representations of the 30S particle are consistent with this representation and each other. That is, the reader may superpose any subset of images from Figure 3 or Figure 4. (continued)

The representation (Figure 3a) includes an mRNA bound to the 30S particle, and depicts an average spacing between the crucial nucleotides of the Shine and Dalgarno domain and the initiation codon (Gold et al., 1981; Stormo et al., 1982a,b). Next we note a crosslinking experiment between acetylvalyl-tRNA and 16S RNA. The acetylvalyl-tRNA was bound into the P site, after which photoactivation

(b)

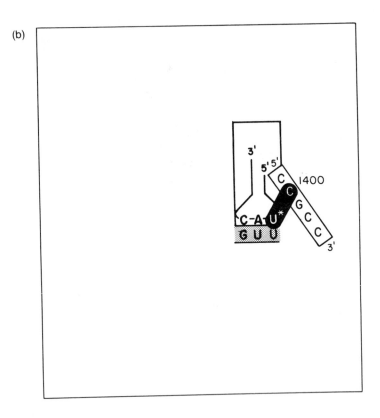

(FIGURE 3, continued)
(b)
We show acetylvalyl-tRNA (anticodon = mo⁵U-A-C) in the ribosomal P
site, base paired with an mRNA (GUU). Photoactivation of the modified
base of the anticodon is performed on 70S ribosome-tRNA-mRNA
complexes. Photoactivation leads to crosslinking between the modified
base of the anticodon and C-1400 of 16S RNA, as is shown in the
figure (Ehresmann et al., 1984; Prince et al., 1982). (continued)

gave a specific crosslink to C-1400 (Prince et al., 1982). The location
of C-1400 (Figure 3b), relative to the P site, has rotational ambiguity,
dependent on the precise details of the photoactivation process (the
tRNA has a modified base at the 5' position of the anticodon, and the
crosslink is between that modified base and C-1400 (Ehresmann et
al., 1984)). We select among the possible orientations of C-1400,
relative to the tRNA in the P site, according to our idea about IF3
translation (see below). In Figures 3c and 3d, we select orientations
and polarities based on the same idea.

    Hearst and his colleagues have perfected the methodology for
intramolecular crosslinking of 16S RNA by psoralen. Most psoralen
crosslinks confirm the 16S RNA secondary structure predicted by the
Woese/Noller phylogenetic arguments. However, a long-range crosslink
between U-1506 and U-956 has been found (Thompson and Hearst,

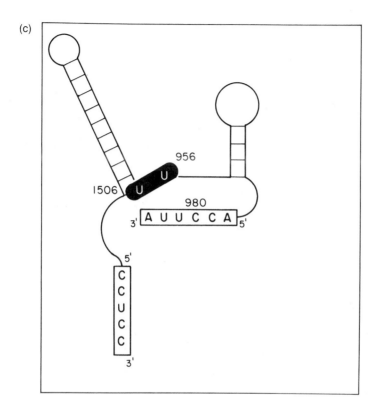

(FIGURE 3, continued)
(c)
We show a long range psoralen crosslink identified by Thompson and
Hearst (1983). The crosslinking was performed on deproteinized 16S
RNA. The same crosslink was identified by electron microscopy
(Wollenzein et al., 1979). Thompson and Hearst proposed a precise
interaction (involving base pairing between the two domains) that
accounts for the capacity of psoralen to crosslink the two U's (see
their Figure 7). Our representation of the two domains maintains the
secondary structure of 16S RNA proposed by Woese et al. (1983),
since we are focused upon the 16S RNA nucleotides 978-983. Our
representation is consistent with the Thompson and Hearst structure.
(continued)

1983). Thus, the C-980 domain of 16S RNA may be brought into the
neighborhood of the mRNA groove, based on Hearst's data (Figure
3c). The orientation of the C-980 domain has rotational flexibility;
the orientation shown is consistent with the proposed secondary
structure of 16S RNA.
    Another crosslink involves the 3' U of a tetranucleotide
messenger and G-462 of 16S RNA (Wagner et al., 1976). The
tetranucleotide mRNA was positioned in the decoding site on the 30S
particle (Figure 3d). The data also implicated the G at position 474,
which is not more than 15 angstroms from G-462 (see Figure 2). We

(d)

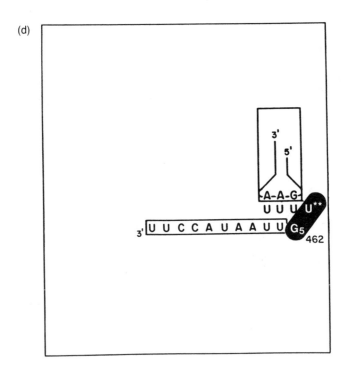

(e)

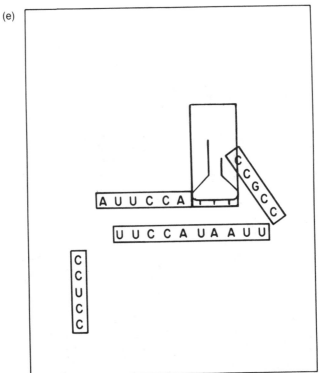

(FIGURE 3, continued)

(d)
We show a tetranucleotide mRNA analog (UUUU**) base paired with phe-tRNA in the ribosomal A site. The modified message contains a reactive adduct on the 2' position of the ribose. After incubation with 30S particles (containing 32P-labeled 16S RNA) the mRNA analogue was found crosslinked to G-462 and G-474 (see Figure 2). The crosslinking to both G-462 and G-474 is consistent with the secondary structure of 16S RNA (Woese et al., 1983). We have shown the nucleotides 3' to G-462 in a manner that excludes the RNA stem involving G-464 ; it is that stem (see Figure 2) that makes plausible G-462 and G-474 as targets for the same crosslinking reaction. The data for this experiment are from Wagner et al. (1976) and Wagner and Gassen (1975). (e) A summary of the 16S RNA locations identified in panels A-D.

show the sequences 3' to G-462, including three nucleotides that are within a stem in the 16S RNA secondary structure (Woese et al., 1983).

In summary, four 16S RNA domains (the 3' end, C-1400, C-980, and G-462) are near each other on the surface of a 30S particle (Figure 3e). The surface we implicate participates in binding of mRNA, the 50S particle (Herr et al., 1979; Vassilenko et al., 1981), and binding of the initiation factor IF3 (Hershey, 1980).

The IF3 mRNA of E. coli

The four 16S RNA domains are shown again (Figure 4a), this time superposed upon a generic mRNA sitting in the 30S particle (exactly as in Figure 3a). One sees that the rotational ambiguity intrinsic to the crosslinking data has been used to potentiate base pairing between the mRNA and the 16S RNA domains. Before going forward, we emphatically note that the three new domains (C-1400, C-980, and G-462) have been identified by crosslinking data which were obtained after beautiful experimentation and without consideration of the specific mRNA we are about to discuss. The power of the model comes from the integration of the crosslinking data with the IF3 mRNA sequence.

IF3 is responsible for dissociating 70S ribosomes that are not making protein, so the 30S subunit can re-engage an mRNA and initiate translation (Hershey, 1980). Purified IF3 is required for 70S ribosome dissociation *in vitro*, and is required for high-level translation on natural mRNAs (Hershey, 1980). The sequence of the IF3 mRNA around the initiation codon is (Sacerdot et al., 1982):

GGAGG̲AAU̲AAGGU *AUU*AAAGGCGG

The Shine-Dalgarno region (boxed) is the only segment of this mRNA that is conventional; the initiation codon is AUU (italic), which is unique among *E. coli* mRNAs. A second potential Shine-Dalgarno sequence (underlined), with improper spacing for the initiation codon, is between the real Shine-Dalgarno region and the AUU; extra potential Shine and Dalgarno regions within ribosome binding sites are very rare (Gold et al., 1981; Stormo et al., 1982a,b). Lastly, a "structurogenic" sequence (doubly underlined) is just 3' to the AUU;

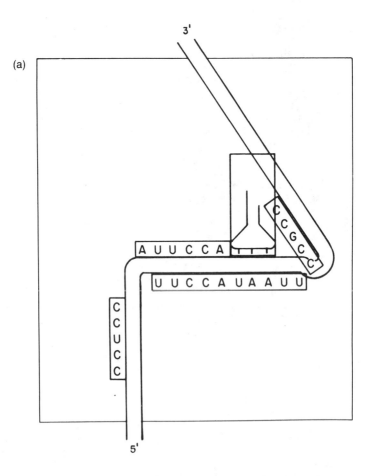

FIGURE 4 Reactions of IF3 mRNA with a 30S particle. (a) We show the summary of 16S RNA locations (from Figure 3e) superposed on to a generic mRNA (as in Figure 3a). (continued)

such G-rich regions are very rare in *E. coli* ribosome binding sites (exclusive of the Shine and Dalgarno region). Thus, this mRNA breaks many rules, at least through a comparison with more than 150 other *E. coli* ribosome binding sites; the IF3 message must be translated at a high rate (Howe and Hershey, 1983). When we first saw this mRNA, we knew something interesting was about to be unveiled, though we did not know what.

Cells must be buffered against minor perturbations in the intracellular levels of key proteins. That is, a "perfect cell" would respond to transiently lowered levels of IF3 by raising the rate of IF3 expression, which is itself needed for the translation of everything in the cell. Similarly, excess IF3 should selectively diminish the relative rate of IF3 production. The IF3 level is adjusted for the ribosome level (Howe and Hershey, 1983). These goals for the cell could be largely accomplished if IF3 was translated in a mode

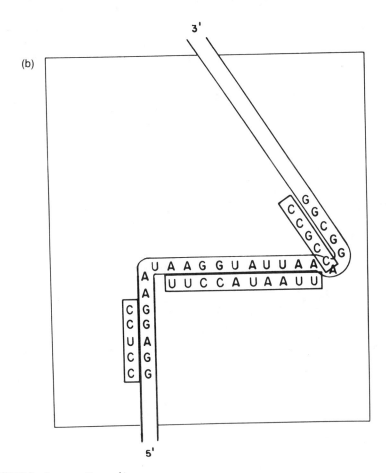

(FIGURE 4, continued)
(b)
The postulated binding reaction between IF3 mRNA and a 30S particle is shown. Base pairing is shown using the Shine and Dalgarno region, the G-462 region, and the C-1400 region.

independent of IF3 function; with mRNA excess, cells would adjust the relative amount of IF3 made as a result of competition between the IF3 mRNA and all others in the cell. High levels of IF3 would stimulate all other mRNAs, leaving few ribosomes for IF3 translation; low levels of IF3 would make available many ribosomes for IF3-independent translation of IF3 itself. (Buffers of this sort would also be sensible for the transcriptional initiation factor sigma.) Although levels of IF3 increase along with the other components of the translational machinery, in response to faster cellular growth rates (Howe and Hershey, 1983), IF3 is not feedback-regulated in the way many of the ribosomal proteins are (Lestienne et al., 1982). Cells carrying plasmids expressing IF3 yield more protein in response to

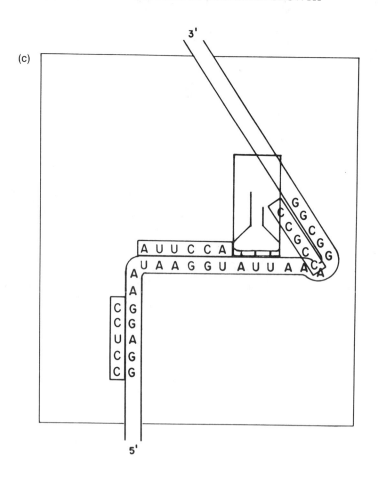

(FIGURE 4, continued)
(c)
The postulated initiation reaction is shown. The G-462 region has been replaced by both the C-980 region and initiator tRNA, allowing the G-462 /G 474 region to return to the hairpin structure shown in Figure 2.

increased plasmid copy number (Lestienne et al., 1982). Obviously we hoped that the odd sequences around the IF3 ribosome binding site would contribute, somehow, to the buffering we thought to exist.
    The IF3 mRNA could utilize its unusual sequence to facilitate translation. In Figure 4b the IF3 mRNA is shown bound to a 30S particle. In this proposed binding mode both the C-1400 and G-462 domains of 16S RNA are paired to IF3 mRNA. The proposed complex rationalizes every odd feature of IF3 mRNA, including the AUU initiation codon. In fact, the AUU of the mRNA participates in a perfect 10-base duplex!
    Once mRNA is bound, the initiator tRNA is recruited (Gold et al., 1981). However, the IF3 initiation codon is part of a helical

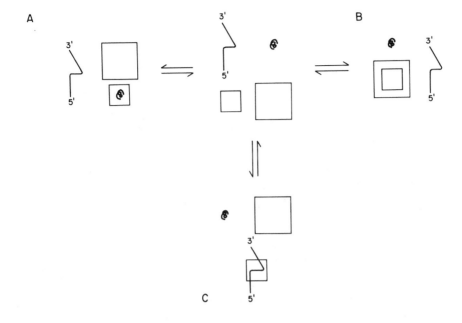

FIGURE 5 Linked equilibria between IF3 mRNA, IF3 protein, 30S particles, and 50S particles. In the top center of the diagram we show the free components (IF3 protein, IF3 mRNA, the 30S particle, and the 50S particle). The 30S and 50S particle can interact to form a 70S particle (top right). The 30S and IF3 protein can interact to prepare the 30S particle for normal translational initiation (top left). The 30S particle can interact directly with IF3 mRNA to form a pre-initiation complex (see Figure 4b) (lower center). The 30S particle with IF3 protein (top left) participates in translational initiation on most E. coli messages, through the Shine and Dalgarno interaction. In this mode IF3 mRNA will be treated as merely another mRNA.

structure in Figure 4b, and hence is unavailable. Fortunately, the C-980 domain of 16S RNA can provide binding energy that along with the initiator tRNA might flip the "binding reaction" into an "initiation reaction" (Figure 4c). The initiator tRNA of *E. coli* can recognize AUG, GUG, AUA, UUG, and AUU (Gold et al., 1981); one experiment suggests that even UUU can be recognized (van der Laken et al., 1980; van der Laken et al., 1979). In an extreme case, *in vitro* an mRNA with a good Shine-Dalgarno domain can lead to initiator tRNA binding even when no nucleotides lie in the P site (Jay et al., 1980).

The proposed interactions between IF3 mRNA and 16S RNA are remarkable. If regions of 16S RNA are simply positioned into locations suggested by biochemical crosslinking data, the IF3 mRNA falls into place, without even being forced to indicate regions to be ignored. The probability that these complementarities arose by chance is close enough to zero to be beneath discussion. Obviously we arrived at this model by first looking at the IF3 mRNA sequence, and building on to the sequence the ideas articulated in the introduction of this paper.

How do the proposed interactions meet the objective of IF3-independent translation? We imagine that translational termination (of all messages) is followed by the release of free 30S and 50S particles (Martin and Webster, 1975). The free 30S particle can react with either a 50S particle, the IF3 protein, or the IF3 mRNA (Figure 5). When the amount of IF3 protein is low relative to the number of 30S particles, IF3-independent translation of the IF3 mRNA will let the IF3 protein level increase; this is the means by which IF3 protein levels respond to increased growth rate (Howe and Hershey, 1983). When the growth rate is lowered, excess IF3 would be diluted. While IF3 is at excess, few 30S particles will be able to initiate translation in the IF3-independent mode.

We would be delighted if the IF3 protein, the 50S particle (through 23S RNA), and the IF3 mRNA competed for the same RNA sequences on the 30S particle (but see Vassilenko et al., 1981 and Herr et al., 1979). All three options utilize the same face of the 30S particle. If the IF3 protein sits on the 30S particle in a way that occludes the G~462, C~980, and C~1400 regions, only the 3' end of 16S RNA will be available for mRNA interactions. Thus, IF3-dependent translation will yield all proteins and the IF3 protein via normal translation (that is, through the Shine-Dalgarno interaction). Probably the IF3 protein is displaced when a 50S particle joins the complex. In this view IF3 protein bound to a 30S particle could provide mRNA selectivity by preventing RNA-RNA interactions that are usually not productive.

We propose further that evolution of the IF3 regulatory sequences happened by a path that is still obvious. The present IF3 mRNA sequence is only two transversions away from a sequence for which conventional translation would yield an IF3 protein missing the amino terminal hexapeptide (Figure 6). In fact, the IF3 protein in the cell is found as a mixture of long and short forms (Suryanarayana and Subramanian, 1977); the short form is fully active and missing exactly six amino acids from the amino terminus (Brauer and Wittmann-Liebold, 1977). The short form begins with Val-Gln-Thr, and thus results from proteolysis rather than translational initiation (through the remnant of the old initiation domain). We imagine that the present amino terminus is important for its contribution to mRNA function rather than protein function. Furthermore, the relationship between nucleotides 3-21 and 22-40 suggests that creation of a small duplication participated in the evolution of the present IF3 mRNA (Figure 6).

CONCLUSIONS

We are happy to have made this presentation as a gift to Ole Maaløe. After all, our model for the mechanism by which the IF3 protein level is regulated is a detailed extension of his own ideas about passive regulation of the synthesis of the main components of the translational apparatus (Ingraham et al., 1983). In the IF3 case, the translational initiation domain is designed to be responsive to the levels of free 30S particles. It has not escaped our attention that free 30S particles are sometimes thought to play a role in diminishing rRNA synthesis

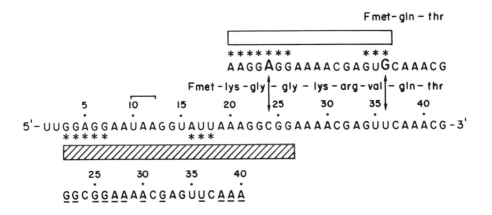

FIGURE 6 Evolution of the IF3 mRNA. The IF3 mRNA sequence is shown (numbered to nucleotide 42). The termination codon for the upstream thrS gene is at position 10-12. The present Shine and Dalgarno domain is at 3 to 7, and the initiating AUU is at 16-18. Our discussion thus far has centered on nucleotides 3-26. The top line of RNA sequence shows the sequence from position 20 to 42 as it might have once looked. The two transversions have created a good Shine and Dalgarno domain (position 20-26) and a good initiation codon (position 34-36), with proper spacing. The hypothetical IF3' protein sequence is also shown. For comparison purposes nucleotides 22-40 of the present sequence are placed under nucleotides 3-21. The sequences share 12 of 19 bases (underlined, bottom RNA sequence), which suggests that a small duplication and subsequent drift led to the present IF3 regulatory sequences.

(Jinks-Robertson et al., 1983). If so, we would imagine that the same surface of the 30S particle (as in Figures 3 to 5) handles that job. Needless to say, we are testing our model and its extensions.

Ole's appreciation of biology includes an idea that has been diluted in recent times. Ole manages somehow to act as though there are answers available from thinking hard about the way systems are integrated. When regulatory loops of unspeakable complexity are under consideration, one should always step back to see if a simple model won't explain the data (or at least the data worthy of such hard thinking). One should hang on to those simple models until they no longer work, and then add as little as possible to the next version. Our model for IF3 translation is not simple, but it does use every interaction (and only those interactions) possible with each 16S RNA region known to be in the approximate location of the mRNA binding cleft on a ribosome. Our model also explains every peculiarity of the IF3 mRNA sequence. "Buffered" gene expression, along with transient fluctuations around the "set-point", are simple concepts contained within the thoughts that have been offered over the years by Ole. Wouldn't it be just great if IF3 expression actually turns out to be regulated in the way we have suggested?

ACKNOWLEDGEMENTS

We thank the Alfred Benzon Foundation for arranging a meeting in June, 1983 at which these ideas began to take serious form. We appreciate the useful conversations we had with Jim Ofengand, Marianne Grunberg-Manago, Mathias Springer and Ed Brody. Sid Shinedling helped us state our ideas clearly, and Kathy Piekarski was crucial to the preparation of the manuscript. This work was supported by NIH Grant GM28685.

REFERENCES

Brauer, D. and Wittmann-Liebold, B. (1977). *FEBS Lett.*, 79, 269.

Ehresmann, C., Ehresmann, B., Millon, R., Ebel, J.-P., Nurse, K., and Ofengand, J. (1984). *Biochem.*, 23, 429.

Gold, L., Pribnow, D., Schneider, T., Shinedling, S., Singer, B. S., and Stormo, G. (1981). *Ann. Rev. Microbiol.*, 35, 365.

Herr, W., Chapman, N. M., and Noller, H. F. (1979). *J. Mol. Biol.*, 130, 433.

Hershey, J. W. B. (1980). *Cell. Biol.*, 4, 1.

Howe, J. G. and Hershey, J. W. B. (1983). *J. Biol. Chem.*, 258, 1954.

Ingraham, J. L., Maaløe, O., and Neidhardt, F. C. (1983). *Growth of the Bacterial Cell*. Sinauer Associates.

Jay, E., Seth, A. K., and Jay, G. (1980). *J. Biol. Chem.*, 255, 3809.

Lestienne, P., Dondon, J., Plumbridge, J. A., Howe, J. G., Mayaux, J.-F., Springer, M., Blanquet, S., Hershey, J. W. B., and Grunberg-Manago, M. (1982). *Eur. J. Biochem.*, 123, 483.

Martin, J. and Webster, R. E. (1975). *J. Biol. Chem.*, 250, 8132.

Prince, J. B., Taylor, B. H., Thurlow, D. L., Ofengand, J., and Zimmermann, R. A. (1982). *Proc. Natl. Acad. Sci. USA,* 79, 5450.

Sacerdot, C., Fayat, G., Dessen, P., Springer, M., Plumbridge, J. A., Grunberg-Manago, M., and Blanquet, S. (1982). *EMBO J.,* 1, 311.

Shine, J. and Dalgarno, L. (1974). *Proc. Natl. Acad. Sci. USA,* 71, 1342.

Singer, B. S. and Gold, L. (1976). *J. Mol. Biol.*, 103, 627.

Steitz, J. A. (1969). *Nature,* 224, 957.

Steitz, J. A. (1973). *J. Mol. Biol.*, 73, 1.

Stormo, G. D., Schneider, T., and Gold, L. (1982). *Nucl. Acids Res.*, 10, 2971.

Stormo, G. D., Schneider, T., Gold, L., and Ehrenfeucht, A. (1982). *Nucl. Acids Res.*, 10, 2997.

Suryanarayana, T., and Subramanian, A. R. (1977). *FEBS Lett.*, 79, 264.

Thompson, J. F. and Hearst, J. E. (1983). *Cell*, 32, 1355.

van der Laken, K., Bakker-Steeneveld, H., Berkhout, B., and van Knippenberg, P. H. (1980). *Eur. J. Biochem.*, 104, 19.

van der Laken, K., Bakker-Steeneveld, H., and van Knippenberg, P. (1979). *FEBS Lett.*, 100, 230.

Vassilenko, S. K., Carbon, P., Ebel, J. P., and Ehresmann, C. (1981). *J. Mol. Biol.*, 152, 699.

Wagner, R. and Gassen, H. G. (1975). *Biochem. Biophys. Res. Comm.*, 65, 519.

Wagner, R. and Gassen, H. G. (1976). *FEBS Lett.*, 67, 312.

Woese, C. R., Gutell, R., Gupta, R., and Noller, H. F. (1983). *Microbiol. Revs.*, 47, 621.

Archaebacteria: Our First Look at Their Ribosome Component Genes

Patrick P. Dennis*, Ivy Hui*, Lawrence Shimmin*,
Joan McPherson*, Chia C. Pao**, and Alastair Matheson†

* *University of British Columbia, Vancouver*
** *Ortho Pharmaceuticals Corp.; Raritan, New Jersey*
† *University of Victoria, British Columbia*

My earliest contacts with Ole Maaløe and the Microbiology Institute were indirect and occurred over a period of several years while I was a postdoctoral student in the laboratories of C. Helmstetter, H. Bremer and M. Nomura. It was only after I had assumed a faculty position at the University of British Columbia that I was able to arrange time to visit Copenhagen and work at the Institute. Those brief visits to Copenhagen between 1976 and 1982 were extremely productive and our collaborative experiments produced some of the earliest evidence in support of what has become known as autogenous translational control of ribosomal protein synthesis (Dennis and Fiil, 1979; Fiil et al., 1980; Johnsen et al., 1982). Our experiments were confined primarily to the L10 operon and the results of those collaborations are described in detail elsewhere in this volume. Briefly, the L10 operon contains two ribosomal protein genes, *rplJ* (L10) and *rplL* (L12) and two RNA polymerase subunit genes *rpoB* (beta subunit) and *rpoC* (beta-prime subunit). Translation of the L10 operon mRNA produces L12 and L10 ribosomal protein in a stoichiometric ratio of 4:1; the two proteins form a tight 4:1 complex that is rapidly assembled into 50S ribosomal subunits. If, however, there is a deficiency in 23S and 5S rRNA production, the free L10 protein or the L10-L12 complex binds to a region in the leader sequence of the mRNA and thereby inhibits further translation of the mRNA.

About two years ago in my laboratory we became interested in archaebacteria and their machinery for protein synthesis. Archaebacteria are a heterogeneous group of organisms that represent

a distinct line of evolution separate from eubacteria and eukaryotes
(Figure 1). Subgroups within the archaebacteria kingdom include the
extreme halophiles and methanogens, the thermoacidophiles and the
sulfur-respiring organisms (Woese, 1981). Our work to date has been
confined largely to the halophilic species *Halobacterium cutirubrum*.
The machinery for protein synthesis in archaebacteria shows features
in common with both eukaryotes and eubacteria. Their ribosomes are
30S and 50S in size, and the 16S rRNA contains a typical pyrimidine-
rich sequence at its 3' end. However, protein synthesis in these
organisms is apparently resistant to the eubacterial antibiotics,
chloramphenicol and kanamycin, and sensitive to the eukaryotic
antibiotic, anisomycin. The archaebacterial elongation factor G is ADP-
ribosylated by diphtheria toxin, and the initiator methionine tRNA is
not formylated. One report suggests the presence of an intervening
sequence in archaebacterial tRNA genes (Kaine et al., 1983). The

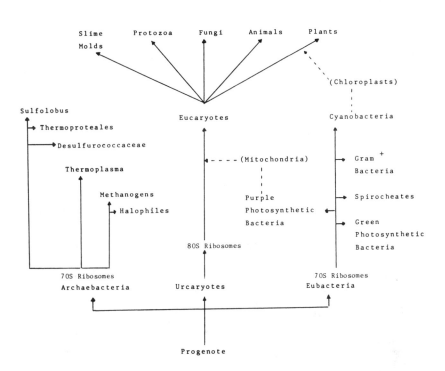

FIGURE 1   Cellular phylogeny of archaebacteria and their relationship
to eukaryotes and eubacteria. Both archaebacteria and eubacteria (true
bacteria) totally lack a nuclear membrane and are classified as
prokaryotic cell types. However, the evolutionary distance between
archaebacteria and eubacteria is no less than the distance between
either group and eukaryotes.

mRNA encoding bacterio-opsin, the only example studied in detail to date, lacks a purine-rich region in front of the AUG codon for initiation of translation; the mRNA leader sequence preceding the AUG codon is only two nucleotides in length (Dunn et al., 1981; Das Sarma et al., 1984).

The ribosomal proteins from these organisms show striking homology with eukaryotic ribosomal proteins and in general much less homology with eubacterial ribosomal proteins (Matheson, 1984). In *Halobacterium cutirubrum* the protein equivalent to L12 (HL20) is a multicopy protein, as in *E. coli,* but in *Halobacterium* it forms a 4:1 complex with a protein, HL11, which is highly related at the N-terminus to the *E. coli* L11 protein. Thus, the HL11 protein of *Halobacterium* may be functionally equivalent to the two proteins, L10 and L11, in the *E. coli* ribosome. Using protein chemistry and recombinant DNA technology we hope to address questions relating to ribosome structure, function and evolution and gene organization and regulation in the archaebacteria.

RIBOSOMAL RNA GENES

The genes encoding ribosomal RNA were isolated from a phage-λ library containing genomic *Halobacterium cutirubrum* DNA sequences. The techniques employed were basically those described by Maniatis et al. (1982). The plaques produced by recombinant phage were screened by hybridization with cDNA prepared from ribosomal RNA and primed with calf thymus DNA fragments and six were identified as carrying ribosomal genes. Two phage, λHc9 and λHc4, were characterized further and found to contain overlapping sequences. Northern blot hybridization of the nick-translated recombinant λ DNAs to 16S and 23S rRNA indicated that λHc4 contained sequences complementary to both RNAs, whereas λHc9 was complementary to only the 16S rRNA; Southern blot hybridization of restricted λ phage DNAs with radioactive cDNA identified the position of the ribosomal DNA sequences within the cloned DNA. Restriction fragments from these positive clones were subcloned into pBR322 and subsequently into pUC or M13 for DNA sequencing.

NUCLEOTIDE SEQUENCES OF THE 16S, 23S, AND 5S RNA GENES

The entire nucleotide sequence of the 16S and 5S rRNA genes and a partial sequence of the 23S gene has been determined (Figure 2). The 5S gene is 123 nucleotides in length and occupies the distal position in the cluster, beginning about 109 nucleotides downstream from the 3' end of the 23S rRNA gene. The sequence we have elucidated is in agreement with the previously published 5S rRNA sequence (Fox et al., 1982; Nazer et al., 1978).

The 16S gene occupies the proximal position in this gene cluster and is 1472 nucleotides in length. The sequence shows 88% homology with the 16S sequence from *H. volcanii* (Gupta et al., 1983); most of the base substitutions occur in regions of helical duplex structure within the RNA and are compensatory; this base substitution profile provides further evidence for evolutionary conservation of a universal structure for 16S rRNA (Woese et al., 1983). The 3' end contains the pyrimidine-rich sequence CCUCCU-OH used in eubacterial ribosomes to identify the translation initiation sites of mRNA (Figure 3).

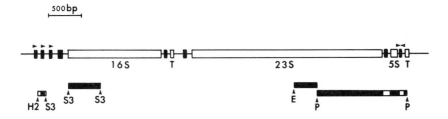

FIGURE 2 The structure of the ribosomal RNA gene cluster of
Halobacterium cutirubrum. The total length of the rRNA gene cluster
is about 6.0 Kb. The 5' flanking sequence consists of three highly
related direct-repeat units of 133 bp in length and are indicated by
the arrows over the solid boxes. The 16S gene is 1472 nucleotides in
length, and its 5' and 3' ends were deduced from sequence comparison
to H. volcanii 16S rRNA. The 23S rRNA gene is estimated by
restriction enzyme analysis to be about 2950 nucleotides in length.
The positions of the 5' and 3' ends of 23S rRNA gene were estimated
within about 4 nucleotides by S1-nuclease mapping. Flanking the 16S
gene and the 23S gene are long inverted-repeat sequences (solid
boxes). The 5S rRNA gene is located about 112 nucleotides downstream
from the 3' end of the 23S rRNA gene and is 123 nucleotides in
length. A putative alanine-tRNA gene is located in the intergenic
space between the 16S and 23S rRNA genes and a putative cysteine-
tRNA gene is located 110 nucleotides downstream from the 3' end of
the 5S gene. The 16S, 23S, 5S, and tRNA genes are shown as open
boxes. A region of inverted-repeat symmetry located between the 5S
gene and the putative cysteine-tRNA is indicated by two arrow heads
above a solid box.

     The 23S gene occupies the central position in this gene cluster
and is about 2950 nucleotides in length. The ends of the gene have
been localized using S1-nuclease mapping and the nucleotide sequence
at the ends has been determined. The 5' and 3' ends contain a 5-bp
complementing region and show some sequence homology with the *E.
coli* 23S rRNA (Brosius et al., 1980).

     In summary, the ribosomal RNA genes in *Halobacterium
cutirubrum* are organized in a manner analogous to that in the typical
prokaryote *E. coli* with the 16S, 23S and 5S gene occupying the
proximal, middle, and distal positions respectively; based on gene
organization, it would seem likely that they are also cotranscribed as
a long primary transcript that is subsequently processed to produce
the mature 16S, 23S and 5S rRNAs.

INVERTED REPEAT SEQUENCES SURROUNDING THE 16S GENE AND THE 23S GENE

The *H. cutirubrum* 16S and 23S genes are surrounded by long, 30-
40 base inverted repeats that are probably utilized for the initial
removal of precursor 16S and 23S sequences from the primary
transcript (Young and Steitz, 1978; Gegenheimer and Apirion, 1981).
The respective complementary regions in the primary transcript can

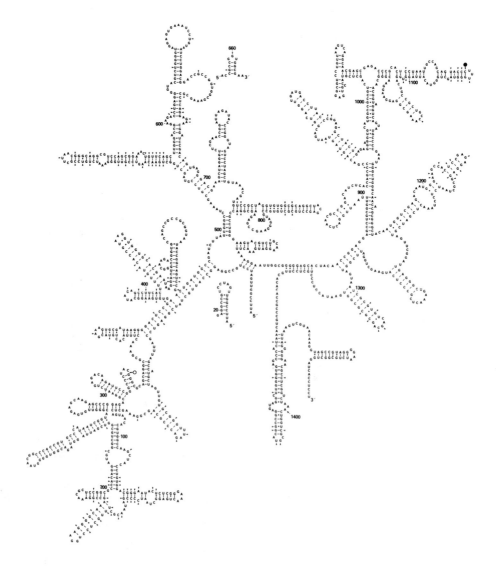

FIGURE 3 The secondary structure map of 16S rRNA of <u>Halobacterium</u> <u>cutirubrum</u>. The secondary structure of the 16S <u>rRNA of H.</u> <u>cutirubrum</u> is adopted from the universal structure proposed by Noller and Woese with several minor modifications. Standard Watson–Crick AU or GC base pairs are indicated by dashes and the nonstandard GU base pairs by dots in helical regions. The H. cutirubrum 16S rRNA is 88% homologous to the 16S rRNA from the related species, <u>H. volcanii</u>. The positions where the sequences show base substitution differences are indicated by triangles. A single base deletion between nucleotides 327-328 and a single base insertion at nucleotide 1078 are indicated in the <u>H. cutirubrum</u> sequence. Alternate structures for the 5' end of the molecule (nucleotides 1-21) and an internal loop region (nucleotides 653-667) are presented.

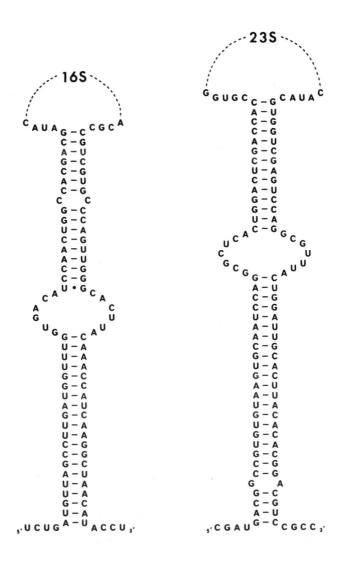

FIGURE 4 Long inverted-repeat sequences flanking the 16S and 23S ribosomal RNA gene. The nucleotide sequence and the potential secondary helical structures of the long inverted repeats surrounding the 16S and 23S rRNA genes of H. cutirubrum are illustrated. The loop length for the 16S structure is 1589 nucleotides and for the 23S structure is about 3070 nucleotides. The positions of the mature 5' and 3' ends of the 16S RNA are 83 and 43 nucleotides from the top of its stem; the mature 5' and 3' ends of the 23S rRNA are about 140 and about 26 nucleotides from the top of its stem.

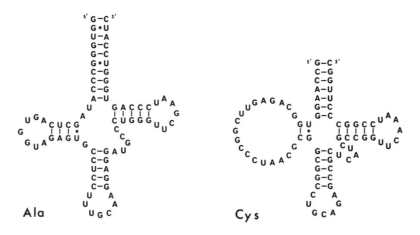

FIGURE 5   Cloverleaf structure of putative alanine and cysteine tRNAs. The nucleotide sequence and universal cloverleaf structure of the two putative tRNA sequences are illustrated. The alanine sequence is located in the 16S-23S intergenic space (nucleotides 2152-2229), and the cysteine sequence is located distal to the 5S rRNA (nucleotide D395 to D469).

form an extended stem structure with the 16S and 23S sequences present in the loops (Figure 4); the helical stem could be a substrate for an RNase III-like processing enzyme. The respective distances between the ends or beginning of inverted repeat sequences and the mature 5' and 3' ends of the 16S and 23S rRNAs are 83, 43, 23, and 140 nucleotides. Thus, if the initial processing sites are within the inverted-repeat sequences, the precursor 16S and 23S rRNAs would require further extensive processing at both the 5' and the 3' ends.

In *E. coli* the 16S and 23S genes are also surrounded by long, 30-40 base, inverted repeats (Young and Steitz, 1978; Gegenheimer and Apirion, 1981). These sequences are utilized by RNase III to remove the precursor 16S and 23S rRNA sequences from the primary transcript.The mature 3' end of the 16S rRNA and the 5' and 3' ends of the 23S rRNA are located very close to or within the inverted repeat sequences; the mature 5' end of the 16S rRNA is 83 nucleotides from the inverted repeat sequences.

TRANSFER RNA-LIKE SEQUENCES

There are two sequences in the vicinity of the rRNA genes that when transcribed, are capable of producing the conventional tRNA cloverleaf structure and contain many of the highly conserved nucleotides at defined positions of a tRNA molecule (Figure 5). It is possible that one or both of these sequences are produced by post-transcriptional processing of a large rRNA primary transcript as in *E. coli* (Lund et al., 1976; Gegenheimer and Apirion, 1981). The first sequence, representing a putative alanine-tRNA gene, is located in the spacer region between the 16S and 23S rRNA genes; its anticodon sequence is UGC. A similar alanine-tRNA gene has been recently located in a much shorter spacer in a rRNA transcription unit from the archaebacterial species, *Methanococcus vannielii* (Jarsch and Böch, 1983).

The second sequence encodes a putative cysteine-tRNA and is located 110 nucleotides downstream from the 3' end of the 5S rRNA gene; its anticodon is GCA. The structure of this tRNA is unusual in that it contains only 2 bp (including a GU base pair) in the D stem and 19 nucleotides in the D loop.

If mature tRNA in *Halobacterium* contains the standard seven base pairs in the acceptor stem, it is clear that neither the alanine nor the cysteine genes contain the unpaired CCA trinucleotide sequence present at the 3' acceptor end of all mature tRNAs. This implies that this sequence is a post-transcriptional addition to the immature tRNA, as in eukaryotic organisms, by a CCA nucleotidyl transferase system.

DIRECT REPEATS IN THE 5' FLANKING REGION

There are three bipartite direct repeats starting at nucleotides -551, -418, and -285 in the 5' sequence flanking the rRNA gene cluster. The two segments within each repeat unit are 27 nucleotides and 8 nucleotides in length (Figure 6). The spacing between these direct repeats is amazingly precise; the short spacers within each bipartite repeat are 13 nucleotides, and the long spacers between the three repeats are 85 nucleotides. Even within the spacer sequences there is a considerable degree of conserved sequence homology. The sequences further upstream also appear to be related to the repeat sequences. These repeat sequences may be an important component of the rRNA promoter.

Typical eukaryotic or eubacterial consensus promoter sequences are not found in regions believed to act as promoter regions on the *Halobacterium* genome (Das Sarma et al., 1984). Comparison of the well-characterized promoter region of the bacterio-opsin gene and the 5' sequence flanking a long open reading frame in two separate transposable sequences from *Halobacteria* has revealed a hexanucleotide sequence, AAGTTA, positioned 28-42 nucleotides upstream from the AUG translation start codon, which is conserved in at least five of the six positions in all three promoters (Das Sarma et al., 1984; Xu and Doolittle, 1983; Simsek et al., 1982; Dunn et al., 1981). The related sequence conserved at five of six positions, AAGTAA, is present in the long segment of the direct-repeat units preceding the rRNA gene cluster and may represent an integral part of the promoter of the ribosomal RNA operon.

TRANSCRIPTION TERMINATION

As in the case of promoter sequences the bacterio-opsin system represents the only well-characterized example of an archaebacterial transcription termination signal (Das Sarma et al., 1984). The sequence surrounding the major site of termination of the bacterio-opsin transcript contains (1) a stretch of 10 consecutive GC base pairs, (2) an imperfect 7-bp inverted repeat and (3) the sequence TTCAACGAC immediately following the inverted repeat. In *E. coli* the second feature is characteristic of rho-independent terminators and the third is related to a sequence found near several rho-dependent

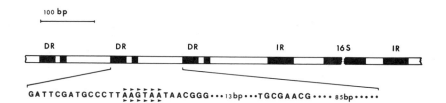

FIGURE 6   Structure of the direct repeat units in the 5' flanking region of the 16S rRNA. The two segments of the bipartite repeat unit are 27 and 8 nucleotides in length; the short interval spacer is 13 nucleotides and the long spacer between the three repeat units is 85 nucleotides. There is extensive conservation of sequences within both the short and the long spacers; the sequence upstream from the first repeat is also closely related to the repeat sequences. The position of the AAGTAA hexanucleotides in the long repeat unit is identified by triangles.

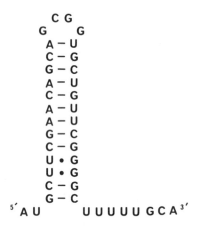

FIGURE 7   Inverted-repeat sequence in the 3' flanking region of the 5S rRNA gene. The 13-nucleotide inverted repeat located in the 110-nucleotide intergenic space between the 5S rRNA gene and the putative cysteine-tRNA gene is capable of forming the secondary helical structure illustrated.

terminator sites. The significance of these three features in *Halobacterium* termination is unknown.

Located in the 110-bp region between the 5S rRNA gene and the putative cysteine–tRNA gene beginning at nucleotide D303 is a 13-bp inverted repeat followed by five T residues (Figure 7). For *Halobacterium* this sequence is not considered to be GC-rich and the stem contains two GU base pairs. If this is an efficient terminater of transcription, the putative cysteine–tRNA would either not be a part of the rRNA transcription unit or would be produced in reduced amounts. The sequence between nucleotide D543 and D589 downstream from the putative cysteine–tRNA sequence contains three blocks of seven, ten, and eight consecutive GC base pairs; the overall sequence in this region is 78% GC and is followed by a 15-bp sequence that is 80% AT. If this is the normal termination site, the cysteine–tRNA gene would be within the rRNA transcription unit. The stoichiometry of tRNA and ribosomes has not been examined in *Halobacterium;* if the rRNA genes are a unique and single copy, and if the alanine and cysteine tRNA genes are part of the rRNA transcription unit and are efficiently processed, then the stoichiometry of each tRNA would be 1:1 with ribosomes. These two tRNA species would be major components of the tRNA pool.

COPY NUMBER OF RIBOSOMAL RNA GENES

Southern blot analysis of genomic DNA using restriction fragments from within or near the rRNA gene cluster indicate that the rRNA genes are present in a single unique copy on the *Halobacterium* genome. The probes include: (1) a fragment from the 5' flanking sequence containing a complete copy of the bipartite direct repeat unit; (2) a fragment from the 5' end of the 16S rRNA gene; (3) a fragment from the middle of the 23S rRNA gene and (4) a fragment spanning the distal end of the 23S rRNA gene, the 5S gene and a portion of the putative cysteine tRNA gene (see Figure 3). Each of these fragments probes only a single band of restricted genomic DNA (in some instances >25 Kbp in size). The results obtained with the 23S gene probe are illustrated in Figure 8. Together these results suggest that the rRNA genes are uniquely represented in the genomic DNA of *H. cutirubrum;* if there are multiple copies of the ribosomal RNA genes in the genome, the unit repeat size must extend for >25 Kbp in both the 5' and 3' directions from the rRNA genes.

The fragment containing the 5' bipartite direct repeat was hybridized at reduced stringency to determine if the repeat unit was present elsewhere in genomic DNA possibly as a component of other promoter sequences. At very low stringency, nonspecific hybridization of the probe to both *Halobacterium* DNA and lambda DNA sequences was observed. At intermediate and high stringency, hybridization was confined to the DNA restriction fragments spanning the 5' flanking sequences of the rRNA gene cluster; this indicates that the direct-repeat sequences are unique and associated only with the ribosomal RNA genes and not with other genes in the *Halobacterium* genome.

THE RIBOSOMAL PROTEINS FROM ARCHAEBACTERIA

An initial characterization of many archaebacterial ribosomal proteins indicates that they are more closely related to their eukaryotic than their eubacterial equivalents (for a review, see Matheson, 1984). The L12-equivalent protein from the *Halobacterium* ribosome (HL20; also generally referred to as the ribosomal A protein) exhibits substantial amino acid sequence homology with the equivalent A protein from eukaryotic ribosomes. The HL20 protein is less closely related to the *E. coli* L12 protein; direct comparison of the amino acid sequence indicates that substantial rearrangements of domains within the protein have probably occurred during evolution.

In the *Halobacterium* ribosome the HL20 protein forms a 4:1 complex with a protein designated HL11 (Matheson, 1984). This protein at the amino terminus is highly related to the *E. coli* L11 protein (and unrelated to L10, the protein which forms the 4:1 complex with L12 in *E. coli*). Thus *Halobacterium* may contain a fusion gene that encodes a multifunctional HL11 protein, equivalent to the *E. coli* L11 at its N-terminal domain and possibly equivalent to L10 at its C-terminal domain. We would like to understand the roles of HL11 and HL20 in 23 S rRNA binding, and in ribosome structure and function. Equally important, we would like to know how genes in archaebacteria, including ribosomal protein genes, are organized on the bacterial genome and how their expression is regulated.

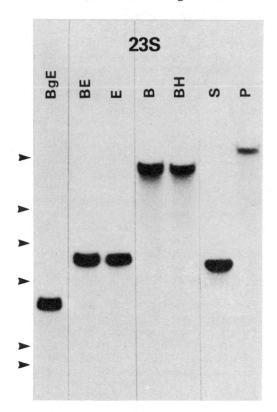

FIGURE 8 Genomic Southern hybridization using a 23S rRNA gene probe. Genomic <u>H. cutirubrum</u> DNA was digested with a variety of restriction endonucleases (BgE, BglII + EcoRI; BE, BamHI + EcoRI; E, EcoRI; B, BamHI; BH, BamHI + HindIII; S, SalI; P, PstI) fractionated by electrophoresis, transferred to nitrocellulose and probed with the ^{32}P-labeled EcoRI-PstI fragment from within the 23S rRNA gene. The size markers (λ DNA cut with HindIII) are indicated by arrows and are from top to bottom: 23.7, 9.5, 6.7, 4.3, 2.3, and 2.0 Kbp.

HL20	N-term.	MET[1]	GLU	TYR	VAL	TYR	ALA[6]	C-term.		
mRNA	5'	AUG	GAA_4	UAU_3	GUU_3	UAU_3	GCU_1	3'		
			G_7	C_8	C_8	C_8	C_8			
					A_3		A_5			
					G_9		G_{16}			
OLIGOS	3'	TAC	CTT_4	ATA_3	CAA_3	ATA_3	CG	5'	17 G/A	
17-MERS			C_7	G_8	G_8	G_8			16 OLIGOS	
	3'	TAC	CTT_4	ATA_3	CAT_3	ATA_3	CG	5'	17 C/T	
			C_7	G_8	C_9	G_8			16 OLIGOS	

FIGURE 9 Oligonucleotide probe for the HL20 gene of Halobacterium cutirubrum. The N-terminal amino acid sequence of the HL20 ribosomal protein is illustrated. Two mixtures, a G/A and a C/T, each containing 16 different 17-mer oligonucleotides were synthesized manually using phosphoramidite chemistry. At high stringency the G/A oligonucleotide hybridizes to a unique sequence in the H. cutirubrum genome. The subscript numbers in the nucleic acid sequences represent the codon-utilization preferences in the bacterio-opsin gene (Dunn et al., 1981).

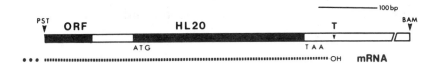

FIGURE 10 The HL20 gene of Halobacterium cutirubrum. The HL20 gene was cloned as a 1.2 Kbp PstI-BamHI fragment in pUC13. The position of the HL20 gene and an open reading frame (ORF) are indicated (shaded). Nuclease S1 mapping indicates that the HL20 mRNA is initiated upstream from the PstI site and terminated about 40 nucleotides beyond the HL20 gene in an AT-rich sequence (T).

CLONING OF THE HL20 GENE FROM *HALOBACTERIA*

From the N-terminal amino acid sequences of HL20, a 17-mer oligonucleotide was synthesized which was complementary to the first six codons of the gene (Figure 9). This oligonucleotide was used to probe restriction digests of genomic H. cutirubrum DNA, and a 1.2-Kb BamHI-PstI fragment that hybridized to the probe at high stringency was identified. The 1.2 Kb fragment was partially purified by acrylamide gel electrophoresis and cloned into the plasmid pUC13. About 7% of the recombinant plasmids were positive when probed with the oligonucleotides. Ten positive clones were chosen for further analysis and all, with one possible exception, contained an identical 1.2 Kb BamHI-PstI insert fragment. A map of the insert fragment illustrating the position and orientation of the HL20 gene is illustrated (Figure 10).

Preliminary base sequence analysis of the clone of the HL20 gene indicates that the AUG translation-initiation codon is not preceded by the usual eubacterial purine-rich Shine-Dalgarno ribosome binding site, which is complementary to the 3' end of the 16S rRNA (Shine and Dalgarno, 1974). However, the sequence GAG does occur at nucleotides 3-5 in the coding region of the gene; this sequence is complementary to the CCUCCU-OH sequence at the 3' end of the *Halobacterium cutirubrum* 16S rRNA. In the bacterio-opsin gene the sequence GGAG occurs at position 6-9 in the coding region, and the mRNA leader sequence in front of the AUG initiation codon is two nucleotides long (Das Sarma et al., 1984; Dunn et al., 1981).

The 183 nucleotides of HL20 5' flanking sequence do not contain a sequence related to the hexanucleotide AAGTTA found within the region 28-42 nucleotides in front of the AUG start codon in bacterio-opsin and two other *Halobacterium* open reading frames. This may indicate that the HL20 gene has either a unique and different promoter structure or that the HL20 gene is part of a multicistronic transcription unit. Our clone, which begins at a PstI site 183 nucleotides upstream from the HL20 coding sequence, contains a potential open reading frame terminating with a UGA stop codon located 89 nucleotides in front of the HL20 AUG translation initiation codon. Preliminary S1 mapping indicates that the HL20 mRNA is initiated somewhere upstream from the PstI site at the beginning of our clone. Transcription termination occurs in an AT-rich (75%) region about 40 nucleotides downstream from the HL20 termination codon; the AT-rich region is preceded by a GC-rich sequence totally lacking dyad symmetry. We are currently completing the sequence of the HL20 gene, recloning to obtain additional 5' flanking sequences and continuing S1 mapping studies to locate the 5' and 3' ends of the HL20 mRNA. Needless to say, we expect to learn a lot more about archaebacterial ribosomal protein genes, their organization and their regulation, and the function of their protein products within the next few years.

ACKNOWLEDGEMENTS

This work was supported by a grant from the Medical Research Council of Canada (MA6340) to P.P.D. We thank Tom Atkinson for help in preparing the HL20 17-mer oligonucleotide probe and Ross MacGillivray for his encouragement and technical advice.

REFERENCES

Brosius, J., Dull, T. J., and Noller, H. (1980). *Proc. Nat. Acad. Sci. USA,* 77, 201.

Das Sarma, S., Raj Bhandary, U., and Khorana, H. G. (1984). *Proc. Natl. Acad. Sci. USA,* 81, 125.

Dennis, P. P. and Fiil, N. P. (1979). *J. Biol. Chem.,* 254, 7540.

Dunn, R., McCoy, J., Shimsek, M., Majumdar, A., Chang, S., RajBhandary, U., and Khorana, H. G. (1981). *Proc. Natl. Acad. Sci. USA,* 78, 6744.

Fiil, N. P., Friesen, J. D., Downing, W. D., and Dennis, P. P. (1980). *Cell,* 19, 837.

Fox, G., Luehrsen, K., and Woese, C. (1982). *Zbl. Bakt. Hyg. Abt. Orig., C3,* 330.

Gegenheimer, P. and Apirion, D. (1981). *Microbiol. Rev.,* 45, 502.

Gupta, R., Lanter, J., and Woese, C. (1983). *Science,* 221, 656.

Johnsen, M., Christensen, T., Dennis, P. P., and Fiil, N. P. (1982). *EMBO Journal,* 1, 999.

Jarsch, M. and Böch, A. (1983). *N.A.R.,* 11, 7537.

Kaine, B., Gupta, R., and Woese, C. (1983). *Proc. Natl. Acad. Sci. USA,* 80, 3309.

Lund, E., Dahlberg, J. E., Lindahl, L., Jashunos, S. R., Dennis, P. P., and Nomura, M. (1976). *Cell,* 7, 165.

Maniatis, T., Fritsch, E., and Sambrook, J. (1982). In *Molecular Cloning,* Cold Spring Harbor Laboratory, Cold Spring Harbor, N.Y.

Matheson, A. (1984). In *The Bacteria,* C. Woese and R. Wolfe (eds.), Academic Press, in press.

Nazar, R., Matheson, A., and Bellemore, G. (1978). *J. Biol. Chem.,* 253, 5464.

Semsek, M., Das Sarma, S., RajBhandary, U., and Khorana, H. G. (1982). *Proc. Natl. Acad. Sci. USA,* 79, 7268.

Shine, J. and Dalgarno, L. (1974). *Proc. Natl. Acad. Sci. USA,* 71, 1342.

Woese, C. (1981). *Sci. Am.,* 244(6), 98.

Woese, C., Gutell, R., Gupta, R., and Noller, H. (1983). *Microbiol. Rev.,* 47, 621.

Xu, W-L. and Doolittle, W. F. (1983). *N.A.R.,* 11, 4195.

Young, R. A. and Steitz, J. A. (1978). *Proc. Natl. Acad. Sci. USA,* 75, 3593.

Some Puzzles of Translational Accuracy

Jonathan Gallant, Robert Weiss, James Murphy, and Megan Brown
University of Washington, Seattle

It was almost a generation ago that Ole Maaløe and his colleagues in Copenhagen established an agenda for the analysis of bacterial growth. The agenda called for definition of the parameters of bacterial growth, e.g., the rates of synthesis of various macromolecules, especially those of the protein synthetic apparatus; due attention to the network of interconnections between different cellular subsystems; and the implicit hope that the formal structure of this network could be elucidated by what we would now call systems analysis.

I like to think of this program in terms of a homely analogy. A master mechanic can often diagnose an automobile engine by listening to it, without needing to take it all apart and examine the separate parts under a microscope. None of us is a master mechanic yet, but Maaløe's program encouraged us to keep this possibility in mind, as a distant goal. This emphasis was visionary, even a bit wild-eyed, at a time when the mainstream of molecular biology was focussed on the approach of taking the machinery apart under the microscope. In those days, terms like "systems analysis" and "holism" were not the buzzwords they have become more recently.

In the event, Onkel Ole's program proved to be fruitful, although sometimes in unexpected ways. The vision is still visionary. To this day, we still lack a general systems analysis of bacterial growth. Nonetheless, the Copenhagen school influenced those of us who count ourselves disciples to pay attention to the integrative features of global control systems and to cross connections between their various levels of control. My group was certainly mindful of this program when we paid attention to the metabolic level of stringent control, a focus which led us directly to the magic spot nucleotides.

The stringent control system is, in essence, *E. coli's* response to the problems posed by limitation for one or more aminoacyl-tRNA species. In recent years, our interests have diverged somewhat from the way the cell copes with these problems to the question of what these problems are. We used to think of them in terms of the coordination of various biosynthetic subsystems (e.g., ribosome proteins, ribosomal RNAs, tRNAs, intermediary metabolism), essentially a matter of economics. More recently, we have realized that another kind of problem is involved: that of the *accuracy* of protein synthesis.

THE POOL BIAS EXPERIMENT

When ribosomes decode a messenger RNA triplet, they perform an act of selection among competing cognate and noncognate aminoacyl-tRNA (aa-tRNA) substrates. The outcome of this competition--which is to say the accuracy of decoding--depends on the kinetic constants for binding and processing of the competing aa-tRNAs, and their concentrations. It is therefore no surprise that if one jiggers the concentrations of aa-tRNAs, one changes the error frequencies at the relevant codons (Edelmann and Gallant, 1977; O'Farrell, 1978; Gallant, 1979; Parker and Friesen, 1980; Wagner and Kurland, 1980; Parker et al., 1980; Parker et al., 1983).

"Pool bias" experiments of this sort have revealed two surprises. One is that the stringent response buffers the translation system somehow against the increase in errors that otherwise would naturally occur at "hungry" codons calling for an aa-tRNA in short supply. Such errors occur infrequently in *rel+* cells which accumulate ppGpp, but much more frequently in *rel* mutants which do not do so. The mechanism evidently involves one or another effect of ppGpp on the ribosome cycle (O'Farrell, 1978; Gallant, 1979; Gallant and Foley, 1980; Wagner and Kurland, 1980; Wagner et al., 1982; Weiss et al., 1984). I will return to this puzzle, which remains to be solved, in the second part of this article. First, I will discuss the other surprising aspect of decoding under conditions of pool bias.

Pool biases elicit not only amino acid substitution errors, but also ribosome frameshifting (Atkins et al., 1979; Gallant and Foley, 1980; Weiss and Gallant, 1983). How does this come about?

We postulate that certain types of noncognate aa-tRNA binding throw translocation out of kilter, and thus shift the ribosome into a new reading frame. To test this interpretation and to inquire into the molecular mechanism, we need three types of information: identification of the sites of ribosome frameshifting, identification of the noncognate aa-tRNAs that do the mischief, and a model that explains why these specificities have anything to do with translocation.

Pool bias by aa-tRNA limitation

A year ago, we reported that limitation for trp-tRNA powerfully suppresses (+) frameshift alleles in the *rIIB* gene located just downstream of the one UGG codon in the early part of the gene (Weiss and Gallant, 1983). We further demonstrated by genetic means that the phenotypic shift into the (-) reading frame responsible for suppression must occur at this specific UGG codon (Weiss and Gallant, 1983). The importance of this demonstration is that it justifies a straightforward *general* method for identifying shifty sites: frameshifts stimulated by limitation for a given aa-tRNA occur at codons calling for that tRNA.

To identify the noncognate aa-tRNA responsible for this shift, we simply sought to redress the pool bias by limiting for a second aa-tRNA as well as trp-tRNA. In most cases, the second limitation either had no effect on the efficiency of phenotypic frameshifting, or reduced it by a small amount, about a factor of two. We think the latter effect results from a modest degree of recharging of trp-tRNA brought about by the second limitation's restriction of demand for trp-tRNA. We have demonstrated by direct assay of trp-tRNA that one secondary limitation does produce such a two-fold recharging (Weiss et al., 1984). However, when the second limitation was for leu-tRNA, then the efficiency of frameshifting was reduced much more drastically, by one to two orders of magnitude (Weiss and Gallant, 1983). The specificity of this result confirms our general hypothesis, and strongly suggests that a leu-tRNA isoacceptor is the culprit.

The difference between the small nonspecific or recharging effect and the large specific effect that identifies the culprit is quantitative rather than qualitative, and hence there is some ambiguity. Nonetheless, the quantitative difference is large, and we will later present genetic evidence confirming the different character of these two effects.

Pool bias by addition and by limitation

John Atkins and his colleagues (1979) followed a different approach, relying on the addition of purified tRNAs to a cell-free protein synthesis system programmed by phage MS2 RNA. They showed that the addition of tRNA-ser$_3$ (AGU/C) stimulated the formation of a frameshifted variant of MS2 coat; and that addition of tRNA-thr$_{maj}$ (ACU/C) stimulated the formation of a variant of MS2 replicase that is very likely a result of frameshifting in that gene. This method directly identifies the culprits, presumably noncognate to the sites at which they act.

They also showed that each of these induced frameshifts was antagonized by the simultaneous addition of one other specific tRNA. The frameshift induced by tRNA-ser$_3$ was antagonized by t-RNA-ala$_{1b}$ (GCA), and that induced by tRNA-thr$_{maj}$ was antagonized by tRNA-pro$_{min}$ (CCU/C). They drew the reasonable inference that the competing tRNAs were the *correct* ones cognate to the positions of the induced frameshifts.

We have tested this inference in a simple way. If it is correct, then limitation for the amino acid carried by the *competing* tRNA species should provoke the very frameshifts they antagonize. And so it does: Figure 1 shows that alanine limitation provokes formation of polypeptide 7, the frameshifted variant of the coat protein, and proline limitation provokes the formation of the putative frameshift variant of replicase.

Limitation for a variety of other amino acids had no such effect, except for two cases: limitation for proline or for serine induced the coat frameshift variant (Figure 1). The implication of this finding is that frameshifts are induced at Ser and Pro codons in the coat gene by other noncognate tRNAs present in the cell-free extract. We tested this inference by searching for the noncognate tRNAs responsible for

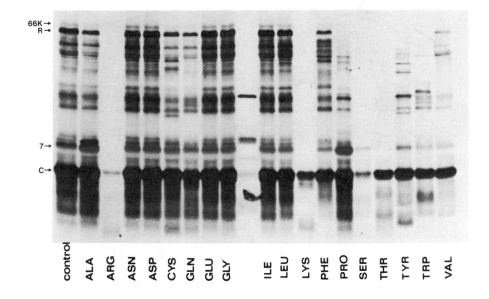

66K→
R→

7→

C→

control ALA ARG ASN ASP CYS GLN GLU GLY ILE LEU LYS PHE PRO SER THR TYR TRP VAL

FIGURE 1 The effect of amino acid limitation on translation of MS2 RNA in vitro. Preparation of MS2 RNA and crude E. coli cell-free extracts, incubation procedures and SDS polyacrylamide (17.5%) gel electrophoresis, and radioautography were done essentially as described by Atkins et al. (1979). Each lane displays an incubation labeled with $[^{35}S]$methionine in the absence of the indicated amino acid, the other 19 being present at 0.2 mM. The central, unmarked lane contains molecular weight standards. Polypeptide 7, a frameshifted variant of coat, and 66K protein, which is probably a frameshifted variant of replicase (see Atkins et al., 1979), are indicated; normal coat is indicated as C and normal replicase as R. Note the increased ratio of polypeptide 7 to normal coat in the incubations starved for Ala, Pro, or Ser; total labeling is much reduced in the Ser-limited incubation, but labeling of polypeptide 7 is about equal to labeling of coat, as compared with the normal ratio of roughly 1:10. Similarly, labeling of replicase is much reduced by limitation for Pro, but the ratio of 66K protein to replicase (both faintly visible) is about 1:1, in contrast to the normal 1:20.

these shifts, and the ser–tRNA and pro–tRNA isoacceptors that antagonized their effect. In both cases, we were able to identify a pair of tRNAs which exhibited the predicted effects. Figure 2 shows that tRNA-thr$_{maj}$ induces the coat frameshift and is antagonized by tRNA-pro$_{min}$ (this is the same pair of tRNAs involved in the replicase shift). Figure 2 also shows that tRNA-leu$_3$ (CUU/C) induces the coat frameshift, and is antagonized by tRNA-ser$_1$ (UCA/G). Note in Figure 2 that the competing effects of the cognate tRNAs are specific, as shown by the fact that tRNA-phe, our negative control, does not antagonize induced frameshifting in either case.

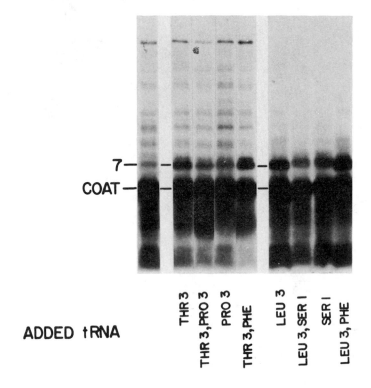

ADDED tRNA

THR 3
THR 3, PRO 3
PRO 3
THR 3, PHE
LEU 3
LEU 3, SER 1
SER 1
LEU 3, PHE

FIGURE 2 Identification of tRNAs involved in polypeptide 7 formation. Procedures were as in Figure 1, with a complete amino acid mixture provided. Purified isoaccepting tRNAs were added at 1 mg/ml. The tRNAs added are indicated beneath the appropriate lanes, and a control incubation with no added tRNAs is shown in the lefthand lane. The tRNAs and cognate codons are as follows: tRNA-thr_3 (ACU/C); tRNA-pro_3 (CCU/C); tRNA-leu_3 (CCU/C); tRNA-ser_1 (UCCA/G).

In short, the methods of tRNA limitation and addition cross-confirm one another in each of four cases, and predictions based on one method have been confirmed by the other, working in both directions.

This concordance between the limitation and addition methods gives some confidence that pool-bias manipulations correctly identify shifty sites and shifter tRNAs. Table 1 presents the information obtained in this way. The puzzle is to discern a pattern in these relationships which explains why translocation is thrown off.

The offset pairing model

For each case listed in Table 1, the first two bases of the shifty site are (or could be) identical to the second two bases normally read by

TABLE 1   Cases of frameshifting by noncognate tRNAs

Case	Shifty codon	Shifter tRNA	Base trans- located	Direction of frameshift	Gene	Ref.*
A	UGG	leu(CUX, UUA/G)	4	(−)	T4 *rIIB*	1
B	GCA	ser_3(AGU/C)	2	(+)	MS2 coat	2
C	CCC/U	thr_{maj}(AGU/C)	2	(+)	MS2 synthe- tase, coat	2,3
D	UCA	leu_3(CUU/C)	2	(+)	MS2 coat	3

The codons normally read by the shifter tRNAs are indicated in parentheses. The direction of the induced frameshift is indicated according to the genetic convention: a (−) frameshift moves the reading frame one base to the right, as when one base is deleted; a (+) frameshift moves the reading frame one base to the left, as when one base is added. (The reference by Atkins et al. unfortunately used the opposite convention.)

*References: (1) Weiss and Gallant, 1983; (2) Atkins et al, 1979; (3) Figure 2.

the shifter tRNA. The probability that this identity would recur at random, in five independent cases, is negligible.

The significance of this relationship is that each shifty codon can pair with a triplet in the anticodon loop of its counterpart shifter tRNA offset by one base to the 5' side of the normal anticodon. In this offset configuration, the U at position 33 of all tRNAs becomes the wobble position, in contrast to the normal configuration in which nucleotide 34 is the wobble base. Figure 3 illustrates the normal and offset pairing configurations for the cases listed in Table 1.

In the offset pairing model, the initial frameshift is thus not on the mRNA *but on the tRNA*. Abnormal pairing of this sort will then naturally lead to incorrect translocation, if mRNA is translocated by virtue of its association with a tRNA molecule, and the tRNA molecule moves. An mRNA paired one base up the anticodon loop from the normal position will be swung to an aberrant position, whatever the detailed mechanism of translocation.

One of us (R.W.) has analyzed space-filling models of the anticodon loop, in order to see what kinds of aberration might occur (Weiss, 1984). In the conventional model of translocation, with peptidyl-tRNA physically moving from the A to the P site, offset pairing appears to lead to two-base translocation. In Woese's "reciprocating ratchet" model of translocation, the whole tRNA molecule does not move, but there is instead an allosteric transition in the

Amino acid	Trp	Ala	Pro	Ser
Shifty codon	U G G	G C A	C C C	U C A
	⋮ ⋮ ⋮	⋮ ⋮ ⋮	⋮ ⋮ ⋮	⋮ ⋮ ⋮
Anticodon	**G A C U**	**U G G U**	**U G G U**	**G A G U**
	⋮ ⋮ ⋮	⋮ ⋮ ⋮	⋮ ⋮ ⋮	⋮ ⋮ ⋮
Normal codon	C U G	A G C/U	A C C/U	C U C/U
Shifty tRNA	tRNA-leu	tRNA-ser$_3$	tRNA-thr$_3$	tRNA-leu$_3$

FIGURE 3 Offset pairing. Nucleotides 36, 35, 34, and 33 in the anticodon loop of each shifty tRNA are shown in boldface, paired in the normal fashion with each cognate codon below, and in the proposed offset configuration with each shifty codon above.

anticodon loop from one stacking configuration to another (Woese, 1970). In this model, it appears that offset pairing might lead to either a two-base or a four-base translocation, depending on how the allosteric transition occurs (Weiss, 1984).

Further tests of the offset pairing model

Certain of the Lys codons (AAA/G) in the *rIIB* message appear to be quite shifty, as judged by strong phenotypic suppression of frameshift mutants induced by limitation for lys-tRNA. Shifts in both directions must occur, for we can suppress certain frameshift alleles of each sign; the suppressible (+) and (-) alleles are located in different regions of the gene, suggesting that Lys codons at different locations are involved.

The offset pairing model predicts that the shifter tRNAs normally read codons in which the second two bases are identical to the first two bases of the Lys codons; hence the suspects are gln-tRNA (CAA/G) and glu-tRNA (GAA/G). We have accordingly constructed the appropriate isogenic pair of temperature–sensitive synthetase mutants, and performed the double limitation test described earlier. The results are shown in Table 2.

It can be seen that gln-tRNA limitation drastically reduces the phenotypic suppression of *FC370*[+] induced by lys-tRNA limitation. As a control, note that simultaneous limitation for ser-tRNA has no such effect. These results suggest that the shifter responsible for suppression of *FC370*[+] is a gln-tRNA, in good agreement with the model.

In the case of *FC151*[-], our results are a little less decisive. Limitation for glu-tRNA does reduce suppression by a factor of about 10, suggesting that this tRNA is the culprit. Limitation for ser-tRNA has little effect, again providing a specificity control. Limitation for gln-tRNA reduces suppression by a factor of two, which provides a second and slightly ambiguous specificity control. We believe that this twofold decrease in suppression reflects the nonspecific recharging of lys-tRNA, an effect alluded to earlier, and the experiments provide internal evidence for this contention. The double limitations involving gln-tRNA were run simultaneously with the two frameshift mutants;

TABLE 2   LX suppression indices

Phage mutant	Host strain	tRNA limiting			
		Lys	Ser,Lys	Gln,Lys	Glu,Lys
370⁺	glnS(Ts)	42.3 ± 8.7 (6)	57.0 ± 21 (3)	3.23 ± 1.4 (8)	
		100%	**135%**	**7.6%**	
151⁻	glnS(Ts)	14.4 ± 3.4 (5)		6.9 ± 1.8 (8)	
		100%		**48%**	
151⁻	gluS(Ts)	27 ± 10.7 (7)	15.2 ± 3.6 (3)		2.8 ± 1.1 (6)
		100%	**56%**		**10%**

Host strains were derivatives of CP79, a lambda lysogen that restricts
rIIB phage development and that is *relA⁻*; temperature-sensitive alleles
of *glnS* and *gluS* (for which we are grateful to Leif Isaksson) were
introduced by cotransduction with nearby Tn10 elements. Phage
infections were carried out essentially as described by Weiss and
Gallant (1983). Briefly, exponential cultures of the host strains,
growing at 30°C in M9-glucose medium, were infected at moi = 0.2;
after 10-12 min, unadsorbed phage were inactivated with antiserum;
after 30 min the infected cells were diluted and plated on a permissive
host (S/6), and the plates were incubated overnight at 30°C. The
resulting plaques enumerate the frequency of rare infected cells that
leak through the block against *rIIB* and produce at least one phage.
This was about 5×10^{-4} for both phage mutants in control infections
and many times higher in infections limited for lysyl-tRNA. The LX
suppression index is the frequency of leaky infections under the latter
condition, divided by that in the control. Suppression indices are
given as mean standard error of the mean, with the number of replicate
experiments in parentheses; the bold numbers indicate the average
LX suppression ratio with single limitation for lys-tRNA alone, times
100; that is, the relative efficiency of Lys-limited phenotypic
suppression in percent. Aminoacyl-tRNA limitations (all during the
first 30 min of the infection cycle) were accomplished as follows: Lys,
100 μg/ml lysine-hydroxamate; Ser, 250 μg/ml serine-hydroxamate;
Gln, *glnS*(Ts) mutant at 39.6°; Glu, *gluS*(Ts) mutant at 40°C.

in each experiment, we infected duplicate aliquots of the *glnS*(Ts)
host culture with each of the phage mutants, and shifted them to the
restrictive temperature side by side. In each of eight experiments,
suppression of *FC370⁺* was reduced much more than suppression of
FC151⁻. Although these experiments are subject to considerable
variability, as the standard errors indicate, the different responses

of the two phage mutants is highly significant ($P < 0.01$ by the t test). Therefore, a secondary limitation for gln-tRNA must restrict suppression of $FC370^+$ and $FC151^-$ by two different mechanisms. This consideration reinforces our conclusion that the large effect on suppression of $FC370^+$ correctly identifies gln-tRNA as the shifter in this case.

Conversely, the weak two-fold effect of gln-tRNA limitation on suppression of $FC151^-$ must be of the nonspecific or recharging variety. This implies that the stronger, ten-fold effect of glu-tRNA limitation identifies glu-tRNA as the shifter here. Thus, we have at least one, and probably two, more confirmations of the general tenet of the offset pairing model--namely, the rule that the first two bases of a shifty codon correspond to the second two bases normally recognized by the shifter tRNA. I submit that we now have a sufficient number of confirmations to be reasonably certain that the offset pairing is at least the first step in the pathway of abnormal translocation.

The data further suggest that the directional specificity of the pathway resides in the structure of the shifter tRNA. Gln-tRNA must promote four-base translocation, a (-) frameshift, so as to suppress $FC370^+$. Atkins et al. (1979) have earlier reported that addition of this tRNA to the *in vitro* system promotes another (-) frameshift in translating MS2 coat. Evidently, gln-tRNA prefers four-base translocation when paired in the postulated offset way. On the other hand, glu-tRNA apparently promotes a two-base translocation, a (+) frameshift, so as to suppress $FC151^-$. Note that tRNA-thr$_{maj}$ appears twice in Table 1—it stimulates frameshifts on both coat and replicase of MS2—and in both cases the shifts are in the (+) direction. This tRNA evidently prefers two-base translocation.

In the Weiss model, the difference between two-base and four-base translocation lies in two different versions of a stacking transition akin to that proposed by Woese (1970), but starting from offset pairing. One might reasonably expect the tendency for the transition to occur in one way or the other to be determined by the sequence of the anticodon loop and stem, and thus to vary from one tRNA to another, as our results suggest.

THE EFFECT OF ppGpp ON ACCURACY

Wagner et al. (1982) have proposed a tricky kinetic model to explain the error-preventing effect of ppGpp. They note that ppGpp binds EF-Tu and thus lowers the effective level of binary complex (EF-Tu-GTP). This will naturally reduce the formation of ternary complex (EF-Tu-GTP-aa-tRNA) for nonlimiting, fully charged aa-tRNA species, including those which can misread hungry codons calling for aa-tRNA species in short supply. They further argue that this reduction in ternary complex will not occur for an aa-tRNA species limited by a prior bottleneck in tRNA charging, because in this case recharging of the tRNA can compensate for the reduction in binary complex. In this way, ppGpp could force nonlimiting ternary complex levels down toward that of the limiting one, and thus even out competition between them.

Several *in vitro* studies indicate that there should be only a slight diminution in ternary complex formation at the concentration of ppGpp and GTP that prevail during the stringent response *in vivo*, about 1 mM each (Miller et al., 1973; Dix et al., 1983; Pingoud et al., 1983). It is therefore not clear that ppGpp could affect accuracy *in vivo* in the way Wagner et al. postulate.

In order to resolve this question *in vivo,* we asked whether a high level of ppGpp significantly reduces the rate of peptide chain growth in whole cells, as the model of Wagner et al. predicts. Ordinarily a high level of ppGpp is maintained in *rel+* cells only during aminoacyl-tRNA limitation, a condition which would complicate the interpretation of the experiment. We avoided this complication by examining a circumstance in which *unstarved* cells retain a high ppGpp level. The *spoT−* mutant allele drastically reduces the rate of ppGpp turnover (Laffler and Gallant, 1974). As a result, the ppGpp accumulated during a brief amino acid starvation in a *spoT−rel+* strain declines very slowly after the amino acid is restored, whereas it plummets to basal level within a minute or two in the *spoT+rel+* genotype (Figure 4). We examined the rates of peptide chain elongation

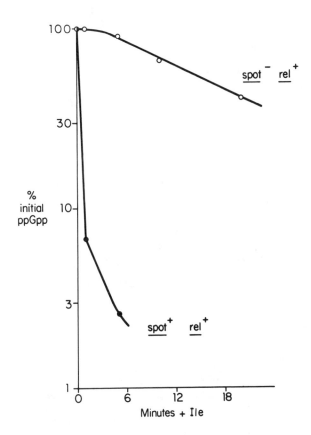

FIGURE 4   Levels of ppGpp. An isogenic $\underline{spoT^+rel^+}$ and $\underline{spoT^-rel^+}$ pair of strains were labeled with $^{32}PO_4$ to equilibrium and subjected to endogenous isoleucine limitation by addition of valine (2.5 mg/ml) for 20 minutes. Isoleucine (100 ug/ml) was then added and samples withdrawn at intervals for assay of ppGpp as described by Gallant et al. (1979). Filled circles: $\underline{spoT^+}$; open circles: $\underline{spoT^-}$.

between five and about fifteen minutes after reversing starvation in this isogenic pair, and in an isogenic *spoT⁻rel⁻* strain that fails to accumulate any ppGpp during the initial starvation. Comparison of these three strains makes the ppGpp level the sole experimental variable.

Our metric of peptide chain growth-rate is the transit time for completion of beta-galactosidase (Schleif et al., 1973). The square root plot of Schleif et al. linearizes the initial induction kinetics, so as to permit accurate extrapolation to the time required for the synthesis of the first beta-galactosidase molecules. Figure 5 shows a representative experiment in which we first starved briefly for isoleucine and then restored it, as in Figure 4. Figure 6 shows the results of a similar experiment involving leucine limitation and

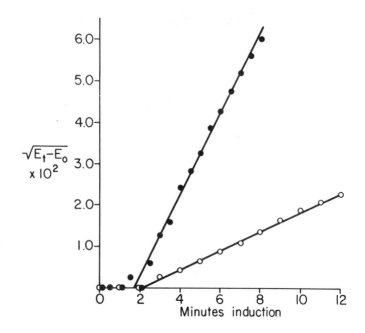

FIGURE 5 Induction kinetics of β-galactosidase. The strains were subjected to endogenous isoleucine limitation through false feedback inhibition of the pathway by valine (2.5 mg/ml). After 20 minutes, isoleucine (100 μg/ml) was added. Five minutes later, the lac operon was induced by the addition of IPTG (2 mM) and cAMP (5 mM). At frequent intervals, 300 μl aliquots were withdrawn to iced tubes containing chloramphenicol (100 μg/ml final concentration). The samples were permeabilized by the method of Putnam and Koch (1975), and assayed for β-galactosidase activity by adding 100 μl of each extract to 400 μl of ONPG (0.8 mg/ml in 0.05 M sodium phosphate, pH 7.0) and following absorbance at 420 nm. One arbitrary unit of enzyme activity corresponds to a change of A_{420} of 1.0 per min in this assay, corrected for the background rate of change in samples taken immediately after induction. The ordinate presents these rates in the square-root plot of Schleif et al. (1973). Closed circles: spoT⁺rel⁺; open circles: spoT⁻rel⁺.

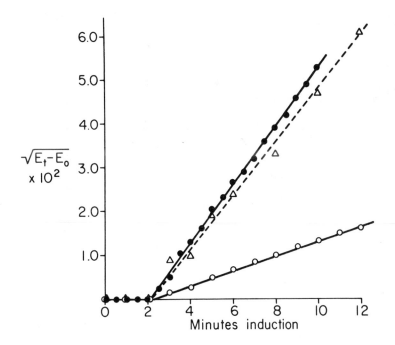

FIGURE 6 More induction kinetics. The experiments were performed as in Figure 3, except that limitation was for leucine rather than isoleucine. The strains, which are leucine auxotrophs, were centrifuged, washed, and incubated for 20 min in leucine-free medium; then, leucine (100 μg/ml) was added back. Closed circles: <u>spoT$^+$rel$^+$</u>; open circles: <u>spoT$^-$rel$^+$</u>; open triangles: <u>spoT$^-$rel$^-$</u>.

restoration. Table 3 summarizes the results of several experiments of each sort.

It is clear that the high level of ppGpp characteristic of the *spoT$^-$rel$^+$* genotype under these conditions does not significantly extend the lag preceding the first appearance of beta-galactosidase activity. We conclude that in whole cells, ppGpp cannot have more than a small effect on ternary complex formation, or any other step that constitutes a sizeable fraction of the step time of protein elongation.

Once beta-galactosidase begins to accumulate, it does so about four times slower in the *spoT$^-$rel$^+$* strain than in its partners (Figure 4 and 5, Table 3). The same is true of the overall rate of protein synthesis, which returns to normal very gradually in the *spoT$^-$rel$^+$* strain, in step with the gradual decline of ppGpp level. During the period that beta-galactosidase began to accumulate in the experiments of Figures 5 and 6, the rate of protein synthesis was about five times lower in this strain than in its two partners. Since the rate of peptide chain growth is not detectably lower, it follows that ppGpp inhibits protein synthesis principally at the level of initiation. The four- to fivefold inhibition we observe *in vivo* is consistent with the effect of ppGpp on IF2 *in vitro* (Legault et al., 1972; Yoshida et al., 1972).

TABLE 3   Summary of peptide chain-growth experiments

Genotype	Lag	Slope
$spoT^+relA^+$	2.17 ± 0.14 (6)	0.81 ± 0.054 (6)
$spoT^-relA^+$	2.28 ± 0.13 (6)	0.19 ± 0.014 (6)
$spoT^-relA^-$	2.4 ± 0.4 (2)	0.68 ± 0.08 (2)

The data for lag (in min) and slope following the lag are given ± standard error of the mean, with the number of replicate experiments in parentheses. Two of the experiments with the $spoT^+rel^+$ and $spoT^-rel^+$ pair were done by isoleucine limitation and restoration, as in Figure 3; four were done by leucine limitation and restoration, as in Figure 4. Both of the experiments with the $spoT^-rel^-$ strains were done by leucine limitation and restoration. The strains are isogenic derivatives of AT2538, constructed as described by Laffler and Gallant (1974a,b). The values for lag (intercept to the x axis) and slope were calculated for each experiment by means of regression analysis.

O'Farrell (1979) has pointed out that ppGpp inhibition of initiation might serve to keep up the charging level of a limiting aa-tRNA, by further reducing the rate of protein synthesis and thus the drain on a limiting aa-tRNA species. This model is refuted by the fact that amino acid-limited rel^+ cells do not exhibit lower rates of residual protein synthesis than their rel^- counterparts (Gallant et al., 1970). Moreover, several studies demonstrate that the charging of a limiting aa-tRNA is not kept at a significantly higher level in rel^+ than rel^- cells (Bock et al., 1967; Yegian and Stent, 1969; Piepersburg et al., 1979; Weiss et al., 1984). Our evidence that ppGpp inhibits initiation (but not elongation) *in vivo* is compatible with both of these observations; computer modelling by Harley et al., (1981) shows that the overall rate of protein synthesis is quite insensitive to the frequency of initiation when the bottleneck is at aa-tRNA charging. (Indeed, these considerations constitute experimental verification of Harley's computer model of the kinetics of protein synthesis.) I should add that the similar charging levels in limited rel^+ and rel^- strains also falsifies the model of Wagner et al., without some very dubious *ad hoc* assumptions about the partitioning of aa-tRNA between the free and ternary complex forms (Weiss et al., 1984; Yarus and Thompson, 1984).

One is thus driven to consider models in which ppGpp affects the intrinsic discrimination specificity of the ribosome, or the operation of a proofreading step, rather than the concentrations of the aa-tRNA substrates. All models of the former type differ from recharging models in the following respect: they predict that a high level of ppGpp increases accuracy at *all* codons, whereas the latter class of models predicts an apparent effect on accuracy only at hungry codons.

We have tested this prediction by examining mistranslation of a mutant UGA codon by trp-tRNA. Other experiments demonstrate that trp-tRNA is normally responsible for UGA readthrough (Hirsh and

TABLE 4   The effect of (p)ppGpp on nonsense suppression, expressed as the ratio of phage produced in a ser-tRNA-limited infection to phage produced in a normal infection

rIIB allele	Genotype of bacteria	
	$relA^+$	$relA^-$
N24 UGA	1.5	1.0
X655 UAG	1.6	1.6

The hosts are an isogenic $relA^+/relA^-$ pair: CP79-111 $relA^+$ and CP79-119 $relA^-$. Ser-tRNA limitation is with 250 µg/ml serine hydroxamate (reversed after 30 min). *N24* UGA indicates UGA leakiness (i.e., readthrough of UGA by normal trp-tRNA), and *X655* UGA indicates amber suppression by *supE44*.

Gold, 1971; Grosjean et al., 1980). We have confirmed this conclusion in the case of the *rIIB* mutant *N24*(UGA) by showing that trp-tRNA limitation decreases readthrough (Weiss and Gallant, 1983). We have now measured readthrough of *N24*(UGA) in an isogenic $relA^+/^-$ pair subjected to limitation for a different aa-tRNA, ser-tRNA. The stringent response to this limitation produces accumulation of ppGpp in the rel^+ but not the $relA^-$ strain. The prediction, then, is that readthrough of *N24*(UGA) under this condition should be decreased in the $relA^+$ strain but not in its $relA^-$ partner.

Unhappily, the prediction was not fulfilled: readthrough of *N24*(UGA) was not significantly affected in either strain, nor was amber suppression by the *supE44* gln-tRNA (Table 4). We used amber suppression as an additional control because ribosome mutations in *rpsL* and *rpsD* that affect discrimination specificity and/or proofreading have corresponding effects both on nonsense suppression (Gorini, 1974) and on normal readthrough by mistranslation (Stringini and Brickman, 1973; Andersson et al., 1982; Kurland and Gallant, 1984). I should add that Jack Parker and his colleagues have obtained similar results with a different and more direct metric of mistranslation: they found that certain amino acid substitution errors in MS2 coat protein were not significantly reduced by ser-tRNA limitation in either member of a different $relA^+/^-$ pair (Parker and Holtz, 1984).

Accordingly, this discussion must end on a note of mystery. *All* plausible models to account for the effect of ppGpp on translational accuracy seem to have been falsified. Perhaps the original designation of ppGpp as "magic spot" was the right one after all. Or, perhaps we need a new idea about stringent control of the ribosome cycle in relation to ribosome accuracy.

ACKNOWLEDGEMENTS

We are grateful to Linda Palmer for expert technical assistance in many of the experiments reported here, and to Dale Lindsley for expert technical assistance in several of the phage experiments of Table 2. This work was supported by grants GM1362 from the National Institutes of Health and NP-279E from the American Cancer Society.

REFERENCES

Andersson, D. I., Bohman, K., Isaksson, L. A., and Kurland, C. G. (1982). *Mol. Gen. Genet.,* 187, 467.

Atkins, J., Gesteland, R., Reid, B., and Anderson, C. (1979). *Cell,* 18, 1119.

Bock, A., Faiman, L. E., and Neidhardt, F. C. (1966). *J. Bacteriol.,* 92, 1076.

Chase, M., and Doermann, A. (1958). *Genetics,* 43, 332.

Dix, D. B., Thompson, R. C., Mackow, E. R., and Chang, F. N. (1983). *Arch. Biochem. Biophys.,* 223, 319.

Edelmann, P., and Gallant, J. (1977). *Cell,* 10, 131.

Gallant et al. (1970). *Cold Spring Harbor Symposium on Quantitative Biology,* Volume XXV.

Gallant, J. (1979). *Ann. Rev. Genet.,* 13, 393.

Gallant, J., and Foley, D. (1980). In *Ribosomes: Structure, Function and Genetics,* G. Chambliss et al. (eds.). University Park Press. pp. 615-638.

Gorini, L. (1974). In *Ribosomes,* M. Nomura, A. Tissieres, and P. Lengyel (eds.). Cold Spring Harbor Laboratory. pp. 791-804.

Grosjean, H. J., DeHenau, S., Houssier, C., and Buckingham, R. H. (1980). *Archs. Int. Physiol. Biochim.,* 88, 168.

Harley, C. B., Pollard, J. W., Stanners, C. P., and Goldstein, S. (1981). *J. Biol. Chem.,* 256, 10786.

Hirsh, D., and Gold, L. (1971). *J. Mol. Biol.,* 58, 459.

Kurland, C. G., and Gallant, J. (1984). In *Accuracy in Biology,* D. Galas (ed.), Mathuen. In press.

Laffler, T., and Gallant, J. (1974). *Cell,* 1, 27.

Legault, L., Jeantet, C., and Gros, G. (1972). *FEBS Letters,* 27, 71.

Miller, D. L., Cashel, M., and Weissbach, H. (1973). *Arch. Biochem. Biophys.,* 154, 675.

O'Farrell, P. (1978). *Cell,* 14, 545.

Parker, J., and Friesen, J. D. (1980). *Mol. Gen. Genet.*, 171, 439.

Parker, J., and Holtz, G. (1984). *Biochem. Biophys. Res. Comm.*, 121, 487.

Parker, J., Johnson, T. C., Borgia, P. T., Holtz, G., and Remaut, E. (1983). *J. Biol. Chem.*, 258, 10007.

Piepersberg, W., Geyl, D., Buckel, P., and Bock, A. (1979). In *Regulation of Macromolecular Synthesis by Low Molecular Weight Mediators*, G. Koch and D. Richter (eds.). Academic Press. pp. 39-52.

Pingoud, A., Gast, F-U., Block, W., and Peters, F. (1983). *J. Biol. Chem.*, 258, 1320.

Putnam, S. L., and Koch, A. L. (1975). *Analyt. Biochem.*, 63, 350.

Schleif, R., Hess, W., Finkelstein, S., and Ellis, D. (1973). *J. Bacteriol.*, 115, 9.

Stringini, P., and Brickman, E. (1973). *J. Mol. Biol.*, 75, 659.

Wagner, G., and Kurland, C. G. (1980). *Mol. Gen. Genet.*, 180, 139.

Wagner, G., Ehrenberg, M., and Kurland, C. G. (1982). *Mol. Gen. Genet.*, 185, 269.

Weiss, R., and Gallant, J. (1983). *Nature,* 302, 389.

Weiss, R., Murphy, J., Wagner, G., and Gallant, J. (1984). *Bentzon Symposium 19,* Munksgaard, Copenhagen. In press.

Weiss, R. (1984). *Proc. Natl. Acad. Sci. USA.* In press.

Woese, C. (1970). *Nature,* 226, 817.

Yarus, M., and Thompson, R. C. (1983). In *Gene Function in Prokaryotes,* J. Beckwith, J. Davies, and J. Gallant (eds.), Cold Spring Harbor Laboratory. pp. 23-64.

Yegian, C. D., and Stent, G. S. (1969). *J. Biol. Biol.*, 39, 45.

Yoshida, M., Travers, A., and Clark, B. F. C. (1972). *FEBS Letters,* 23, 163.

Tuning the Ribosome

Charles G. Kurland
The Biomedical Center; Uppsala, Sweden

INTRODUCTION

The quintessential question posed by the Copenhagen School begins with the phrase "How many...." This quantitative setting is the natural habitat for anyone interested in the parameters which determine the frequency of translational errors. It turns out, however, that there is a still deeper connection between the biology of translational accuracy, and the mission of Ole Maaløe's cohort.

Until recently, a principle concern of students of translational accuracy was to determine how the physical properties of codons and anticodons would limit the precision of protein synthesis. Now, however, we believe that there is in principle no physical limit on the accuracy of tRNA selection. Rather, we believe that the accuracy of translation is set at a biologically optimal level, which is determined by the requirements for maximum growth rates.

Maaløe's view of bacterial growth is based in part on the intuitive notion that gene expression must be carried out at maximum efficiency. Hence, he would expect that a bacterium that accumulates more ribosomes than are needed to support its protein synthesis will have to pay a price for such superfluity. That is, this extra baggage would in Maaløe's view keep the bacterium from attaining the maximum growth rate supported by a given medium. This notion of optimality can be formally defined and extended to all of the components of exponentially growing bacteria (Ehrenberg and Kurland, 1984, Kurland and Ehrenberg, 1984). Among its many, sometimes surprising consequences are those concerning the optimal accuracy of gene expression.

We can introduce the issue of the optimality of error frequencies by considering first the influence of a given average missense frequency per codon on macromolecular devices of different size. Thus, it is easily shown that for the class of small enzymes with amino acid sequence lengths of the order of 300, an average missense frequency

108

close to 5×10^{-4} means that roughly 90% of the sequences can be produced without errors. On the other hand, the amino acid sequences incorporated into a single ribosome are close to an aggregate length of 10,000. Consequently, there will be fewer than 5% error-free ribosomes at this error frequency. In order to accumulate ribosomes that are 90% error free the missense frequency would have to be close to 10^{-5} per codon. How does the bacterium choose between an error frequency of 5×10^{-4} and one closer to 10^{-5}? Clearly an assessment of the optimal error level would depend on the weight given to the relative costs associated with the incorporation of errors into protein structures versus those associated with avoiding these errors.

The consequences of errors

Until recently the principle methodology for assessing the consequences of amino acid substitutions in proteins has been the analysis of missense mutants. Unfortunately, a systematic bias is introduced by this sort of approach. For example, when a few hundred mutants with altered forms of beta-galactosidase were screened, only five percent were identified as missense mutants, while the vast majority were found to be nonsense mutants (Langridge, 1968a). This distribution is the mirror image of what would be expected from the relevant distribution of codon frequencies, and it suggests that most of the missense mutations are silent. In other words, contrary to a common prejudice, it would appear that the consequences of many missense events are difficult to detect.

Indeed, if nonsense mutations expressed in the genes coding for beta-galactosidase (Langridge, 1968b, Langridge and Campbell, 1969) the *lac* repressor (Miller et al., 1982), and RNA polymerase (Nene and Glass, 1982) are suppressed by batteries of different tRNA-nonsense suppressors, a large collection of amino acid-substituted proteins is obtained, and these can then be screened with respect to their activities. The resulting data show that most of the altered forms retain their original activities to a large degree, though aberrations such as enhanced temperature sensitivity are often observed. Even the amino acid substitutions in so important a protein as RNA polymerase most often (135/190) lead to a protein with sufficient activity to sustain bacterial growth (Nene and Glass, 1982).

Such results are underscored by recalling that most tRNA-nonsense suppressors have significantly lowered affinity for their corresponding charging enzymes (Smith, 1979). In addition, the charged suppressor tRNA species are most often processed with a lower efficiency than are wildtype tRNA by the ribosome (Gorini, 1971). Both of these kinetic limitations on the activities of suppressor tRNAs would tend to lower the expression level of the suppressed gene products. This means that some of the amino acid-substituted versions of either beta-galactosidase or RNA polymerase that fail to support growth, may have done so because the concentration of the suppressed enzymes was too low and not because the specific activity of the altered protein was too low. In summary, if the above three proteins are as representative as they seem to be, the inescapable conclusion is that proteins with a single amino acid substitution retain most often a major portion of their wildtype activity.

It is also possible to compare the amino acid sequences of the same nominal enzyme isolated from different organisms. Here, we most often find multiple amino acid differences, the number of which is often taken as an index of the evolutionary distance separating the molecules. Likewise, the suppression of frameshift mutations by multiple reading-frame shifts often involves strings of amino acid substitutions between the compatible frameshifts (Crick et al., 1961). Nevertheless, the resulting gene products are often nearly as effective as the wildtype proteins (Crick et al., 1961).

Thus, while major or minor changes in the amino acid sequences may influence kinetic parameters, temperature sensitivity, ionic response and the like, the fact is that the amino acid sequences that support particular activities are never found to be unique; they are always, to some extent, degenerate. Of course, a minor fraction of the amino acid substitutions may virtually inactivate a protein. Yet another small fraction of aberrant events may be required to produce small amounts of a critical activity (Bennoun, 1982; Kastelien et al., 1982). Nevertheless, we would expect most often that missense errors in translation influence the kinetic characteristics of the substituted enzymes so that they function somewhat less than optimally. Likewise, noncatalytic proteins would be most often only partially inactivated by amino acid substitutions (see below). How then can the suboptimality of the missense-substituted proteins be evaluated?

The kinetic setting

If bacteria are growing exponentially in a medium of constant composition, those with the fastest growth rates will eventually dominate the growing population. Although an invariant exponential-growth constraint by no means realistically describes the extra-laboratory conditions of bacterial growth, most of the bacteria with which we work have been selected for decades under precisely such conditions. Accordingly, we have developed a growth kinetic theory in which the efficiency of all the bacterial components can be optimized to yield a maximum growth rate. In what follows we use this kinetic optimization to describe the laboratory behavior of the bacteria in the exponential growth state (Ehrenberg and Kurland, 1984, Kurland and Ehrenberg, 1984).

One important optimal property of the catalytic devices of the growing bacteria that emerges from our theory is that the flow of products normalized to the mass of each device (enzyme, ribosome, polymerase, etc.) should be maximized. This condition very obviously provides a quantitative criterion for the evaluation of the effects of amino acid substitutions on the "fitness" of protein structures. Accordingly, we will take some time to inspect this criterion of optimal construction.

An important novelty in our theory of growth optimization is that the mass of intermediate substrate molecules is included in the cost analysis (Ehrenberg and Kurland, 1984). This is particularly important for the translation apparatus for which the immediate substrate for ribosome-mediated peptide-bond formation is the ternary complex containing aminoacyl-tRNA, EF-Tu, and GTP. Thus, we might require the concentration of ternary complex to be high in order to keep the rates of protein synthesis high, and thereby, keep the growth rate high. However, this tendency will be opposed by the need to minimize the fraction of the metabolic flow invested in the accumulation of ternary complexes. Here, the relevant constraint is the upper limit on the flow of substrate into the bacterium, and in this limit the total metabolic activity of the bacterium is fixed.

Under the least complicated circumstances, this limit of the substrate inflow may be described as an optimization of the surface area. Thus, at a fixed and presumably maximized pump density at the surface, the only way to increase the substrate uptake would be to increase the surface area. However, to the extent that this would lead to an increased internal volume, it would decrease the concentration of bacterial devices, and therefore, it would tend to lower the growth rate. In effect, an optimal inflow can be obtained only when these opposing tendencies are balanced against each other.

According to this view, the rates of substrate utilization and the growth rates are related to each other by density terms (concentrations) and rate constants. Clearly, if the density terms are fixed within certain limits, the rate constants should be maximized in order to maximize the growth rate. Indeed, this argument is reflexive in the sense that by maximizing the rate constants of devices, their concentrations can be minimized, which would maximize the growth rates. Thus, the substrate flows are in general limited, and this creates a kinetic competition between each metabolic pathway for limited resources, i.e., for both the substrates and the catalytic devices that are the products of these substrates. Hence, not only is there a premium on high rates of function, but the relevant criterion is high rate of function per mass of protein device, because the mass is a measure of the metabolic investment.

This criterion of optimal design has important consequences (Ehrenberg and Kurland, 1984). Chief among these is that it provides a welcome antidote to a persistent affliction of the literature, which consists of amazed reports that this component or that partial sequence is not absolutely required to mediate a given chemical reaction.

Since enzymes are catalysts that can only speed up the course of spontaneous reactions, they are never absolutely required for their cognate reactions. Our criterion of optimal design implies that increasing the mass of a bacterial device will have a favorable effect on the growth rate if the flows of substrates and products per mass of the device increase. Therefore, the large size of enzymes may be accounted for in a variety of ways in addition to the direct effects of the structure of enzymatic sites on the catalytic rates (Ehrenberg and Kurland, 1984). For example, an increase of protein mass that reduces the occupation time of the device in nonfunctional states could improve the rate/mass index. Likewise, a mass increase that decreases the folding or assembly time of newly synthesized devices could improve the performance of the total mass invested in a particular enzymatic pathway.

Indeed, the bacterial ribosome provides a very good, if complex, example of the kinetic tuning functions of protein mass. Although these ribosomes contain 50 or more proteins that make up more than one-third of its mass, it has been very difficult to associate individual proteins with unique partial reactions in protein synthesis. For example, the *in vitro* reconstitution of ribosomes lacking one or another protein revealed a diffuse pattern of more or less depressed activity for the defective particles (Nomura et al., 1969). Similarly, a number of bacterial mutants lacking one or another ribosomal protein have been identified more recently; to varying degrees these grow less rapidly than do wildtype bacteria, but they are perfectly viable

(Dabbs, 1979; Dabbs et al., 1983). One extreme interpretation of the dispensability of ribosomal proteins is that their primary function is to facilitate the folding of ribosomal RNA into functional forms that provide the essential working sites of the ribosome (Kurland, 1974). This view is supported by more recent functional studies of ribosomes (Garrett, 1983a) as well as the observation of a remarkable evolutionary conservation of ribosomal RNA sequences and structure (Garrett, 1983b). Accordingly, the contrasting evolutionary variability as well as dispensability of ribosomal proteins is consistent with the idea that they function primarily as local tuning devices that maximize the kinetics of ribosome function and assembly.

Ribosomal proofreading

It would be a mistake to underestimate the importance of local tuning functions. Indeed, a major theme of the present contribution is that the level at which the accuracy of translation is tuned will influence the maximal allowable growth rate of bacteria. In the next section we will consider the tuning effects of individual ribosomal proteins as well as the costs of accurate translation. However, in order to make the connections between these different topics we will need first to make a digression.

It has long been considered axiomatic that the codon–anticodon interaction determines the course of aminoacyl-tRNA selection on mRNA-programmed ribosomes. Nevertheless, the influence of ribosome structure as well as tRNA structure at sites distant from the anticodon has suggested that the selection mechanism might be more complex (Gorini, 1971; Kurland et al., 1975; Buckingham and Kurland, 1976; Yarus, 1983). Furthermore, studies of the stabilities of oligonucleotide interactions, tRNA-tRNA interactions, and the like in solution have suggested that these might not be adequately specific to account by themselves for the specificity of tRNA selection by the codon (Kurland, 1970; Uhlenbeck et al., 1970; Eisinger et al., 1971). One solution to the problem of an intrinsic limitation of the specificity of the codon–anticodon interaction is to repeat the limited elementary selection step one or more times and to build up a great final accuracy to tRNA selection from two or more modest selection steps. This sort of selection is called kinetic proofreading (Hopfield, 1974; Ninio, 1975).

A proofreading selection differs in at least one important way from a conventional selection: in order to amplify the accuracy of the selection beyond the maximum allowed in a single, elementary binding step, an extra driving force is needed (Kurland, 1978). This extra driving force can be obtained by coupling the flow of the selected substrate to the flow of nonselected substrate. For example, the flow of aminoacyl-tRNA on the codon–programmed ribosome can be coupled to that of GTP. When the flows of these two substrates are coupled in a proofreading system, the accuracy enhancement is associated with the discard from the ribosomes of aminoacyl-tRNAs following GTP hydrolysis dependent on EF-Tu. This excess dissipative loss will be greater for mismatched tRNAs than for correctly matched tRNAs. It is these dissipative characteristics that provide the experimentalist with a signal that the ribosome proofreads.

Although the initial attempts to identify a proofreading signal in a binding system dependent on aminoacyl-tRNA and EF-Tu were positive (Thompson and Stone, 1977), experiments of this sort are open to alternative explanations (Kurland, 1978). Therefore, we developed a steady state system that synthesizes *in vitro* polypeptides at accuracies and rates comparable to those of growing bacteria (Jelenc and Kurland, 1978; Wagner et al., 1982; Andersson et al., 1984). It has been possible to exploit this system to detect the excess dissipation of ternary complexes characteristic of proofreading flows (Ruusala et al., 1982).

It could be shown in this system during steady state protein synthesis that the stoichiometric ratio of ternary complexes dissipated per peptide bond formed by poly (U)-programmed ribosomes with Phe-tRNA, is about 1.1 while that for Leu-tRNA$_2$, Leu-tRNA$_3$, and Leu-tRNA$_4$ corresponds to about 100, 75 and 50, respectively. Furthermore, antibiotics such as streptomycin and kanamycin, which depress the proofreading flows, also depress the accuracy of translation (Kurland et al., 1984; Jelenc and Kurland, 1974). Likewise, mutations that influence the structure of the ribosome and that also influence translational fidelity, increase or decrease the dissipative proofreading flows (Andersson and Kurland, 1983; Kurland et al., 1984). The diversity of these different correlations between proofreading flows and the accuracy of translation in an *in vitro* system that matches reasonably well the performance of living bacteria strongly suggest that the bacterial ribosome attains its accuracy of function *in vivo* by kinetic proofreading. This in turn greatly influences the ways that we would expect the accuracy of gene expression to be optimized to support maximum bacterial growth rates.

Tuning the ribosome

The excess dissipation of ternary complex over a proofreading ribosome is a rough measure of one of the costs of accuracy amplification. Three parameters can in principle influence the cost (Blomberg et al., 1981): (1) The intrinsic discrimination between cognate and noncognate tRNA species at each proofreading step is inversely related to the dissipative cost; the minimum cost is paid when the intrinsic discrimination is maximized. However, this maximum is most likely set by the physics of the codon-anticodon interaction, and not by the structure of the ribosome. (2) When all the other relevant parameters are fixed, the dissipative cost can in principle be decreased by increasing the number of proofreading steps up to a certain point. Unfortunately, our present understanding of the physical limits on the number of proofreading steps is not well developed. All that can be said about this now is that increasing the number of proofreading steps will also tend to have the deleterious effect of decreasing the maximum rate of ribosome function. (3) There is a discard parameter that might be thought of as determining the "stickiness" of the tRNA association with the codon-programmed ribosome; this parameter is very clearly controlled by the structure of the ribosome. Thus, the intrinsic discrimination between correct and incorrect tRNA interactions at the codon-programmed site will set the limits on the maximum relative discard ratio of the different tRNA species. However, the degree to which this maximum ratio is expressed is determined by the stickiness parameter, which sets the absolute scale of the discard flows by determining the fraction of correct ternary complex species that is dissipated. Indeed, for any reasonably efficient ribosomal proofreading flow the principle dissipative loss will be expressed in the cognate flows.

Now, the discard flow of cognate species will present the system with at least two kinetic costs. One will be the excess GTP flow that will tend to lower the saturation level of the EF-Tu cycle. The other will be that the excess ternary complex flow will tend to lower the efficiency with which peptide bonds are catalyzed by ribosomes. In other words, any substantial increase of the ribosomal proofreading flows will increase the accuracy of translation but will also tend to lower the rate and the efficiency of the process. In contrast, if the system is optimally designed, the discard flows of cognate species can be small, for example, of the order of ten percent. Accordingly, relatively large decreases of the proofreading accuracy will be obtained with little or no appreciable increase in the rate or efficiency of translation. Recent experimental studies of ribosome mutants have confirmed in surprising detail the relevance of these expectations.

We find that mutant (Ram) bacteria with lower accuracy of translation than wildtype *in vivo* produce ribosomes that translate poly(U) with higher missense error rates and lower proofreading flows *in vitro* (Andersson et al., 1982; Andersson and Kurland, 1983). The rates of protein elongation by Ram ribosomes both *in vivo* and *in vitro* are indistinguishable from wildtype. A contrasting picture is obtained with streptomycin-resistant mutants that restrict translational errors (Bohman et al., 1984). We observe *in vivo* a reciprocal correlation between the degree of restriction and the rate of polypeptide elongation: the more accurate ribosomes work least quickly. Similarly, we observe *in vitro* that the most accurate mutants have the largest proofreading flows and these in turn are associated with an increase in the phenomonological K_m for ternary complex. Thus, the more actively the ternary complexes are discarded by proofreading, the greater is the apparent concentration of ternary complex required to saturate the ribosomes.

The relevance of the kinetic effects associated with enhanced proofreading to the growth optimization of the bacteria has been established in two independent ways: (1) The growth rates of the different ribosome mutants have been compared in batch cultures. It is found that the more accurate, aggressive proofreaders grow less rapidly than do wildtype bacteria. Within the group of restrictive streptomycin-resistant mutants, the growth rate decreases with degree of restriction (Andersson and Kurland, unpublished data). (2) An analysis of many streptomycin-dependent and pseudo-dependent mutants reveals a direct trade-off relationship between translational accuracy and growth rate. Briefly, streptomycin (Str) is a fairly specific inhibitor of the proofreading flows; it has in contrast only a small effect on the initial selection steps for tRNA on the ribosome (Ruusala et al., 1984). The antibiotic seems to stimulate the growth of some mutant bacteria by suppressing excessively high proofreading flows. Most significant is the finding that the growth stimulatory effect of Str on these particular mutants is accompanied by a dramatic reduction of the accuracy of translation. Since these results are so crucial to the present argument, we shall consider these mutants in more detail.

Zengel et al. (1977) have described a mutant strain that grows at roughly one half the wildtype rate in the absence of Str, and at close to the wildtype rate in the presence of high antibiotic concentrations (100 microgram/ml). This pseudo-dependent mutant elongates proteins at about one half the normal rate in the absence of Str and at almost the normal rate in the presence of antibiotic.

We have studied the ribosomes of the pseudo-dependent mutant *in vitro* and found that they are unusually aggressive proofreaders in the absence of the drug (Ruusala et al., 1984). For example, when poly(U)-dependent poly(Phe) synthesis is carried out by the pseudo-dependent ribosomes, it takes an average of ca. 1.7 ternary-complex cycles to insert one peptide in the absence of Str, compared to close to one ternary complex per peptide bond in the presence of Str. In contrast, the wildtype ribosome has stoichiometric ratios of ternary-complex cycles per peptide bond close to one in both the presence and absence of Str (1.1 vs. 1.0).

The antibiotic has only a small (30%) inhibitory effect on the k_{cat} of both pseudo-dependent and wildtype ribosome. Instead, the effect of the drug on the proofreading flows shows up as a ca. twofold reduction of the K_m of a pseudo-dependent ribosome, which is to be compared to a 50% increase of the K_m of a wildtype ribosome. In other words, at nonsaturating concentrations of the ternary complex the efficiency of usage of the ternary complex is lower for the pseudo-dependent mutant than for the wildtype in the absence of the drug. However, the efficiency is increased by the drug, which may account for the stimulation of the rate of protein synthesis *in vivo*. Most important, by studying both rates of suppression of nonsense mutations and the electrophoretic heterogeneity of proteins we have shown that the growth stimulation of the pseudo-dependent mutant by Str is accompanied by a dramatic increase of the error frequency of translation (Ruusala et al., 1984).

Thus, the pseudo-dependent mutant is a bacterium with ribosomes that are far more accurate and, therefore, far less efficient than those of wildtype bacteria in the absence of Str. The kinetic efficiency of the pseudo-dependent ribosome is improved so much by the presence of Str that it more than compensates for the loss of efficiency caused by the enhanced error frequency attending the inhibition of proofreading by the drug. The classical Str-dependent mutants respond in a comparable way to Str; however, these mutants are far more stringent than the pseudo-dependent ones, with the consequence that they cannot grow in the absence of the drug.

Clearly, the tuning functions of the proteins that control the accuracy of translation are not negligible. Structural change in proteins such as S4 and S12 can increase or decrease the error frequency, dissipative losses and rates of translation. Accordingly these proteins tune one of the principal mass flows of the bacterium. Although these and related proteins under laboratory conditions may seem to be dispensable, in reality, their seemingly marginal functions can determine the selective fitness of the bacteria.

Looping the loop

The data discussed in the previous section persuade us that bacteria set the error frequencies of translation far above the minimum attainable level. That is, the accuracy level is set at an optimum rather than at a maximum. This fits nicely with our earlier suggestion that the consequences of missense substitutions in proteins will in general be more marginal than dramatic. In other words, while there is not much to gain in the kinetic efficiency of the protein products, there is much to lose in the efficiency of ribosomes as well as ternary complexes by raising the accuracy of translation too precipitously.

There have been a number of estimates made for the global missense frequency of translation in bacteria and these usually place this rate close to 10^{-4} per codon (Ellis and Gallant, 1982). In contrast, direct measurements of the missense frequencies have been made for only two codons, and these set the error rate closer to 10^{-3} (Bouadloun et al., 1983). By measuring the differences in the error frequencies obtained at the codons in a wildtype and a restrictive Str-r mutant, a minimum estimate can be made for the contribution of the wildtype ribosome to the missense events. In the case of both codons, most of the error (> 70%) is incurred on the wildtype ribosome. Therefore, it seems reasonable to conclude, at least tentatively, that the error level of gene expression is tuned primarily at the level of the ribosome. This brings us to our final issue.

We have seen that the accuracy and the efficiency of ribosome function can be influenced by structural modifications of ribosomal proteins that alter the proofreading flows. Clearly, the missense substitutions that occur during the translation of ribosomal proteins also should lead to structural changes that influence the accuracy of their function. What will be the aggregate effect of the accumulation of errors of translation in ribosomes on the accuracy of ribosome function?

This problem was first introduced by Orgel (1963) and has been treated in detail in a number of theoretical studies that will not be discussed at any length here. Instead, we will simply note that these studies assume that a ribosome or a polymerase molecule containing an error of construction would be expected to be less accurate in its function than a canonically perfect device. In other words, beginning with Orgel (1963) it has been assumed without inspection that the devices responsible for gene expression work at a maximum accuracy level. In contrast, we have emphasized both here and elsewhere (Kurland and Ehrenberg, 1984; Ehrenberg and Kurland, 1984) that the accuracy of gene expression is demonstrably optimized at levels far below any physical maximum.

It follows that some errors of construction may increase the accuracy of function, just as others may decrease the accuracy. What will the average effect be? Stated more colloquially, what is the nature of the ribosomal error loop?

A straightforward answer is not possible at this time. However, it should be noted that the proofreading flows over the ribosome are controlled by the "stickiness" of the tRNA at its binding site. In general, the more sticky, the less accurate and the less costly the polypeptide product will be. Furthermore, we know that mutational change of ribosomal proteins influences the accuracy of translation by changing the stickiness parameters, as shown in the previous section. It would seem to be an advantage to evolve ribosome structures in which the average consequence of an amino acid missense substitution would be to decrease the stickiness of tRNA to the ribosome slightly. This would tend to increase the accuracy of function at a small increased cost of proofreading dissipation by error-containing ribosomes. What is important in such a construction is that the error loop would be damped in the sense that the errors of translation would not be able to generate an autocatalytic error catastrophe. Such a principle of ribosome construction would seem to have great selective value in light of the data suggesting that an average ribosome produced in the laboratory by a nominally wildtype bacterium contains roughly ten amino acid missense errors (Bouadloun et al., 1983).

In principle, any selection system that operates far from its maximum attainable accuracy, as does the bacterial ribosome, can respond to errors in its construction by working more accurately. However, what is special about proofreading systems is that they are virtually unlimited in their maximum accuracy as long as the number of proofreading steps can be increased. Only the cost of proofreading places a practical limit on its accuracy of function. It is this property along with the natural damping of the error loop that would make the proofreading mechanism of the ribosome an endlessly adaptable mechanism through which evolution can operate. Here, the kinetic parameters determining the speed, accuracy and cost of translation can be tuned in small steps by correspondingly small adjustments of the structures of the ribosomal proteins.

SYNOPSIS

1. The existence of mutant bacteria with ribosomes more accurate than those of nominally wild type bacteria indicates that the accuracy of translation is a parameter that is optimized rather than maximized. How are the costs to the bacterium of errors in protein construction balanced against the costs of reducing these errors?

2. Data from systematic studies of mutationally altered enzymes suggest that missense substitutions most often create kinetically suboptimal rather than inactive proteins. Therefore, the costs of errors of protein construction to bacteria are appropriately evaluated in terms of their influence on the growth kinetics.

3. Likewise, the costs of reducing errors can be evaluated in the growth-kinetic context. A major cost of gene expression that increases and decreases with accuracy of translation is that associated with ribosomal proofreading. This cost is expressed in the decreased efficiency of ribosome and EF-Tu function.

4. Excessive accuracy of ribosome function in some mutants leads to loss of growth rate, and antibiotics that lower the accuracy as well as the proofreading costs of translation can stimulate mutant growth rates. This behavior is seen with streptomycin-dependent mutants.

5. Optimized selection systems such as proofreading ribosomes can be constructed so that errors in their construction do not lead to enhanced error rates in their function. In other words, there is no necessary positive feed back of errors of gene expression, and, accordingly, the so-called error catastrophe is not the problem it was thought to be.

ACKNOWLEDGEMENTS

I wish to thank my colleagues at the department for helpful discussions and criticism. I am especially indebted to Måns Ehrenberg for critically reading this text. My work is supported by funds supplied by the Swedish Cancer Society and Natural Sciences Research Council.

REFERENCES

Andersson, D. and Kurland, C. G. (1983). *Mol. Gen. Genet.*, 191, 378.

Andersson, D. I., Bohman, K., Isaksson, L. A., and Kurland, C. G. (1982). *Mol. Gen. Genet.*, 187, 467.

Andersson, D. I., Ruusala, T. Ehrenberg, M., and Kurland, C. G. (1984). In preparation.

Bennoun, M. (1982). *FEBS Lett.*, 149, 167.

Blomberg, C., Ehrenberg, M., and Kurland, C. G. (1980). *Q. Rev. Biophys.*, 13, 231.

Bohman, K., Ruusala, T., and Kurland, C. G. (1984). In preparation.

Bouadloun, F., Donner, D., and Kurland, C. G. (1983). *EMBO J.*, 2, 1351.

Buckingham, R. H. and Kurland, C. G. (1977). *Proc. Natl. Acad. Sci. USA*, 74, 5496.

Crick, F. H. C., Barnet, L., Brenner, S., and Watts-Tobin, R. J. (1961). *Nature,* 192, 1227.

Dabbs, E. R. (1979). *J. Bacteriol.*, 140, 734.

Dabbs, E. R., Hasenbank, R., Kastner, B., Rak, K-H., Wartusch, B., and Stoffler, G. (1983). *Mol. Gen. Genet.*, 192, 301.

Ellis, N. and Gallant, J. (1982). *Mol. Gen. Genet.*, 188, 169.

Ehrenberg, M. and Kurland, C. G. (1984). *Q. Rev. Biophys.*, in press.

Eisinger, J., Feur, B., and Yamane, T. (1971). *Nature New Biol.*, 231, 126.

Garrett, R. (1983a). *TIBS,* 8, 75.

Garrett, R. (1983). *TIBS,* 8, 189.

Gorini, L. (1971). *Nature New Biol.*, 234, 261.

Hopfield, J. J. (1974). *Proc. Natl. Acad. Sci. USA*, 71, 4135.

Jelenc, P. C. and Kurland, C. G. (1979). *Proc. Natl. Acad. Sci. USA*, 76, 3174.

Katelein, R. A., Remaut, E., Fiers, W., and van Duin, J. (1982). *Nature,* 295, 35.

Kurland, G. C. (1970). *Science,* 169, 1171.

Kurland, C. G. (1974). In *Ribosomes,* M. Nomura et al. (eds.), Cold Spring Harbor Laboratory, pp. 309-332.

Kurland, C. G. (1978). *Biophys. J.,* 22, 373.

Kurland, C. G. and Ehrenberg, M. (1984). *Progr. Nuc. Acid. Res. and Mol. Biol.,* in press.

Kurland, C. G., Rigler, R., Ehrenberg, M., and Blomberg, C. (1975). *Proc. Natl. Acad. Sci. USA,* 72, 4248.

Langridge, J. (1968a). *Proc. Natl. Acad. Sci. USA,* 60, 1260.

Langridge, J. (1968b). *J. Bacteriol.,* 96, 1711.

Langridge, J. and Campbell, J. H. (1969). *Mol. Gen. Genet.,* 103, 339.

Miller, J. H., Coulondre, C., Hofer, M., Schmeissner, U., Sommer, H., and Schmitz, A. (1979). *J. Mol. Biol.,* 131, 191.

Nene, V. and Glass, R. E. (1982). *Mol. Gen. Genet.,* 188, 399.

Ninio, J. (1975). *Biochemie,* 57, 587.

Nomura, M., Mizushima, S., Ozaki, M., Traub, P., and Lowry, C. V. (1969). *Cold Spr. Hbr. Symp.,* 34, 49.

Orgel, L. (1963). *Proc. Natl. Acad. Sci. USA,* 49, 517.

Ruusala, T., Ehrenberg, M., and Kurland, C. G. (1982). *EMBO J.,* 1, 75.

Ruusala, T., Andersson, D. I., Ehrenberg, M., and Kurland, C. G. (1984). In preparation.

Smith, J. D. (1979) In *Nonsense Mutations and tRNA Suppressors,* J. E. Celis and J. D. Smith (eds.), Academic Press, London, pp. 109-126.

Thompson, R. C. and Stone, P. J. (1977). *Proc. Natl. Acad. Sci. USA,* 74, 198.

Uhlenbech, O. C., Baker, J. C., and Doty, P. (1970). *Nature,* 225, 508.

Wagner, E. G. H., Jelenc, P. C., Ehrenberg, M., and Kurland, C. G. (1982). *Eur. J. Biochem.,* 122, 193.

Zengel, J M., Young, R., Dennis, P. P., and Nomura, M. (1977). *J. of Bact.,* 136, 1320.

Labeling of a Nucleoside-Transporting Protein with β-Galactosidase by Gene Fusion

Agnete Munch-Petersen and Nina Jensen
University Institute of Biological Chemistry B;
Copenhagen, Denmark

ABSTRACT

The *nupG* region on the *E. coli* chromosome encodes the components of a nucleoside transport system. Protein fusions have been constructed in which the promoter and part of a structural gene from the *nupG* operon is fused to the *lacZ* gene. Cells containing the protein fusions produce hybrid proteins with beta – galactosidase activity. The synthesis of the hybrid proteins is regulated by the *cytR* and the *deoR* control systems, which are known to regulate the synthesis of the *nupG* nucleoside–transport system.

Two such protein fusions have been cloned on resistance–carrying plasmids and introduced into a minicell–producing strain, and protein synthesis has been examined in the minicells. The hybrid proteins carrying the beta–galactosidase activity were found in the membrane fraction of the minicells.

INTRODUCTION

In the 1960s when our small group (Jan Neuhard and I) started work in the building where Ole Maaløe's famous Institute of Microbiology was located, the concept of thymineless death was an important subject of work and discussion at the Institute. We were encouraged to participate in these discussions and soon became sufficiently interested in the field to initiate our own project concerning thymineless death. Our approach was a biochemical investigation of the metabolism of deoxyribose compounds in *E. coli*, during thymineless death as well as during normal growth. This project turned out to be a very fruitful one, which during the past years has been the basis for extensive studies of pyrimidine and purine metabolism in *E. coli* and *S.*

121

typhimurium (for references, see J. Neuhard, this volume, and P. Nygaard, 1983).

In a thymine-requiring mutant the low or high requirement for exogenous thymine is determined by the cellular pools of deoxyribose 1-phosphate, and these pools again depend on a balance between the activities of the nucleoside-catabolizing enzymes. Thus, we were specifically led to a study of this group of enzymes and of the metabolism of nucleosides in general. Besides purification and characterization of these enzymes we have investigated the cellular control systems that govern their synthesis. These control systems, all acting at the transcriptional level, are highly complex and are now being explored and resolved on the DNA level (for references see Valentin-Hansen et al., 1984).

The same control mechanisms have been shown to regulate the synthesis of transport systems that are necessary for the access of exogenous nucleosides to the cells, but while the nucleoside-catabolizing enzymes have been characterized in detail, the transport systems are less explored. It is known that *Escherichia coli* contains at least two different, active nucleoside-transport systems (Komatsu, 1973; Doskocil, 1974; Leung and Visser, 1977), designated the *nupC* and the *nupG* systems (Munch-Petersen et al., 1979). The two systems, which can be separated genetically, differ in their specificity towards the nucleoside substrates: the *nupG* system facilitates the transport of all nucleosides, whereas the *nupC* system mediates transport of nucleosides other than guanosine and deoxyguanosine (Komatsu and Tanaka, 1972; Roy-Burman and Visser, 1972). The *nupG* system may be the more complex of the two systems; there are indications that overlapping as well as separate components are functioning in the transport of purine and pyrimidine nucleosides (Peterson and Koch, 1966; Petersen et al., 1967; Munch-Petersen and Pihl, 1980). The *nupG* region covers 2.5-3.0 kb on the chromosome (Westh Hansen, in preparation) and is believed to constitute an operon. This transport system is known to be regulated by the *cytR* and *deoR* genes (Munch-Petersen and Mygind, 1976), the two repressor genes, which have been previously characterized as the genes controlling the synthesis of the nucleoside catabolizing enzymes (Munch-Petersen et al., 1972). The *nupC* system seems to be regulated by the *cytR* gene alone. In connection with the *cytR* control, nucleoside transport mediated by either system is subject to catabolite repression through the cyclic AMP-CRP system (Mygind and Munch-Petersen, 1975).

An outer membrane protein, the phage T6 receptor, encoded by the *tsx* gene (Hantke, 1976; Krieger-Brauer and Braun, 1980) has been shown to stimulate the transport of certain nucleosides at low (less than 10^{-6} M) concentrations, but at higher substrate concentrations this protein is not required for transport. Apart from the *tsx*-encoded protein, no single protein connected with nucleoside transport has been isolated or characterized.

This paper describes the selection and investigation of strains containing protein fusions in which part of a gene, involved in the *nupG* transport system, is fused to the *lacZ* gene. The fusion-containing strains produce hybrid proteins in which the N-terminus, encoded by the *nupG* gene, is fused to an enzymatically active beta-galactosidase, in such a way that the synthesis of this enzyme is controlled by the *cytR* and *deoR* repressors. Moreover, the beta-galactosidase is converted to a membrane-bound state by the protein fusion.

RESULTS

Selection of nupG-lacZ fusions

A λ phage placMu1, constructed by Bremer et al. (1984) specifically for isolation of protein fusions, was used. The phage is plaque-forming and carries both ends of the Mu phage as well as a lacZ gene and a lacY gene. The lacZ gene is deleted for its transcription- and translation-initiation signals, and the Mu sequence in front of the lacZ gene is an open reading frame fused in frame to codon 8 in lacZ. Infection of a Δlac strain and selection for growth on lactose yields strains in which the synthesis of the lacZ and the lacY gene products is controlled by exogenous promoters.

Strain MP614, nupC Δlac was used as a recipient. This strain contains the nupG nucleoside-transport system only (Table 1). After infection with placMu1 and a helper phage, λpMu507, Lac⁺ colonies were selected and screened for growth on nucleosides and for resistance to the toxic nucleoside analog, fluorodeoxyuridine (FUdR), 5 µg/ml. Phage P1 lysates were prepared from possible candidates and used for transduction of MP619Δlac nupC metC, selecting for met⁺ on minimal plates containing Xgal (20 µg/ml) and glucose as carbon source. The metC and nupG loci have been shown to cotransduce with a frequency of 20 percent (B. Mygind, unpublished). Blue transductant colonies were picked, purified, and tested for inability to grow on nucleosides and for resistance to FUdR. A second selection of a nupG-lacZ fusion was made in a similar way, this time making use of λplacMu3 (Table 1). From each selection a fusion strain was chosen for further analysis. These two strains, MP733 and MP750, both failed to grow on nucleosides, were resistant to FUdR, and produced enough beta-galactosidase for growth on lactose. To verify that the failure of these strains to use nucleosides as sole carbon sources was in fact due to a blocked transport system and not to a spontaneous mutation in one of the nucleoside-catabolizing enzymes, cell suspensions prepared from exponentially growing cells were directly assayed for transport of nucleosides (see Methods). In both cases the fusion strain had lost the capacity for significant nucleoside uptake (results not shown).

Properties of the nupG-lacZ fusion strains

The two fusion strains, MP733 and MP750, were transduced to tetracycline resistance with a P1 lysate of S01025 glc::Tn10. [The glc gene, specifying the enzyme malate synthetase G, which is necessary for growth on glycollate as carbon source (Vanderwinkel and DeVlieghere, 1968) maps between metC and nupG (Bachmann and Brooks Low, 1980).] In both cases a number of transductants were obtained that simultaneously had become Lac⁻ and had regained the ability to use nucleosides as carbon sources. Thus, the fusion strains contain only one copy of placMu1, tightly linked to the nupG gene.

Cells with a nupC nupG genotype are unable to grow on nucleosides (1 mg/ml) as sole carbon source. However, a limited uptake

TABLE 1   Characteristics and sources of strains and plasmids used in the present work

Strain	Relevant genotype	Origin or reference
E. coli		
MC4100	Δ*lac(U169)*	Casadaban, 1975
SO758	*metC*	Taylor and Trotter, 1967
SO1025	*glc*::Tn10	Mygind, this lab
MP614	Δ*lac nupC*	From MC4100 by resistance to showdomycin
MP619	Δ*lac nupC metC*	This paper
MP733	Δ*lac nupC ϕ(nupG-lacZ)* hyb1(λp*lac*Mu1)*	This paper
MP750	Δ*lac nupC ϕ(nupG-lacZ)* hyb2(λp*lac*Mu3)*	This paper
BD1854	*minA minB his*	Diderichsen, 1980
Phage		
λp*lac*Mu1	*cI⁺*	Bremer et al., 1984
λp*lac*Mu3	*cI⁺ imm21*	Bremer etal., 1984
λpMu507	*cI857 S7 MuA⁺B⁺*	Bremer et al., 1984
λpMu507.3	*imm21 S7 MuA⁺B⁺*	Bremer et al., 1984
Plasmid		
pMLB524	*bla⁺*	Silhavy and Beckwith, 1983
pMP5	*nupG⁺bla⁺*	Munch-Petersen and Jensen, 1984
pMP12	pMLB524 ϕ(*nupG-lacZ*)*hyb1*	This paper
pMP14	pMLB524 ϕ(*nupG-lacZ*)*hyb2*	This paper

*The abbreviation *hyb* indicates that the gene fusion encodes a hybrid protein.

of exogenous nucleosides does take place, sufficient to cause some induction of the nucleoside-catabolizing enzymes. The synthesis of one of these, cytidine deaminase, is induced by cytidine, the low molecular effector in the *cytR* control system. The synthesis of thymidine phosphorylase is controlled by both the *cytR* and the *deoR* systems and in the latter system deoxyribose 5-phosphate, formed from exogenously added thymidine, acts as inducer. Table 2 shows the induction pattern in the fusion strains after addition of cytidine or thymidine. Clearly the response of beta-galactosidase activity to the two inducers follows the response of thymidine phosphorylase, i.e., the synthesis of beta-galactosidase is regulated by both the *cytR* and the *deoR* control systems in the fusion strains. It has not yet been possible to construct fusion strains with a *cytR* or a *deoR* mutation (see later).

TABLE 2   Induction of enzyme activities in strains containing
nupG-lacZ fusions

Strain	Inducer added (0.5 mg/ml)	Enzyme activities (U/mg protein)		
		β-Galacto-sidase	Cytidine deaminase	Thymidine phosphorylase
MP733, Δ*lac*	None	20	20	20
φ(*nupG-lacZ*)	Cytidine	95(5)	260(13)	150(7.5)
hyb1	Thymidine	165(8)	30(1.5)	190(9.5)
MP750, Δ*lac*	None	50	20	30
φ(*nupG-lacZ*)	Cytidine	210(4)	330(16.5)	120(4)
hyb2	Thymidine	380(7.5)	15(1)	180(6)
MP619	None	0	35	55
Δ*lac nupG*+	Cytidine	0	990(13)	400(7)
	Thymidine	0	40(1)	570(10)

Cells were grown exponentially in glycerol-salts medium. Nucleosides were added as indicated, and growth was continued for 90
min. Cells were harvested and cell extracts were prepared and
assayed for enzymes as described (Hammer-Jespersen et al., 1971).
Figures in parentheses indicate the factor of induction.

Cloning of the nupG-lacZ fusions

The fusion strains, MP733 and MP750, were induced by UV-irradiation,
and the lysates used to infect MP614Δ*lac*. Plating was done in the
presence of Xgal in order to detect *lac*+ phages. For each fusion a
few *lac*+ lysogens appear, which were picked and purified, and by
subsequent UV-induction and reinfection shown to represent specialized
transducing phages, carrying the protein fusions. Two of these *lac*+
lysogens were isolated and used for preparation of λ DNA, containing
either of the two fusions.

As a cloning vehicle, pMLB524 (Silhavy and Beckwith, 1983)
was used. The two DNAs were digested with EcoRI, each DNA was
mixed with EcoRI-digested pMLB524 and ligated, and the mixtures
were used to transform MP614Δ*lac*. Selection was made on LB plates
containing ampicillin (50 μg/ml) and Xgal (20 μg/ml). Blue colonies
were picked and purified. Two transformants, MP754 and MP756,
obtained in this way and containing the cloned gene fusions from
MP733 and MP750, respectively, were used for large batch preparation
of the fusion-containing plasmids pMP12 and pMP14 (see Table 1).

The two plasmids were digested with a variety of restriction
endonucleases, both singly and in combination. The restriction maps
are shown in Figure 1 and compared with that of plasmid pMP5, onto
which the total *nupG* region is cloned (Munch-Petersen and Jensen,
1984; Westh Hansen, in preparation). The relative location of the
various restriction endonuclease sites serves to localize the insertion
points of the p*lac*Mu phage in the two fusions. As indicated by the

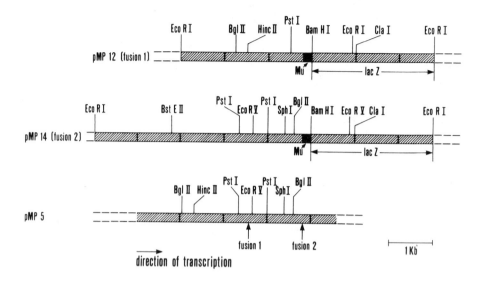

FIGURE 1 Partial restriction endonuclease maps of the <u>nupG</u> operon, cloned on pMP5, and the two fusion-carrying plasmids pMP12 and pMP14. The maps were used to determine the insertion points of the <u>placMu1</u> phage in the two fusion plasmids by comparing the restriction patterns with that of the <u>nupG</u> operon on pMP5. The insertion points are indicated by arrows.

arrows, the insertion point of fusion 1 is located approximately 1200 bp upstream of that of fusion 2.

The pattern of the restriction endonuclease sites also suggests different excision points when the specialized transducing phages, carrying fusion 1 and fusion 2, respectively, were formed. The proximal BglII site is retained in pMP12, but not in pMP 14, and the excision therefore must have taken place on either side of the BglII site in the two plasmids.

Synthesis of nupG–lacZ hybrid proteins

The plasmids containing the cloned gene fusions were introduced into a minicell-forming strain. The minicells were isolated (see Methods) and allowed to synthesize proteins in the presence of [^{35}S]methionine. Polyacrylamide gel electrophoresis of the extracts showed the presence of hybrid protein bands, moving in the region of beta–galactosidase on the gel, but with somewhat larger apparent molecular weights than wildtype beta–galactosidase (Figure 2). The two hybrid proteins produced by the minicells are of different size. This is in accordance with the finding (Figure 1) that the insertion points of the placMu1 phage in the two fusions are some distance apart from each other. Plasmid pMP12 repeatedly yielded two protein bands in the beta-. galactosidase region, suggesting the formation of a precursor protein in these minicells (lane 2, Figure 2).

(a)

(b)

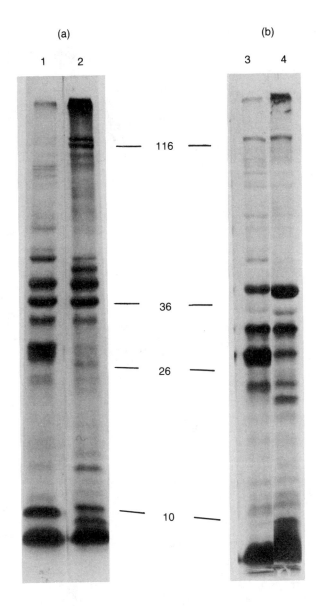

FIGURE 2 Autoradiogram of proteins, synthesized by minicells carrying pMP12(hybl) and pMP14(hyb2). The numbers between panel (a) and panel (b) are the molecular weights (in Kd) of marker proteins with the indicated electrophoretic mobilities. (a) Lane 1: pMP12, total protein; Lane 2: pMP12, membrane fraction. (b) Lane 3: pMP14, total protein; Lane 4: pMP14, membrane fraction.

Electrophoretic analyses of the membrane fraction, prepared by fractionation of the crude minicell-protein extract, indicated that the hybrid proteins specified by the plasmids were mainly located in the cell membrane (lanes 2 and 4, Figure 2). This membrane location of the hybrid proteins may be the reason why all attempts to transduce the protein fusions into Δlac strains with cytR or deoR mutations have failed. When P1 lysates of MP733 and MP750 were employed and selection was made for metC on plates containing Xgal (20 μg/ml), a few blue transductant colonies were obtained that on subsequent purification rapidly segregated to lac⁻ colonies. A derepression of the synthesis of a membrane-located hybrid protein with an unusually high molecular weight is likely to cause this instability of the cells.

DISCUSSION

The transposable phage λplacMu1 (Bremer et al., 1984) was used to isolate lysogenic strains containing protein fusions, in which the promoter and the 5' end of a nupG gene is fused to the lacZ gene. The fusions were identified in the following way: (1) The lysogenic strains containing the fusions are unable to grow on nucleosides as sole carbon sources. (2) The strains produce hybrid proteins with beta-galactosidase activity. (3) The synthesis of the hybrid proteins is controlled by the cytR and deoR repressor genes, as evidenced by the induction of beta-galactosidase activity when cytidine or thymidine is added to the growth medium of the lysogenic strains (Table 2).

The lacZ fusions have been cloned on a multicopy plasmid, pMLB524, and the fusion-carrying plasmids introduced into a minicell-forming strain in order to detect the hybrid proteins and to determine their localization in the cell. Both plasmids encode proteins with molecular weights higher than wildtype beta-galactosidase. Comparison of the restriction endonuclease maps (Figure 2) indicates that the insertion points in the two fusions are approximately 1200 bp apart, but since it is not known how many proteins are encoded by the nupG operon, it cannot be decided whether the two fusions have occurred in one and the same gene or in two adjacent genes. A signal sequence that directs the polypeptide chain to the membrane may be present in the N-terminal end of the fused proteins, since in both cases the beta-galactosidase is found associated to the membrane fraction of the minicells. The finding indicates that in the intact state the protein(s), in which the fusions have occurred, must be located in the cell envelope, in accordance with a possible function in a transport system. The information, which determines the ultimate location of a protein in the cell envelope, is believed to reside in the structural part of a gene (Tommassen et al., 1983), and, hence, the present results cannot be used to identify the compartment of the cell envelope in which the target protein is located.

The regulation of the nupG transport system by the cytR and the deoR control systems was previously shown qualitatively by transport assays on strains harboring cytR or deoR mutations (Munch-Petersen and Mygind, 1976). The fusion of nupG genes to the lacZ gene, described in the present paper, allows a more quantitative estimation of the effect of these control systems. In an induction experiment (Table 2), the response (extent of induction) of the fused beta-galactosidase to the inducers resembles that of thymidine phosphorylase, which is one of the enzymes encoded by the deo operon. It is possible that the regulatory regions of the two operons will show certain similarities, and work is in progress to characterize the various parts of the *nupG* operon at the DNA level.

The protein fusions described above were constructed in an attempt to resolve the *nupG* transport system in its components. The hybrid proteins labeled with enzymatically active beta-galactosidase are now being purified and used to raise antibodies that will recognize the N-terminal end of the *nupG* protein(s) and thus may serve to identify them in a purification procedure.

MATERIALS AND METHODS

Bacterial strains and plasmids. The *E. coli* strains are listed in Table 1 together with phages and plasmids used in the present work. Unless otherwise specified, the strains were grown in minimal AB medium (Monod et al., 1951) with appropriate requirements and carbon sources as indicated. Growth was measured as an increase in absorbancy at 436 nm.

Materials. Nucleosides, showdomycin, and other fine chemicals were purchased from Sigma Chemical Co., St. Louis. ^{14}C-labeled nucleosides were from the Radiochemical Centre, Amersham, England; ^{35}S-methionine was from New England Nuclear. Restriction endonucleases and T4 ligase were obtained from Boehringer Mannheim. Xgal (5-bromo-4-chloro-3-indolyl-β-d-galactoside) was from the Bethesda Research Laboratories.

Enzyme assays. Determinations of beta-galactosidase activity were carried out according to Miller (1972). The nucleoside-catabolizing enzymes were assayed in extracts from glycerol-grown cells, as described previously (Hammer-Jespersen et al., 1971).

Transport assays were performed as described previously (Mygind and Munch-Petersen, 1975) using suspensions of glycerol-grown cells harvested in the exponential phase. Concentrations of the labeled nucleosides were $0.2-0.5 \times 10^{-6}$ M, and specific activities ranged from 50-90 Ci per mol.

Genetic techniques. Transductions and conjugations were carried out according to Miller (1972). Transformation procedures were those described by Kushner (1978).

Phage manipulations. The λplacMu phages (Table 1) were constructed by Bremer et al. (1984). Lysates of these phages were a gift from Dr. W. Boos, University of Konstanz, FRG.

λ lysogens were grown in Luria broth. UV-induction was performed as described by Silhavy et al. (1984). Lac$^+$ phages were detected by plating on tryptone broth agar plates containing Xgal (20 μg/ml). Stocks of λplacMu phages were prepared on MP614.

Preparation of minicells. The minicell-producing strains were grown for 15 h in 100 ml cultures. To derepress the synthesis of transport proteins, the cells were grown in AB minimal medium with glycerol (0.2%) as carbon source. The minicell fractions were separated from normal cells by two consecutive sucrose gradients as described by Inselburg (1972). Before incubation for protein synthesis, the minicells were suspended in AB medium and the cell density adjusted to an A_{436} of approximately 3.0.

Protein synthesis in minicells was carried out as follows: A minicell suspension, usually 1-1.5 ml, was preincubated at 37 C with glycerol (0.2%) biotin (5 μg/ml), thiamine (5 μg/ml), and amino acids

(0.01 mM) other than methionine. After 20 min of equilibration, [^{35}S]methionine (25 µCi) was added and incubation was continued for 30 min.

Fractionation of minicell-synthesized proteins was carried out using the method of Weiner et al. (1978). After incubation with [^{35}S]methionine, the minicell suspension (1.5 ml) was divided into two parts: (1) *Total protein:* 250 µl were centrifuged, washed with minimal medium, and frozen at -70 C. (2) *Crude membrane fraction:* 1.2 ml was digested with lysozyme (10 mg/ml) for 60 min at 4 C in the presence of 10 mM EDTA. A 750 µl volume of Tris-HCl (63.5 mM, pH 6.8) was added and the suspension was sonicated, followed by centrifugation for 15 min at 3000 g. The supernatant was then centrifuged for 12 h at 110,000 g. The resulting membrane pellet was stored at -70 C.

Isolation of plasmid DNA was carried out according to the procedure of Celwell (1972) and Jorgensen et al. (1977). Large-scale isolation of DNA was performed as described by Silhavy et al. (1984).

Polyacrylamide gel electrophoresis was performed on sodium dodecyl sulfate polyacrylamide slab gels (12.5%) according to Laemmli (1970) and Ames (1974). Before electrophoresis the minicell fractions were suspended in 100 µl sample buffer and boiled for two min. Samples (15-20 µl) were applied to the gel. The protein bands were located by staining with Coomassie brilliant blue and by auto-radiography.

ACKNOWLEDGEMENTS

Our thanks are due to Dr. Winfried Boos, University of Konstanz, FRG, for sending us lambda phages and appropriate information before this information was published. Economic support from the Danish Natural Research Council is gratefully acknowledged.

REFERENCES

Amer, G. (1974). *J. Biol. Chem.,* 249, 634.

Bachmann, B. and Low, K. B. (1980). *Microbiol. Rev.,* 44, 1.

Bremer, E., Silhavy, T., Weisemann, J., and Weinstock, G. M. (1984). *J. Bacteriol.,* 158, 1084.

Casadaban, M. J. (1976). *Proc. Natl. Acad. Sci. USA,* 72, 809.

Clewell, D. (1972). *J. Bacteriol.,* 110, 667.

Diderichsen, B. (1980). Dissertation, University of Copenhagen.

Doskocil, J. (1974). *Biochem. Biophys. Res. Commun.,* 56, 997.

Hammer-Jespersen, K., Munch-Petersen, A., Nygaard, P., and Schwartz, M. (1971) *Eur. J. Biochem.,* 19, 533.

Hantke, K. (1976). *FEBS Letters,* 70, 109.

Inselburg, J. (1970). *J. Bacteriol.,* 102, 642.

Jørgensen, P., Collins, J., and Valentin-Hansen, P. (1977). *Molec. Gen. Genet.,* 155, 93.

Komatsu, Y. and Tanaka, K. (1972). *Biochim. Biophys. Acta,* 288, 390.

Krieger-Brauer, H. J. and Braun, V. (1980). *Arch. Microbiology,* 124, 233.

Kushner, S. R. (1978). In *Genetic Engineering,* H. B. Boyer and S. Nicosia (eds.), Elsevier. p. 17.

Laemmli, V. K. (1970) *Nature,* 227, 680.

Leung, K. K. and Visser, D. W. (1977) *J. Biol. Chem.,* 252, 2492.

Miller, J. H. (1972). In *Experiments in Molecular Genetics,* Cold Spring Harbor Laboratory, Cold Spring Harbor, N.Y.

Monod, J., Cohen-Bazire, G., and Cohn, M. (1951). *Biochim. Biophys. Acta,* 7, 585.

Munch-Petersen, A., Nygaard, P., Hammer-Jespersen, K., and Fiil, N. (1972). *Eur. J. Biochem.,* 27, 208.

Munch-Petersen, A. and Mygind, B. (1976). *J. Cellular Physiology,* 89, 551.

Munch-Petersen, A., Mygind, B., Nicolaisen, A., and Pihl, N. J. (1979) *J. Biol. Chem.,* 254, 3730.

Munch-Petersen, A. and Pihl, N. J. (1980). *Proc. Natl. Acad. Sci. USA,* 77, 2519.

Munch-Petersen, A. and Jensen, N. (1984). In *The Cell Membrane: Its Role in Interaction with the Outside World,* Plenum, p. 80.

Mygind, B. and Munch-Petersen, A. (1975). *Eur. J. Biochem.,* 59, 365.

Nygaard, P. (1983). In *Metabolism of Nucleotides, Nucleosides and Nucleobases in Microorganisms,* Academic Press, p. 27.

Peterson, R. N. and Koch, A. L. (1966). *Biochim. Biophys. Acta,* 126, 129.

Peterson, R. N., Boniface, J., and Koch, A. L. (1967). *Biochim. Biophys. Acta,* 135, 771.

Roy-Burman, S. and Visser, D. W. (1972). *Biochim. Biophys. Acta,* 282, 383.

Silhavy, T. and Beckwith, J. (1983). In *Methods in Enzymology,* 97, 21.

Taylor, A. L. and Trotter, C. D. (1967). *Bact. Rev.,* 31, 332.

Tommassen, J., Huub van Tol, and Lugtenberg, B. (1983) *EMBO J.,* 2, 1275.

Valentin-Hansen, P., Hammer, K., Larsen, J. E. L., and Svendsen, I. (1984). *Nucleic Acids Res.,* 12, 5211.

Vanderwinkel, E. and DeVlieghere, M. (1968). *Eur. J. Biochem.,* 5, 81.

Weiner, J. H., Lohmeier, E., and Schryvers, A. (1978). *Can. J. Biochem.,* 56, 611.

Growth of the Graminaceous Leaf

Kirsten Gausing and Rosa Barkardottir
University of Aarhus, Denmark

INTRODUCTION

The leaves of the members of the grass family (Graminea) have a relatively simple pattern of growth: only the cells in the basal meristematic zone are dividing and the cells become progressively older with increasing distance from the meristem. Thus, samples of cells of different, defined age can be obtained by sectioning of uniformly grown leaves. At the Microbiological Institute in Copenhagen we learned to appreciate the virtues of *E. coli* cultures grown under reproducible steady state conditions (Maaløe and Kjeldgaard, 1966). The question that poses itself after a transition to higher plants is whether the leaves of grass plants can be grown and harvested under sufficiently reproducible conditions to warrant a detailed compositional analysis, which was such a fruitful approach in *E. coli* 20 years ago, or whether the knowledge about prokaryotes and eukaryotes alike has now progressed to a stage where efforts to elucidate leaf cell development are better carried out with other approaches. After several years with bacterial cultures in steady states of growth (e.g. Gausing, 1977) it certainly seems a negligence not to attempt to arrive at some degree of understanding of macromolecular synthesis on a whole leaf basis. In the following some characteristics of the growth of the first 7-day-old barley leaf are described, and the limitations of the system are discussed.

As rapid and efficient growth is the hallmark of bacteria (Ingraham et al., 1983), so unquestionably is photosynthesis in higher plants. Chloroplast biogenesis therefore assumes a central role in plant molecular biology, and happily the chloroplast has retained so many prokaryotic characteristics, some of which will be described, that it seems like an old friend. On the other hand, most of the chloroplast genes are located in the nucleus, and the control of chloroplast development (and the development of other plastid forms) within the cells that harbor them raises a number of questions that are specific for eukaryotic cells.

PARAMETERS OF GROWTH

The young grass leaf can be divided into three zones: (1) at the base, the meristematic zone where cell divisions take place, (2) above that, a zone where the cells are no longer dividing but are still elongating, and (3) above that, a zone where the cells have reached their final length and where the cells move up only as a result of the activity in the two lower zones. In order to be able to convert distance from the leaf base to cell age, the growth of the leaf tip was followed from the time it emerges from the seed until harvest (Figure 1a). Next the length of the zone of cell elongation was determined by slicing windows in the coleoptiles and marking the leaves with ink dots (Boffey et al., 1980). Leaves were dotted on days 4, 5, and 6 at the base or at 5 mm and at 5-mm intervals further up the leaves. The position of the dots were recorded over the next days. The results showed that more than 90% of the length increment of the whole leaf takes place in the basal 5 mm. The distance between the dots at 5 and 10 mm increased at most to 7 mm and the distance between dots further up the leaf remained constant. Since the zones of cell division and elongation are confined to the basal 10 mm, and the leaf grows about 30 mm per day during the last three

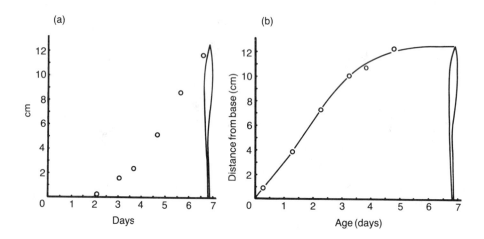

FIGURE 1 Growth of the first barley leaf. Seeds were imbibed at time zero, sowed under cover 15 hrs later, and the cover removed at 2.5 days. The plants were grown with 16 hr light periods at 22°C and 8 hr dark periods at 18°C. 750 leaves were measured on day 4, 5, and 6, and the length of leaves at harvest is the average of 1500 measurements (12.5 ± 1.1 cm). (a) Growth of the leaf tip. (b) After determination of the length of the cell division/elongation zones the results of Figure 1(a) were converted to give the correlation between ages of cells at different points along the leaf.

days before harvest, the correlation between cell age and distance from the base is almost linear up to about 10 cm from the leaf base (Figure 1b). The simple correlation found in Figure 1b was unexpected, because in wheat the zone of cell elongation extends up to 30 mm from the leaf base, and the correlation between age and distance is quite complex (Boffrey et al., 1980).

The advantages of the regular growth of grass leaves are obvious to a bacterial physiologist, but such systems have been explored surprisingly little. The only comprehensive analysis has been carried out on wheat leaves by Dr. Leech's group (e.g., Dean and Leech, 1982a). Since our work with barley is still in its initial phases,

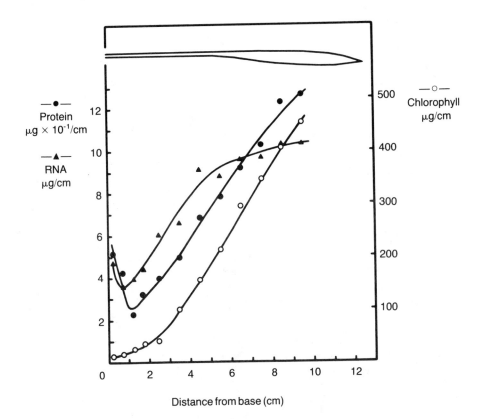

FIGURE 2 Chlorophyll, protein, and RNA in sections of the first barley leaf. 7-day-old barley leaves measuring between 11.5 and 13.5 cm were cut close to the seed, and the coleoptile and the second leaf were removed. Leaves were aligned at the base, and 5 or 10 mm sections were cut to 10 cm from the base. Chlorophyll was measured on 50 leaf sections according to Arnon (1949). Protein was determined on 30 leaf sections by the method of Lowry et al. (1951). RNA was estimated after PCA extraction and alkaline hydrolysis of 20 leaf sections by the absorption at 260 nm.

we shall refer frequently to the wheat results, even though they can be applied to barley only in a general way, as already indicated with respect to the growth of the leaves. At this point, we have measured amounts of chlorophyll, protein, and RNA per cm of leaf (Figure 2). The chlorophyll and protein determinations are fairly straightforward. RNA was determined by UV absorption on alkaline hydrolysates, but we are not satisfied with the purity of the samples, and we consider the results to be preliminary. The barley values for both RNA and protein are significantly lower than in the (hexaploid) wheat leaves, but the ratios of protein to RNA are similar (Dean and Leech, 1982a). In Figure 2 the drop in RNA and protein in the basal sections of the leaf does not mean that these macromolecules are breaking down; protein turnover is in fact very limited in young leaf cells with a halflife of 6-7 days (Davies, 1982). Instead the drop in RNA and protein reflects the higher cell density in the meristem and cell elongation zones. Maximal rates of RNA and protein accumulation continues for about 36 hrs up to 4-5 cm from the leaf base, and then RNA accumulation already declines. Protein accumulation continues almost linearly for another 36 hr, and during this period massive synthesis of chloroplast proteins takes place, in particular, of ribulose biphosphate carboxylase.

In addition to the protein, RNA, and DNA measurements in the wheat system, the composition of RNA and of DNA (nuclear or plastid), as well as chloroplast numbers and distribution of different cell types were determined. Also, other parameters can be measured, especially related to gene expression using sensitive nucleic acid hybridization methods. Although the resolution of the experiments with respect to time is limited (to work with sections less than 5 mm in length, corresponding to 4 hrs of growth is not possible for most experiments), the grass-leaf system compares favorably to other higher plant systems. Unfortunately, different batches of seeds will not give rise to identical leaves, much less seeds from different cultivars or different species, as seen here for wheat and barley (the differences between the K12 and B strains of *E. coli* are trivial in comparison). Thus, not surprisingly, we come to the conclusion that analysis of higher plants should not follow the *E. coli* model — this is amply illustrated in the next section.

AN ATTEMPT TO ESTIMATE THE AVERAGE PEPTIDE CHAIN GROWTH RATE IN YOUNG LEAF CELLS

In the present context it seems appropriate to try to use the RNA and protein measurements to estimate the efficiency of the ribosome in protein synthesis in the most active sections of the barley leaf. The estimate we arrive at is very approximate indeed, but the exercise serves to pinpoint the missing pieces of information, some of which would be very difficult to obtain.

The plant cell synthesizes protein in three separate compartments, the cytoplasm, the chloroplast, and the mitochondria. The mitochondria probably make a limited contribution, at least the fraction of stable RNA that is in the mitochondria is very low (Walbot, 1979). In the region of the leaf where the rate of protein accumulation per weight unit of RNA is maximal (2-4 cm from the base) there are nearly as many chloroplast ribosomes as ribosomes in the cytoplasm (in the wheat leaf, Dean and Leech, 1982a), and for the purpose of

the present calculation it is assumed that they are equal in number.
A priori there is no reason to assume that the chloroplast and
cytoplasmic ribosomes function with the same efficiency; there are no
simple means of ascertaining this point, and the figure we arrive at
thus represents the average activity of their combined activities. There
are three major types of leaf cells, mesophyll cells, epidermal cells,
and cells in the vascular system. Only the mesophyll cells contain
chloroplasts, and probably also the highest number of cytoplasmic
ribosomes, but that is not known. The uneven distribution of
ribosomes among the different cell types will affect our calculations
only if the ribosome efficiency is significantly different in different
cells.

The maximal rate of protein synthesis in the 2-4 cm region of
the barley leaf is 300 ng protein per microgram RNA per hr. The
average molecular weight of an amino acid is taken to be 110 dalton,
and the average molecular weight of the rRNA complements of the
large and small ribosome is 1.7×10^6 dalton. The fraction of total
RNA that is rRNA and how much of that is in mature ribosomes is
not really known, but estimates from electrophoretic fractionations of
total RNA (e.g. Dean and Leech, 1982a) indicate that 85% is probably
a reasonable value. From these numbers the average ribosome
efficiency in barley leaf cells is calculated to be 1.5 amino acid per
ribosome per sec at 21 C. In *E. coli* the growth rate varies as a
simple function of temperature, and the growth rate at 21 C is 47
percent of the growth rate at 30 C (Herendeen et al., 1979). Cell
composition does not vary significantly as a function of temperature
(Schaechter et al., 1958), and we assume that the average peptide
chain growth rate changes with the same factor between 21 C and 30
C as the cell growth rate. It means that the ribosome efficiency in
barley corresponds to 3.2 amino acids per ribosome per sec at 30 C.
This value is significantly lower than the average efficiency in *E. coli*
(Gausing, 1981), yeast (Lacroute, 1973), and *Neurospora* (Alberghina
et al., 1975), which is 7.5-8 amino acids per ribosome per sec at 30
C (and at medium to high growth rates). However, even if the
determinations of RNA and protein in barley and the assumption made
above are substantially correct, we do not at present know to what
extent diurnal variations in the rate of protein synthesis may
contribute to the low average. In addition, neither eukaryotic
(Alberghina et al., 1975) nor bacterial ribosomes (Gausing, 1980) keep
the pace in slow-growing cultures, and with the limited information
available it would be wrong to conclude that the intrinsic ability of
the two types of plant ribosomes to polymerize amino acids is poorer
than that of the ribosomes of lower eukaryotes and prokaryotes.

CHLOROPLAST BIOGENESIS: COOPERATION BETWEEN NUCLEAR
GENES AND THE PLASTID GENOME

From classical genetic studies it has long been known that plastid
development requires the activity of both nuclear and plastid genes,
but only recently has it become clear just how tightly the activities
in the nucleus and in the plastids are interwoven. The progress
results from the assignment of a growing list of genes to either the
nucleus or the plastid genome. The tight linkage between nuclear and
plastid genes is strikingly illustrated by the fact that every major
multimeric structure or enzyme analyzed has components that are

products of the nuclear genome and of the plastid genome. This is the case for the large subunit (plastid-coded) and the small subunit (nuclear-coded) of ribulose biphosphate carboxylase (Ellis, 1981) of the ATPase subunits (Krebbers et al., 1982; Watanabe and Price, 1982) and for the ribosomal proteins (Boynton et al., 1980). One elongation factor, Tu, is encoded in the plastid genome (Cifferi et al., 1979) and Ts in the nucleus (Fox et al., 1980). EF-G is nuclear-encoded in *Euglena* (Breitenberger et al., 1979), but appears to be coded by the chloroplast genome in spinach (Cifferi et al., 1979). The genes for the subunits of the RNA polymerase have not been mapped with certainty (Watson and Surzycki, 1983), and the DNA polymerase, whose structure is not known, is entirely coded in the nucleus, since nuclear mutants, such as *iojap* in maize, which are completely devoid of chloroplast ribosomes, still maintain plastids with normal DNA (Walbot, 1979).

Considering how few protein genes are left on the plastid genome (about 100), it does not seem plausible that the strict sharing of genes for multimeric components could have evolved by chance. Instead, this division may be a prerequisite for coordinating cellular and plastid development. It is believed that plastid development is governed by the nucleus, because the state of differentiation of the cell, supposedly a reflection of nuclear gene activity, determines the fate of plastid development: root cells contain amyloplasts, flowers and fruits contain chromoplasts, and leaf mesophyll cells contain chloroplasts. However, at present it cannot be excluded that regulatory signals may pass from the plastids to the nucleus.

With respect to coordinating activities of genes for defined components of a multimeric enzymes, research is still in an early stage, and practically all information comes from studies of synthesis of ribulose biphosphate carboxylase. In the wheat leaf the synthesis of the two subunits (measured by 3-hr pulse-labeling of whole leaves) and the levels of their mRNAs (measured by *in vitro* translation) were analyzed in sections of the leaf (Dean and Leech, 1982b). In both cases, the ratios of the activities for the large and small subunit remained fairly constant in all leaf sections, indicating that their synthesis is indeed well coordinated. The experiment did not permit a calculation of the stoichiometry of the synthesis of the two subunits. However, there does not appear to be an obligatory linkage in the biosynthesis. Thus, when the synthesis of the large subunit is inhibited in *Chlamydomonas* with chloramphenicol, the synthesis of the small subunit continues; in addition, the cells possess a mechanism for selective degradation of the excess of the small subunit, probably located inside the chloroplast, since the excess small subunit is processed and transported normally into the chloroplasts (Schmidt and Mishkind, 1983).

An especially intriguing cooperation problem is presented by the synthesis and assembly of chloroplast ribosomes. The rRNA genes and the genes for about one third of the r-proteins are located on the plastid genome (Eneas-Filho et al., 1981). The rest of the r-protein genes are in the nucleus, together with the two other sets of r-protein genes that the plant cell possesses. Clearly the elegant coupling between rRNA and r-protein synthesis found in *E. coli* (and described in numerous papers in this volume) cannot operate for the nuclear-coded plastid r-protein genes. It is a very complex problem to investigate experimentally, but we have made a start. As part of a general search for nuclear coded plastid genes, we hope to identify nuclear plastid r-protein genes and, as described in the next section, we have also initiated more specific experiments that may lead to the isolation of such genes.

THE PROKARYOTIC NATURE OF PLASTIDS

Heated discussions on the endosymbiont hypothesis for the origin of plastids still prevail in spite of the rapidly mounting evidence that the plastid ancestors of higher plants belonged to oxygenic-photosynthetic prokaryotes (the plastid ancestors of red algae were of cyanobacterial type) (Gray and Doolittle, 1982). Comparisons of the translation and transcription machineries in plastids and bacteria have supplied particularly compelling evidence for the eubacterial origin of plastids. Much of the evidence has been reviewed recently (Gray and Doolittle, 1982), and Table 1 provides a short summary of newer results that reflect the very rapid progress in the elucidation of plastid gene structure that has taken place during the last few years. The plastid protein sequences listed in Table 1 were derived from the DNA sequences of plastid genes with the exception of r-protein L12, and it is not known if L12 is coded in the nucleus or by the plastid genome. So far, nuclear gene sequences for r-proteins or translation factors have not been isolated, but the nuclear gene for yeast mitochondrial EF-Tu was recently cloned (Nagata et al., 1983). It is 66 percent homologous to *E. coli* EF-Tu and was isolated via hybridization to *tufB*.

TABLE 1 Plastid and *E. coli* sequences

Molecule	% homology	References*	
		E. coli	Plastid
16S RNA	74	2	12
23S RNA	71	2	3
ATPase β	63	10	6
ATPase ε	23	10	6
r-protein S4	39	11	13
r-protein S7	35	9	8
r-protein S12	68	4	8
r-protein S19	55	16	14
r-protein L12	50	15	1
EF-Tu	70	5	7

*(1) Bartsch et al., 1982; (2) Brosius et al. 1978; (3) Edwards and Kossel, 1981; (4) Funatsu et al., 1977; (5) Jones et al., 1980; (6) Krebs et al., 1982; (7) Montandon and Stutz, 1983; (8) Montandon and Stutz, 1984; (9) Reinholt et al., 1978; (10) Saraste et al., 1981; (11) Schlitz and Reinholt, 1975; (12) Schwarz and Kossel, 1980; (13) Subramanian et al., 1983; (14) Sugita and Sugiura, 1983; (15) Terhorst et al., 1973; (16) Yaguchigi and Wittman, 1978.

(a)

λ spc1

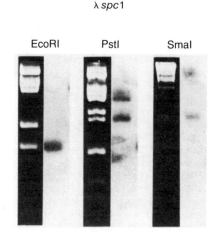

(b)

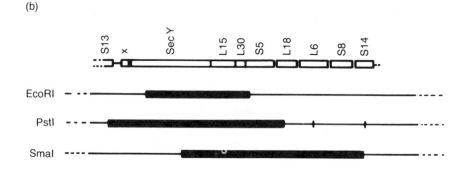

FIGURE 3 Hybridization of barley leaf cDNA to restriction fragments of λ spc1 DNA. (a) λ spc1 DNA was digested with EcoRI, PstI, and Smal and fractionated on a 2% agarose gel (5 μg DNA/lane). The DNA fragments were transferred to a nitrocellulose filter and hybridized to 10^7 cpm of cDNA synthesized by reverse transcription of 1 μg poly(A) mRNA purified from the basal 4 cm of 7-day-old barley leaves. The hybridization was carried out according to standard procedures, but at low stringency (25% formamide at 42°). Left lanes show DNA stained with ethidium bromide, and right lanes show autoradiograms of the hybridized nitrocellulose filter. (b) Map of the region of λ spc1 that contains the DNA fragments hybridizing to barley cDNA (Cerretti et al., 1983). The DNA fragments that hybridize are indicated with heavy bars. In the autoradiogram of the PstI digest of λspc1, two bands are detected; the lower band comes from the spc operon, and the upper band comes from the cI region of λ; the same λ region has been seen to hybridize to EcoRI and Smal digests of λspc1 in other experiments.

We have initiated a search for homology between *E. coli* genes for proteins of the transcription and translation machinery and barley nuclear genes. The yeast genome is sufficiently small that the mitochondrial EF-Tu gene could be isolated by direct cloning of nuclear DNA. This approach is clearly impossible with the genome of a higher plant. Instead we asked whether the poly(A) mRNA population of young barley leaf cells contains detectable amounts of mRNA homologous to *E. coli* genes carried by λ*spc*. Hybridization with cDNA also has the advantage that we could be reasonably sure that positive hybridization signal(s) originated from active genes. The recently described cases of mitochondrial and plastidic sequences in nuclear DNA (Farelly and Butow, 1983; Timmis and Scott, 1983), as well as the detection of sequences homologous to the *E. coli rpoB* gene in both the nuclear and plastid genome of *Chlamydomonas* (Watson and Surzycki, 1983), indicate that heterologous hybridization experiments are not without pitfalls. The results of hybridizing cDNA from barley leaf poly(A) mRNA to λ*spc* DNA is shown in Figure 3. Only a single small region of λ*spc* DNA showed homology with barley cDNA and the hybridization could only be detected at low stringency (25% formamide at 42 C), but the signal is strong enough to suggest that the nuclear plastid genes for r-proteins L30 and/or L15 may be isolated with the aid of the *E. coli* genes. Homology detected between the same region of λ*spc* DNA (the 3% EcoRI fragment) and *Chlamydomonas* nuclear DNA corroborates the nuclear location of one or both of these plastid r-protein genes, albeit the λ*spc* fragment also hybridized to *Chlamydomonas* plastid DNA (Watson and Surzycki, 1983). Formally it can not be concluded that our hybridization is due to a plastid r-protein gene. However, eukaryotic r-proteins are less likely to share sufficient homology with prokaryotic r-proteins and, as discussed earlier, the mitochondrial counterparts of these two r-proteins are probably expressed at a very low level.

ACKNOWLEDGEMENTS

We thank Inger Bjørndal and Bent Sørensen for skillful technical assistance and Aase Sørensen for help in preparing the manuscript. The work was supported by the Danish Natural Research Council.

REFERENCES

Alberghina, F. A. M. and Sturani, E. (1975). *J. Biol. Chem.*, 250, 4381.

Arnon, D. I. (1949). *Plant Physiol.*, 24, 1.

Bratsch, M., Kimura, M. and Subramanian, A. R. (1982). *Proc. Natl. Acad. Sci. USA*, 79, 6871.

Boffey, S. A., Sellden, G., and Leech, R. M. (1980). *Plant Physiol.*, 65, 680.

Boynton, J. E., Gillham, N. W., and Lambowitz, A. M. (1980). In *Ribosomes, Structure, Function, and Genetics*, G. Chambliss et al. (eds.), University Park Press. pp. 903-950.

Breitenberger, C. A., Graves, M. C., and Spremulli, L. L. (1979). *Arch. Biochem. Biophys.*, 194, 265.

Brosius, J., Palmer, M. L., Kennedy, P. J., and Noller, H. F. (1978). *Proc. Natl. Acad. Sci. USA*, 75, 4801.

Brosius, J., Dull, T. J., and Noller, H. F. (1980). *Proc. Natl. Acad. Sci. USA*, 77, 201.

Cerretti, D. P., Dean, D., Davis, G. R., Bedwell, D. M., and Nomura, M. (1983). *Nucleic Acids Res.*, 11, 2599.

Ciferri, O., Tiboni, O., Di Pasquale, G., and Carbonera, D. (1979). In *Genome Organization and Expression in Plants*, C.J. Leaver (ed.). Plenum Press, pp. 373–379.

Davies, D. D. (1982). In *Encyclopedia of Plant Physiology*, Vol. 14A, A. Pirson and M. H. Zimmermann, (eds.), Springer-Verlag, pp. 189–228.

Dean, C. and Leech, R. M. (1982a). *Plant Physiol.*, 69, 904.

Dean, C. and Leech, R. M. (1982b). *FEBS Lett.*, 140, 113.

Edwards, K. and Kossel, H. (1981). *Nucleic Acids Res.*, 9, 2853.

Ellis, R. J. (1981). *Ann. Rev. Plant Physiol.*, 32, 111.

Eneas-Filho, J., Hartley, M. R., and Mache, R. (1981). *Mol. Gen. Genet.*, 184, 484.

Farrelly, F. and Butow, R. A. (1983). *Nature*, 301, 296.

Fox, L., Erion, J., Tarnowski, J., Spremulli, L., Brot, N., and Weissbach, H. (1980). *J. Biol. Chem.*, 255, 6018.

Funatsu, G., Yaguchi, M., and Wittmann-Liebold, B. (1977). *FEBS Lett.*, 73, 12.

Gausing, K. (1977). *J. Mol. Biol.*, 115, 335.

Gausing, K. (1980). In *Ribosomes, Structure, Function, and Genetics*, G. Chambliss et al. (eds.). University Park Press, pp. 693–718.

Gausing, K. (1981). *Mol. Gen. Genet.*, 184, 272.

Gray, M. W. and Doolittle, W. F. (1982). *Microbiol. Rev.*, 46, 1.

Herendeen, S. L., Van Bogelen, R. A, and Neidhardt, F. C. (1979). *J. Bacteriol.*, 139, 185.

Ingraham, J. L., Maaløe, O., and Neidhardt, F. C. (1983). *Growth of the Bacterial Cell*, Sinauer Associates.

Jones, M. D., Petersen, T. E., Nielsen, K. M., Magnusson, S., Sottrup-Jensen, L., Gausing, K., and Clark, B. F. C. (1980). *Eur. J. Biochem.*, 108, 507.

Krebbers, E. T., Larrinua, I. M., McIntosh, L., and Bogorad, L. (1982). *Nucleic Acids Res.*, 10, 4985.

Lacroute, F. (1973). *Mol. Gen. Genet.*, 125, 319.

Lowry, O. H., Rosebrough, N. J., Farr, A. L., and Randall, R. J. (1951). *J. Biol. Chem.*, 193, 265.

Maaløe, O. and Kjeldgaard, N. O. (1966). *Control of Macromolecular Synthesis*, W.A. Benjamin, New York.

Montandon, P.-E. and Stutz, E. (1983). *Nucleic Acids Res.*, 11, 5877.

Montandon, P.-E. and Stutz, E. (1984). *Nucleic Acids Res.*, 12, 2851.

Nagata, S., Tsunetsugu-Yokota, Y., Naito, A., and Kaziro, Y. (1983). *Proc. Natl. Acad. Sci. USA*, 80, 6192.

Reinbolt, J., Tritsch, D., and Wittmann-Liebold, B. (1978). *FEBS Lett.*, 91, 297.

Saraste, M., Gay, N. J., Eberle, A., Runswick, M. J., and Walker, J. E. (1981). *Nucleic Acids Res.*, 9, 5287.

Schaechter, M., Maaløe, O., and Kjeldgaard, N. O. (1958). *J. Gen. Microbiol.*, 19, 592.

Schiltz, E. and Reinbolt, J. (1975). *Eur. J. Biochem.*, 56, 467.

Schmidt, G. W., and Mishkind, M. L. (1983). *Proc. Natl. Acad. Sci. USA*, 80, 2632.

Schwarz, Z. and Kossel, H. (1980). *Nature*, 283, 739.

Subramanian, A. R., Steinmetz, A., and Bogorad, L. (1983). *Nucleic Acids Res.*, 11, 5277.

Sugita, M. and Sugiura, M. (1983). *Nucleic Acids Res.*, 11, 1913.

Terhorst, C., Moller, W., Laursen, R., and Wittmann-Liebold, B. (1973). *Eur. J. Biochem.*, 34, 138.

Timmis, J. N. and Scott, N. S. (1983). *Nature*, 305, 65.

Walbot, V. and Coe, E. H. (1979). *Proc. Natl. Acad. Sci. USA*, 76, 2760.

Watanabe, A. and Price, C. A. (1982). *Proc. Natl. Acad. Sci. USA*, 79, 6304.

Watson, J. C. and Surzycki, S. J. (1983). *Current Genetics*, 7, 201.

Yaguchi, M. and Wittmann, H. G. (1978). *FEBS Lett.*, 88, 227.

Selection of Somatic Variation in Plants

Karl G. Lark
University of Utah, Salt Lake City

To Ole and Taj in memory of 1954

FORWARD

In September of 1953 I began a postdoctoral stay in Copenhagen, at the Statens Serum Institut, in the laboratory of Ole Maaløe. Ole was 39 and I was 22. Both of us were recent entries to the phage group, Ole at Cal Tech with Max Delbrück, myself at Cold Spring Harbor, with Mark Adams and Max Delbrück. This postdoctoral period changed my professional life: I left the study of phage per se, but remained part of the "Phage Group." In exchange I began a study of cell physiology and cell division, which was later to focus on DNA replication. The intellectual approach was that of the phage group: to study the cell holistically as a black box and to divine from the output what went on within.

While I was in Copenhagen, none of the exciting work for which Ole is now known had begun, but it was a privilege to be present at the sowing of the seeds that were going to lead to that work. In studying synchronous cultures of *Salmonella*, an interest began to develop for both of us in DNA replication and in the different responses due to rapid growth at higher temperatures or in richer media. My reward was to work intimately with Ole in the laboratory, hands on. Ole was a master experimentalist, and during that first year he gave me a legacy in technique and in experimental design for which I will always be grateful.

Intellectually, the most exciting development that occurred during my stay in Copenhagen was the proposal by Niels Jerne (Taj) of the first selective model for antibody synthesis. During my first year in Copenhagen, that is, 1953-54, there was a great deal of discussion of all aspects of science, usually between Ole and Taj. These occurred briefly after Taj got in at 10-11 in the morning and resumed after 4 in the afternoon when others in the lab had gone

144

home and we had remained (Ole, Taj, Perla Avengo—a post-doc with Taj—and I). Taj left for Cal Tech during the fall of 1954 and sent us the first draft of his selective model. For weeks, Ole and I discussed the model in all of its aspects. These discussions of selection and variation made a deep impression.

INTRODUCTION

The influence of these ideas played an important role in the decision that I made ten years ago: to develop plant tissue culture as a model system for studying growth and regulation in eukaryotes. An important aspect of that decision was an attempt to develop plant tissue culture as a system of somatic cell genetics. In attempting to isolate mutants, we soon learned that selective isolation of variants (mutants?) led with high frequency to the isolation of variant cell lines that were not stable. In contrast, variants selected by nonselective screening of duplicate colonies were infrequent but stable. The problem that I shall discuss is related to this aberration.

Although this work is less developed than other research in our laboratory, there is a clear relation to the ideas and thinking of my post-doctoral days in Copenhagen. It therefore seems appropriate that the problem (and a possible solution) should make its debut here. In what follows, I shall argue that one of the most important differences between higher plants and higher animals is that the gamete cell line is sequestered in animals, whereas in plants it is not. As a result, plants can adapt heritably to their environment, because genetically altered somatic cells, which are selected, can produce gametes and thus seed.

BACKGROUND

Since the 1920s, there have been continuing reports of inherited changes in plants that have occurred in response to changes in their environment. My attention was drawn to this phenomenon by the following data from our laboratory: soybean cells contain about 2200 rRNA genes, though it is not known whether all of these are transcribed (Jackson, 1981). When soybean cells are grown in suspension culture with glucose or sucrose as an energy source, they divide every 24 hr. However, when maltose is substituted as a carbon-energy source, they grow slowly, dividing every 200 hr (Limberg, Cress, and Lark, 1979). The ribosome content is decreased about 8-fold in the slowly growing cells, and the rate of rRNA synthesis is reduced about 70-fold (Jackson and Lark, 1982). This is similar to the dependence of ribosome content on growth rate that has been observed in bacteria. A surprising result was that approximately a third of the rDNA genes (about 750 copies) were lost when cells were grown for several generations in maltose (Jackson, 1980, 1981). This was accompanied by a change in the sites at which the genomic rDNA could be cut by restriction endonucleases: i.e., high molecular weight restriction fragments were obtained after digestion of DNA extracted from rapidly growing cells, but such large fragments could not be observed when DNA extracted from maltose cells was digested. Thus, cells grown for only a few generations in a medium with apoorly utilized carbon source had (1) lost a large fraction of rDNA genes and (2) presented an altered restriction pattern of the remaining rDNA genes. As yet, we have not been able to demonstrate that the missing rDNA genes represent a particular class of rDNA distinguished by restriction polymorphism. However, we are tempted to conclude this from the data.

These observations are similar to ones reported by Cullis (1979; Cullis and Charlton, 1981; Durant, 1958) for flax. In this plant, growth in a "poor" soil leads to stunted or small plants that contain a reduced number of ribosomal RNA genes. This reduction in rDNA is inherited for many generations.

LAMARCKIAN INDUCTION OR DARWINIAN SELECTION

Because of our own results in tissue culture, I felt that the findings in flax were real and had to be explained. In tissue culture, the remarkable finding was the rapid loss of ribosomal RNA genes, but to a microbiologist or cell physiologist there was nothing surprising about the inheritance of this condition during vegetative growth. It is surprising, however, that in an intact plant the phenotype can be transmitted to the progeny. Such Lamarckian phenomena could be the result of artifacts. However, it is possible to reconcile these results with Darwinian selection in the following way. In plants, the gametic line is not sequestered. Almost any living cell can give rise to a gamete (Reinert, Bajaj, and Zbell, 1977). (This has been shown by innumerable tissue culture studies in which cells from roots, stem, leaves, etc., have been regenerated into complete plants.) As a result, selection operating on somatic cells is, in theory, capable of being transmitted to the germ plasm. Genetic variants can arise during vegetative growth. Those dividing cells that are selectively favored will be preferentially multiplied during somatic growth, and gametes arising from these will transmit the genetic change.

If this can happen, why does it not occur more frequently? The answer may lie in the nature of the reproductive process. Each stage in the formation of a new plant by sexual reproduction invokes the transcription and translation of genes not ordinarily used in vegetative growth (Goldberg et al., 1981; Jackson, 1980; Kamalay and Goldberg, 1980). Thus, flower and gamete formation, embryogenesis (seed formation), and seed germination are processes that are informed from special genes. A genetic change that is advantageous during vegetative growth may not be compatible with stages in the reproductive process. Compatability can then only be achieved through further genetic change. This is difficult to achieve within the life cycle of an annual plant that must germinate, grow, flower, and set seed before it dies within the time span of less than a year. If the necessary changes have not occurred within the limited cell population produced during this cycle, changes that are selected during somatic growth will be lost.

THE PERENNIAL PLANT (DO TREES HAVE MEMORIES?)

Perennial plants maintain and amplify their somatic cell populations during large numbers of successive growing seasons. Variation selected during somatic growth can be further modified to become compatible with the various stages of sexual reproduction. In such plants a population of variant cells could be maintained or even amplified during successive years, and attempts at gamete formation and embryogenesis might eventually meet with success owing to the occurrence of second-site suppressors. In a manner similar to the immune system, a perennial plant can become the repository of genetic information selected to cope with the environmental changes experienced within the plant's lifetime.

Regulation of somatic variation is related to meristematic growth. Meristems are groups of cells that divide to provide the growth of the plant. Usually, the apical, or topmost, meristem inhibits growth of meristems below it, to produce the "christmas tree" appearance seen on so many plants or plant branches. As growth proceeds, new groups of meristems are formed, and hence the population of meristems expands.

In an annual plant, such as tobacco or corn, the number of such meristems in the mature flowering plant approaches about 100 to 200 above or below ground. Each meristem is comprised of about 10^5 to 10^6 dividing cells (calculated from data in references (Poethig, 1984; Sax and Erickson, 1956). Thus, the total population of cells from which variants can be formed is limited. Since the population of meristematic cells is usually small, because there are few nodes, selective enrichment of a particular meristem population would have to begin at an early developmental stage to show any effect. (Indeed, such a rapid and early change is exactly what occurs in the reduction of rDNA genes in tissue culture (Jackson, 1981) and in the flax plant (Cullis and Charlton, 1981)).

In contrast, the node population in a perennial is very large. A mature aspen or poplar tree may have as many as 10^5 nodes and thus contain a population of 10^{10} cells from which variation can be selected. Moreover, because of the nature of the apical inhibition discussed above, selection imposed by the environment, which inhibits most meristems (including the apical meristem), will remove the inhibition due to apical growth, thus releasing those variant meristems capable of growth under the selective conditions. The variants will then be amplified relative to the remainder of the meristem population. At the end of such a growing season, these cells will form a topmost, or apical, meristem and will thus maintain their predominance over the rest of the branch or shoot from which they arose.

For plants most environmental selection is quantitative, or a question of degree (e.g., cold or drought). In both microorganisms and eukaryotes, intensity of phenotype often is related to gene copy number (Schimke, 1982). Thus, gene amplification and deamplification could provide a quantitative form of regulation that is inherited. It is interesting to note that plants, far more than animals, contain large amounts of repeated DNA and are often polyploid or heteroploid, especially when growing in variable environments. Another unique form of genetic information in plants is their cytoplasmic DNA. The information content of such DNA can range from 500 to 3500 Kb (Bendich, 1982). Since the selective process that I am proposing occurs at a cellular level, variation in extrachromosomal DNA also will be selected. Because of the large number of copies of such DNA in each cell, it is possible to obtain mutations, rearrangements, and duplications of this type of DNA by recombination (Belliard et al., 1979). Moreover, intracellular regulation of the proportion of such aberrant organelles could result in quantitative variation of the cell phenotype.

Another form of variation which may occur frequently is the transposition of regulatory elements, which appears to be especially high in plants.

CONCLUSION

The concepts presented here are based on two fundamental sets of observations: (1) that plant cells do not sequester a germline and, hence, selection can operate on a cellular level, as it does in microorganisms; and (2) that the maintenance of large cell populations in perennial plants leads to the accumulation within selected cells of genetic rearrangements that are the result of several successive events. Because plant cells are in direct contact with their environment and because they can be maintained in tissue culture, the effects of these changes can be studied under controlled conditions.

In the discussion just presented, I have described certain attributes of plants that conceptually are very similar to the behavior of microbial cell populations. In this sense the ideas and approaches developed by the Copenhagen group can be applied to the study of plants and plant tissue culture. What makes this an exciting new field of endeavor is that the basic unit, the plant cell, is quite different from the bacterial cell and therefore poses new questions, which offer new opportunities for the scientist interested in cell regulation and physiology.

ACKNOWLEDGEMENTS

Portions of this work were supported by grant NIES 01498. A more detailed exposition of these ideas will appear elsewhere. I am grateful to Dr. T. Whitham for his comments on these ideas. Similar ideas developed by Dr. Whitham based on ecological studies appear in an excellent paper by Whitham and Slobodchikoff (1981): "The adaptive significance of somatic mutations in plants." *Oecologia,* 49, 287.

REFERENCES

Belliard, G., Vedel, F., and Pelletier, G. (1979). *Nature,* 281, 401.

Bendich, A. J. (1982). In *Mitochondrial Genes,* Cold Spring Harbor Laboratory, p. 477.

Cullis, C. A. (1979) *Heredity,* 42, 237.

Cullis, C. A. and Charlton, L. (1981). *Plant Science Letters,* 20, 213.

Durrant, A. (1958). *Nature,* 181, 928.

Goldberg, R. B., Hoschek, G., Tam, S. H., Ditta, G. S., and Breidenbach, R.W. (1981). *Develop. Biol.,* 83, 201.

Goldberg, R. B., Hoschek, G., Ditta, G. S. and Breidenbach, R. W. (1981). *Develop. Biol.,* 83, 218.

Jackson, P. J. (1980). *Fed. Proc.,* 39, 1878.

Jackson, P. J. (1981). Ph. D. Dissertation, University of Utah.

Jackson, P. J. and Lark, K. G. (1982). *Plant Physiol.*, 69, 234.

Kamalay, J. C. and Goldberg, R. B. (1980). *Cell*, 19, 934.

Limberg, M., Cress, D., and Lark, K. G. (1979). *Plant Physiol.*, 63, 718.

Poethig, S. (1984). In *Contemporary Problems in Plant Anatomy*. R. A. White and W. C. Dickison (eds.) Academic Press. p. 234.

Reinert, J., Bajaj, Y. P. F., and Zbell, B. (1977). In *Plant Tissue and Cell Culture*. H. E.Street (ed.), University of California Press, p. 389.

Sax, K. B. and Erickson, R. O. (1956). *Proc. Amer. Phil. Soc.*, 100, 499.

Schimke, R.T. (ed.) (1982). *Gene Amplification*. Cold Spring Harbor Laboratory.

PART TWO

CONTROL OF GENES
AND REGULONS

INTRODUCTION

The popularity and influence of the book *Control of Macromolecular Synthesis* (O. Maaløe and N. O. Kjeldgaard, 1966. W. A. Benjamin, Inc.) have contributed to the readiness with which we associate Ole Maaløe and the Copenhagen school with two dominant themes of growth physiology: adjustment of the size of the protein-synthesizing machinery to achieve efficiency, and coordination of DNA synthesis and cell division during growth at different rates. These themes recur in Parts I, III, and IV of the present volume.

The contributions in this part expand our view of the Copenhagen influence. What follows is a collection of studies on the regulation of individual genes and sets of operons (regulons), primarily, but not exclusively, in *E. coli*. The titles of these presentations indicate something of their diversity, but even an initial scan reveals shared elements. These papers all bear a Copenhagen stamp; their family resemblance is shown, at least collectively, in the nature of the questions they pose, their experimental approach, and the depth of their molecular analysis.

The questions addressed are in no instance far removed from the immediate business of cell growth and survival. In each case, the molecular analysis is driven by the goal of understanding coordination of *in vivo* function. Molecular details are sought not in the abstract but in the context of the growing cell. Studies on individual operons are presented in the metabolic domains of transport, catabolism, central pathways, and biosynthesis; they are illustrative of the classical modes of regulation by induction and repression and by the less well-understood operation of endogenous growth-rate-related signals. Molecular mechanisms are probed in great depth (for example, *ara, lac, pyr,* and *gnd*), and one study tackles the difficult task of assessing the relative contributions of different control mechanism operating on the same system (*trp*) under different growth conditions.

Three of the chapters deal explicitly with multigenic responses of bacterial cells to environmental stresses: temperature shift, radiation, and phosphate deprivation, and the theme of global control networks appears in many of the other papers through consideration of catabolite repression and the stringent response. Indeed, the analysis of molecular mechanisms that tie together unlinked operons that are subject to individual controls provides a strong thread connecting these studies to those in Parts I and III.

Frederick C. Neidhardt

Studying Regulation by Using Gene Fusions

Michael L. Berman
Litton Institute of Applied Biotechnology;
Rockville, Maryland

INTRODUCTION

Spring 1972 was an exciting year to begin graduate school. I had just moved to Boston to become a student in the Department of Microbiology and Molecular Genetics at the Harvard Medical School. One of my first classes was taught by Dr. Jon Beckwith and concentrated on the genetics of the *lac* operon. A scheduling coincidence juxtaposed two particular lectures. The first described a beautiful genetic analysis of transcription signals in the *lac* operon by using *trp-lac* gene fusions (Ippen et al., 1968); the second lecture described the phenomenon of "stable RNA regulation" and the stringent response (Ingraham et al., 1983). There was little to be said about the genetics of expression of stable RNA at that time. It occurred to me that regulation might be analyzed by gene fusions to *lac,* but several years passed before I returned to this idea.

My laboratory work during the next two years armed me with the tools of modern genetics. Under the tutelage of Dr. Luigi Gorini, I began to appreciate the power and subtlety of *E. coli* genetics. Meanwhile the idea of studying regulation that was dependent on growth rate by using gene fusions slowly matured. Two critical observations were provided before I could initiate such a study. The first, by Paul Primakoff (then a postdoctoral fellow in Beckwith's lab), was that the suppressor tRNA gene *supIII* is subject to stringent control (Primakoff and Berg, 1970). This meant that I had found a stable RNA with an easily selectable phenotype for constructing gene fusions genetically. The second important event was the development of a new *in vivo* method for isolating gene fusions by a fellow graduate student Malcolm Casadaban (Casadaban, 1976). Casadaban's original method had immediate impact on a number of ongoing studies (Bassford et al., 1978), and, most important for me, it provided the technique to construct stable-RNA gene fusions *in vivo.*

154

My original idea was to use *lac* gene fusions to study expression of stable RNA in *E. coli*. I conceived of both *cis*-acting, as well as *trans*-acting mutations that could be isolated using *lac* genetics. One obvious class of mutations to analyze would be those affecting the promoter region of a stable RNA gene. Since the DNA sequence of the promoter region of *tyrT* (*supIII*) had been analyzed by Khorana and colleagues (Sekiya and Khorana, 1974), I concentrated my efforts on this target. Slowly I began to appreciate the phenomenon of regulation of stable RNA. I spent many hours with the 1966 book by Maaløe and Kjeldgaard in order to convince myself that a threefold induction of synthesis of stable RNA (with increasing growth rate) is "as dramatic" as the 1000-fold induction of the *lac* operon. In fact, two classic experiments, namely, the coordinate regulation of beta-galactosidase synthesis with growth rate and the effect of a shift of carbon source on beta-galactosidase synthesis, convinced me (and my thesis committee) that I had isolated *tyrT-lacZ* gene fusions (Berman and Beckwith, 1979a).

TyrT-PROMOTER MUTANTS AND THE ROAD TO COPENHAGEN

The *tyrT-lac* fusions had allowed me to isolated promoter-defective mutations in the *tyrT* gene (Berman and Beckwith, 1979b). The DNA sequences of these mutations, which are summarized in Table 1, helped to define critical sequences within the promoter (Berman and Landy, 1979). Fortunately, I had asked Dr. Niels Fiil, who was visiting Beckwith's lab at that time, to review these manuscripts prior to submitting them. This resulted in a 6-month visit to the Institute of Microbiology at the University of Copenhagen. Armed with my newly acquired skill in DNA sequencing, I left Landy's lab at Brown

TABLE 1 Promoter mutants of *tyrT* as gene fusions

Allele	Nucleotide position	Mutation	β-Galactosidase, units	
tyrT	—	Wildtype	522	(100%)
tyrTp20	−8	T → A	7	(1%)
tyrTp27			13	(3%)
tyrTp45	−8	T → C	8	(2%)
tyrTp51			8	(2%)
tyrTp20-3			8	(2%)
tyrTp119	−13	T → C	9	(2%)
tyrTp9-6			9	(2%)
tyrTp74	−16	C → G	46	(11%)
tyrTp11-4	−26/27	ΔG	10	(2%)

Data is summarized from Berman and Beckwith (1979) and Berman and Landy (1979). The nucleotide position of the mutations corresponds to the numbering shown in Figure 2.

University and arrived in Copenhagen in the fall of 1979. There I began to construct recombinant clones on high-copy-number plasmid vectors and to study the *tyrT-lac* fusions that I had isolated.

MANIPULATING *lacZ* FUSIONS *IN VITRO*

The critical regions for the *tyrT* promoter were defined by the defective mutations already sequenced. In order to manipulate the promoter region *in vitro,* it was necessary to delineate a suitable DNA fragment. For instance, the site of the fusion joint in *tyrT* should define a point within the structural gene after which no regulatory signals would be found. (These important signals are defined by the regulation seen to be shared by the tRNA gene and *lacZ* in the fusion.) The results of the DNA sequencing showed that in the single gene fusion analyzed the last intact base of *tyrT* was codon 19 of the structural gene (Figure 1), which is followed by 48 bp from the end of phage Mu. The presence of Mu phage DNA, which had been predicted by CsCl buoyant density analysis of transducing phages, is the result of *in vivo* manipulations involved in constructing the fusion. Following the Mu DNA are sequences from the *trpB* gene. Once again, the presence of these sequences was expected because of the method used to construct the original *tyrT-lacZ* fusion. This result shows that the promoter and its regulatory signals could be mobilized on a fragment with one end at the naturally occurring AvaI site within the structural gene.

The extent of the promoter-regulatory region preceding the structural gene was not as easy to establish. Once again, the promoter mutations defined a certain region. Khorana and his colleagues had shown that a synthetic *tyrT* gene can be transcribed from a fragment including sequences through -60 (Sekiya et al., 1976; Khorana et al., 1979). However, our DNA sequence analysis had revealed a homologous region that may serve as a second-degree promoter site located at -244 (Ross et al., 1980). This "promoter" probably does not function *in vivo,* because the point mutations isolated previously in the -10 region are defective for transcription. However, this upstream region may play a role in regulating expression, so we chose the rather large 306-bp DNA fragment shown in Figure 2 as a portable *tyrT* promoter.

```
        tyrT'    Δ      Mu      Δ    'trpB
    ...CCCGAGCGGC|TGAAGC...GCTTCA|TTCACCGTTC...
```

FIGURE 1 DNA sequence of the tyrT-lacZ DNA fusion joint of fusion 6 (Berman and Beckwith, 1979a). The unique joint in tyrT occurs at base 19 of the structural gene adjacent to a naturally occurring AvaI site (underlined). The tyrT sequences are followed by 48 nucleotides from the ends of bacteriophage Mu fused to the trpB gene at a site 42 nucleotides from its end. This arrangement is subsequently fused to lacZ via the W209 fusion joint originally isolated by Mitchell et al. (1975). Details of this gene-fusion analysis will be published elsewhere.

```
            10        20        30        40        50        60        70        80        90       100
          Mbo I                      Hha I                                                          Hha I
-248   GATCATACCTACCTACACAGCTGAAGAGATATGATGCGCGCAGGTCGTGACGTCGAGAAAAACGTCTTAAGTCGTGCACTATACAAAGTACTGGCACAGGCGGCGTCT
                                        Hha I

-148   TTGTTTACGGTAATCGAACGATTATTCTTTAATCGCCAGCAAAAATAACTGGTTACCTTTACGGATGAAAATTACGCAACCAGTTCATTTTT

                             Hha I              Hha I
-48    CTCAACGTAACACTTTACAGCGGCGGCGTCATTTGATATGATGCGCCCCGCTTCCCGATAAGGGAGCAGGCCAGTAAAAAGCATTACCCCGTGGTGGGGTT
                                                  ●                                      o
                                          <------------- precursor ------------->

       Ava I
+53    CCCGAGATC
         Mbo I
```

FIGURE 2 A portable 306-nucleotide promoter fragment from tyrT. The MboI site at -248 is naturally occurring, while the MboI site at +53 is engineered. Various restriction enzyme sites are listed. The start of transcription (solid circle) is following by a precursor sequence as indicated. There are two additional bases, an A at position +27 and a G at position +38; these give a precursor length of 43 compared with 41 bases, as determined from the RNA sequence (Altman and Smith, 1971).

Once an appropriate promoter fragment had been chosen, it was necessary to test its function *in vivo*. To do so, we used the cloning vector MLB1010 (Chattoraj et al., 1984), a derivative of the plasmid pBR3232 carrying the *lacZ* gene in a clockwise orientation within the *tet* region. The *lac* promoter region has been replaced by a segment of a *trp-lac* operon fusion (W205) isolated *in vivo*, whose nucleotide sequence is known (Xian-Ming et al., 1984). Data from our laboratory (R. Zagursky, unpublished) established the fusion points between a synthetic BamHI linker and the *trpA* gene in the W205 *trp-lac* fusion. The details of this arrangement are shown in Figure 3. Basically, the *trpA-lacZ* deletion removes the *lac* promoter and leaves the ribosome binding site intact. Expression of *lacZ* from this plasmid is dependent on clockwise transcription originating from within or crossing into the

```
                                                   (W205)
                   Δ                                Δ
     Sma I       |←'trpA                          |←'lacO          lacZ
GAATTCCCGGGGATCCGG|ATT GAG CAG...AGT TAA...TCTTTA|TCACACAGGAAACAGCT ATG ACC...
  EcoRI    BamHI  |30  31  32    268  *  ..44 bp..|      S.D.        1
```

FIGURE 3   Nucleotide sequence of the operon-fusion cloning vector pMLB1010 (Chattoraj et al., 1984). Fusion joints are marked with vertical bars and triangles. A synthetic DNA linker was joined to trpA sequences by Casadaban (1980). The fusion joints have been confirmed by DNA sequence analysis in our lab. The DNA sequence of the trpA-lacZ fusion joint is published (Xian-Ming et al., 1984). Relevant genes are indicated. Codons are underlined and numbered according to the position of the corresponding amino acid residue in the mature gene product. bp = base pair; S.D. = Shine-Dalgarno sequence.

TABLE 2   β-Galacosidase levels of cloned operons

Promoter (fusion)	β-Galactosidase, units
None (W205)	700
P_{tet} (W205)	8,900
P_{tyrT} (W205)	16,700
P_{lac} (wildtype)	6,931
P_{lacL8} (wildtype)	443

All clones are derivatives of pBR322. Assays were according to Miller (1972) on cells grown in TYE medium at a cell density of ca. 5 x 10^8 cells/ml The plasmids labeled W205 are derived from the operon-fusion cloning vector pMLB1010 (Chattoraj et al., 1984). The cloned *lac* promoters are measured from a derivative of pMLB1010 that carries all of *lacI* and *lacZ* in the wildtype configuration. These clones were induced with 1 mM IPTG for 4-5 generations prior to assays.

trpA region. A comparison of the levels of beta-galactosidase synthesis supported by insertion of the *tyrT* promoter fragment compared with other promoter regions is shown in Table 2. Clearly the *tyrT* promoter fragment functions *in vivo*.

A defined promoter fragment plus defined promoter mutations constitute the perfect substrates for biochemical analysis. By a simple recombination with an existing promoter point mutation, *tyrTp27*, (Berman and Beckwith, 1979b) we could obtain large amounts of DNA from both wildtype and mutant promoters for *in vitro* transcription experiments. Analysis of these promoters, as well as additional deletions of the wildtype promoter, provided evidence for the extensive interaction of RNA polymerase in the distant upstream region (-85 to -135; Travers et al., 1983).

What have we learned about the regulation of the *tyrT* gene? Mutations and deletion have defined critical sequences responsible for expression, though the DNA sites for recognizing the regulatory signals (for regulation of growth rate as well as for the stringent response) remain elusive. A simple experiment that holds to the tenets expressed in a review by Beckwith (1981) shows how elusive these signals may be. The lessons from the *lac* operon show that promoter-defective mutations may or may not alter the site of action of the components of the *crp-cya* positive regulatory system (Beckwith et al., 1972). In fact, such measurements were central in establishing the two-site promoter model for *lac*. In a similar way we have measured the growth-rate-dependent expression of the various *tyrT* promoter mutants in the *tyrT-lacZ* fusion background. The results show that this response is unaltered in the promoter mutants (S. Brown, unpublished data). We have yet to define the site of action of these signals *in vivo*.

Recently we have found another opportunity to use gene fusions to explore regulation of synthesis of stable RNA. Careful biochemical measurements have led Nomura and his colleagues to propose a feedback model for global regulation of ribosome biosynthesis (Jinks-Robertson et al., 1983). This study used a 10.3-kb plasmid (derived from pBR322) that carries an intact copy of the *rrnB* clone dependent

TABLE 3   Effect of a plasmid carrying *rrnB* on expression of a *tyrT-lacZ* operon fusion

Plasmid	β-Galactosidase, units	
	Expt. 1	Expt. 2
pBR322	1845	1270
pN01301	719	979

Assays of β-galactosidase activity were performed according to Miller (1972). The plasmids are in strain MBM7014 (Berman and Beckwith, 1976a) that has been lysogenized with a transducing phage λp1041 (Berman and Jackson, 1984). This phage carries the operon fusion constructed *in vitro* using the *tyrT* promoter fragment and the *lacZ* vector pMLB1010 (see Table 2).

on expression of the *tyrT-lacZ* gene fusion. The previous results had shown a 40% decrease in expression of the *tyrU* gene, whereas our results show a similar effect on expression of *tyrT* (Table 3). Presently, we are trying to devise selective conditions using the *tyrT-lacZ* fusion that will allow us to select mutations that relieve this effect.

MORE FUSION GENETICS

The utility of *lac* gene fusions goes beyond the study of regulation and isolation of mutants: for example, identifying a gene product. Recently, we set about to identify the gene corresponding to a component of the machinery for protein export, *prlA*. Studies had shown that *prlA* mutations map in the region of the *spc* operon (Emr et al., 1981). These mutations suppress defects in the signal sequence of the outer membrane protein LamB and thereby restore its localization to the outer bacterial membrane.

Characterization of the *prlA* locus once again has relied on gene fusions. Precise mapping had shown that the *prlA* mutations were located in the distal region of the *spc* operon. The DNA sequence of this region had been published by Post et al. (1980) and revealed an open reading frame containing more than 100 codons. In order to determine if this open reading frame could be expressed *in vivo* we systematically fused the 5' end of this putative gene to a large segment of *lacZ*. Regulated transcription was provided by the *lacPO* region (Shultz et a., 1982). This clone did produce a large hybrid beta-galactosidase protein carrying at least 300 residues from the *prlA* open reading frame. This hybrid is now being used with immunological methods (Weinstock et al, 1983) to detect the synthesis and cellular localization of PrlA.

REGULATION OF PORIN EXPRESSION IN *E. coli*

My association with the Silhavy laboratory over the past few years has provided me with an opportunity to study another regulatory system in *E. coli*. Previous experiments had shown that the two major outer membrane porins of *E. coli*, OmpF and OmpC, are regulated by a third gene product, OmpR (Hall and Silhavy, 1979). In addition to the positive regulatory protein OmpR a second locus, designated *envZ*, was shown to affect the relative levels and to balance synthesis of the two porins (Hall and Silhavy, 1981). Although the DNA sequence of the *ompR* gene had been reported (Wurtzel et al, 1982), we undertook a detailed gene-fusion study to explore the mechanism of action of this positive regulator.

Selection of Lac+ gene fusions *in vivo* requires the ability to utilize lactose as a carbon source (Miller et al , 1970). The original *trp-lac* fusion studies defined two critical conditions for selection of gene fusions: (1) the target gene must be adjacent to, expressed, and transcribed in the same orientation as *lacZ*, and (2) the *lacZ* gene must be "cryptic," that is, it must require activation by a deletion or insertion of active signals (Figure 4). In order to utilize lactose two activities are required: beta-galactosidase and lactose permease. Although we had developed simple and rapid techniques for gene cloning of *lacZ*, unfortunately we had neglected the permease. However, once again the *tyrT* gene came to the rescue. We found

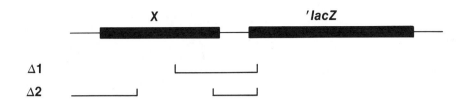

FIGURE 4 General scheme for selection of <u>lacZ</u> gene fusion in a LacY[+] host. Spontaneous deletions that activate a cryptic <u>lacZ</u> are most commonly fusions to the target gene <u>X</u>. The target in <u>lacZ</u> is limited, while the target in gene <u>X</u> is random. For details, see Berman and Jackson (1984).

that a *tyrT-lacY* fusion could supply permease and allow us to select *lacZ* gene fusions with a cloned gene (Berman and Jackson, 1984). In the construction of plasmids and phage clones to analyze the *tyrT* promoter region we had isolated an *in vitro* deletion that created a *tyrT-lacY* fusion. This arrangement was crossed into a lambda transducing phage and could be used to provide lactose permease to any cell. Finally, we could apply true selective pressure to isolate arrangements of cloned genes adjacent to a cloned cryptic *lacZ* segment.

    The results of this simple experiment are presented in Table 4. By allowing Lac[+] papillae to form on lactose MacConkey media, within one week hundreds of independent *ompR-lacZ* hybrid genes can be isolated. This selection has been applied to a number of target genes and seems to depend on expression of the *tyrT-lacY* fusion.

TABLE 4   OmpR-LacZ protein fusions

Allele number	Omp residue	First LacZ residue	Unexpected phenotype
103	96	19	
107	134	17	
25	134	4	OmpR[-d]
133	174	30	Deepest *lacZ* fusion
122	190	16	
108	203	12	
118	210	11	
109	213	15	
120	213	4	
104	231	25	
11	237	4	OmpR2 phenotype*

*This phenotype is OmpF[+]OmpC[-] (Hall and Silhavy, 1981). For a complete description of the selection of these fusions, see Berman and Jackson (1984).

Once again, among the OmpR-LacZ hybrids there came a few kernels of information. Two of the fusions listed in Table 4 provided the course for future studies of OmpR function. The majority of the fusions do not express OmpR function and, therefore, if cloned in an OmpR⁻ cell, will not turn on expression of OmpF or OmpC. One fusion (#11) will make such a cell OmpF⁺OmpC⁻. This phenotypic classification, designated OmpR2, had been described previously (Hall and Silhavy, 1981). Comparison of fusion alleles 11 and 120 shows that the protein domain responsible for activating *ompF* expression must include critical residues from OmpR between amino acids 213 and 237. The second notable fusion (#25) will make a genotypically *omp⁺* cell phenotypically OmpR⁻. The existence of such a classical dominant negative phenotype is taken as evidence for subunit mixing in OmpR. This phenotype will allow us to study protein-protein interactions of OmpR.

All good gene fusion studies should not neglect *lacZ*. Table 4 shows one unusual OmpR-LacZ fusion. Allele #133 replaces the first 30 residues of *lacZ*, which removes amino acids thought to be critical in forming the active oligomeric form of beta-galactosidase (Zabin, 1982). Whether the same protein domain responsible for the OmpR⁻ phenotype is involved in holding this beta-galactosidase together remains to be seen. Clearly a comprehensive set of random gene fusions provides new genetic handles for studying biological phenomena.

SUMMARY AND POSTSCRIPT

On the occasion of this meeting I have taken the opportunity to trace my study of gene regulation in *E. coli*. In doing so, I have not tried to review the many other recent gene fusion experiments that have been central in developing models of global regulation. However, it is clear that the full power of gene fusion analysis has yet to be appreciated. Hopefully, the appearance of recent reviews will make this technology more accessible to new investigators (Weinstock et al., 1983; Silhavy et al., 1984).

It is always difficult to credit the few individuals who have opened their laboratories and become true friends within the scientific community. However, it is clear to me that I owe thanks to my Danish colleagues, in particular, Ole Maaløe, for stimulating my interest in the regulation of bacterial growth.

ACKNOWLEDGEMENTS

Research was sponsored in part by the National Cancer Institute, DHHS, under contract No. 1-CO-23909 with Litton Bionetics, Inc.

REFERENCES

Altman, S. and Smith, J. D. (1971). *Nature New Biol.*, 233, 35.

Bassford, P., Beckwith, J., Berman, M., Brickman, E., Casadaban, M., Guarente, L., Saint-Girons, I., Sarathy, A., Schwartz, M., Shuman, H., and Silhavy, T. (1978). In *The Operon*, J. H. Miller and W. S. Reznikoff (eds.), Cold Spring Harbor Lab., p. 245.

Beckwith, J., Grodzicker, T., and Arditti, R. (1972). *J. Mol. Biol.,* 69, 155.

Beckwith, J. (1981). *Cell,* 23, 307.

Berman, M. L. and Landy, A. (1979). *Proc. Nat. Acad. Sci. USA,* 76, 4303.

Berman, M. L. and Beckwith, J. (1979a). *J. Mol. Biol.,* 130, 285.

Berman, M. L. and Beckwith, J. (1979b). *J. Mol. Biol.,* 130, 303.

Berman, M. L. and Jackson, D. E. (1984). *J. Bact.,* 159, 750.

Casadaban, M. J. (1976). *J. Mol. Biol.,* 104, 541.

Casadaban, M. J. (1980). *J. Mol. Biol.,* 138, 179.

Chattoraj, D. K., Cordes, K., Berman, M. L., and Das, A. (1984). *Gene,* 27, 213.

Emr, S. D., Hanley-Way, S., and Silhavy, T. J. (1981). *Cell,* 23, 79.

Hall, M. N. and Silhavy, T. J. (1979). *J. Bact.,* 140, 342.

Hall, M. N. and Silhavy, T. J. (1981). *J. Mol. Biol.,* 151, 1.

Ingraham, J. L., Maaløe, O., and Neidhardt, F. C. (1983). *Growth of the Bacterial Cell.* Sinauer, Chapter 8.

Ippen, K., Miller, J. H., Scaife, J, and Beckwith, J. (1968). *Nature,* 217, 825.

Jinks-Robertson, S., Gourse, R. L., and Nomura, M. (1983). *Cell,* 33, 865.

Khorana, H. G. (1979). *Science,* 203, 614.

Miller, J.H., Reznikoff, W.S., Silverstone, A.E., Ippen, K., Signer, E. R., and Beckwith, J. R. (1970). *J. Bact.,* 104, 1273.

Miller, J. H. (ed.) (1972). *Experiments in Molecular Genetics.* Cold Spring Harbor Lab.

Mitchell, D. H., Reznikoff, W. S., and Beckwith, J. R. (1975). *J. Mol. Biol.,* 93, 331.

Post, L. E., Arfsten, A. E., Davis, G. R., and Nomura, M. (1980). *J. Biol. Chem.,* 255, 4653.

Primakoff, P. and Berg, P. (1970). *Cold Spring Harb. Symp. Quant. Biol.,* 35, 391.

Rossi, J. J., Egan, J., Berman, M. L., and Landy, A. (1980). In *Genetics and Evolution of RNA Polymerase, tRNA, and Ribosome,* S. Osawa, H. Ozeki, H. Uchida, and T. Yura (eds.). Univ. of Tokyo Press.

Schultz, J., Silhavy, T. J., Berman, M. L., Fiil, N., and Emr, S. D. (1982). *Cell,* 31, 227.

Sekiya, T. and Khorana, H. G. (1974). *Proc. Nat. Acad. Sci. USA,* 71, 2978.

Sekiya, T., Gait, M. J., Noris, K., Ramamoorthy, B., and Khorana, H. G. (1976). *J. Biol. Chem.,* 251, 4481.

Silhavy, T. J., Berman, M. L., and Enquist, L. W. (1984). *Experiments with Gene Fusions.* Cold Spring Harbor Lab.

Travers, A. A., Lamond, A. I., Mace, H. A. F., and Berman, M. L. (1983). Cell, 35, 265.

Wurtzel, E. T., Chou, M.-Y., and Inouye, M. (1982). *J. Biol. Chem.,* 22, 13685.

Weinstock, G.M., Berman, M.L., and Silhavy, T.J. (1983). In *Expression of Cloned Genes in Procaryotic and Eucaryotic Vectors,* T.S. Pappas, M. Rosenberg, and J.G. Chirikjian (eds.) Elsevier/North-Holland.

Xian-Ming, Y., Munson, L. M., and Reznikoff, W. S. (1984). *J. Mol. Biol.,* 172, 355.

Zabin, I. (1982). *Mol. Cell Biochem.,* 49, 87.

In vivo Regulation of the *trp* Operon of *E. coli* by Repression and Attenuation: The Effects of Tryptophan Starvation and the Temperature of Growth

Charles Yanofsky
Stanford University, California

INTRODUCTION

The Copenhagen school of bacterial physiologists, led for so many years by Ole Maaløe, is largely responsible for our recognition of the importance of *in vivo* quantitation of macromolecular synthesis to the analysis of bacterial regulatory phenomena (Maaløe and Kjeldgaard, 1966; Maaløe, 1979; Ingraham, Maaløe and Neidhardt, 1983). Knowledge of the detailed mechanisms of gene control is of course fundamental to the understanding of gene regulation. However, only by quantifying the use of these mechanisms can one appreciate their significance to the growing bacterium. The Copenhagen group--by their attention to rates of synthesis of the major classes of macromolecules and the cell components involved in transcription and translation--have introduced a way of thinking about the growing cell that places regulatory studies in a true-to-life perspective. Thus, investigations in my laboratory, in addition to being concerned with mechanisms of repression and attenuation control (Kelley and Yanofsky, 1982; Yanofsky, 1981), have examined the *in vivo* use of these regulatory mechanism in establishing the level of *trp* operon expression (Yanofsky, Kelley and Horn, 1984).

The product of the enzymes specified by the *trp* operon, tryptophan, is an expensive amino acid to synthesize. E. coli has the regulatory capacity to adjust that synthesis as the need for tryptophan varies. How are repression and attenuation used to control *trp* operon expression as tryptophan starvation becomes increasingly severe? How does the bacterium adjust to different growth temperatures by altering its capacity to synthesize tryptophan? These questions are addressed in this article.

Analysis of the separate contributions of repression and attenuation

Transcription of the five structural genes of the *trp* operon of E. coli is regulated by repression and attenuation (Kelley and Yanofsky, 1982; Yanofsky, 1981). We wished to devise a means of measuring the contribution of each of these regulatory processes in controlling transcription of the structural genes of the operon. To realize this objective we constructed a *trpPOL'-lacZ'* fusion that could be used to measure repression independently of attenuation (Yanofsky, Kelley and Horn, 1984). We fused a DNA segment containing the *trp* promoter/operator and much of the coding region for the *trp* leader peptide to *lacZ* such that beta-galactosidase formation is regulated by *trp* repressor. The attenuator was deleted in this construct. This *trpL-lacZ* fusion was introduced into the bacteriophage lambda genome (phage λTL) which was subsequently inserted into the chromosome of a *lac*-deletion strain by lysogenization (Kelley and Yanofsky, 1982). We then measured expression of the *trp* operon as an indicator of the combined effects of repression and attenuation and beta-galactosidase formation as an indicator of repression (Yanofsky, Kelley and Horn, 1984). We calculated the contribution attributable to attenuation by subtracting the observed regulatory effect of repression. An advantage of using the intact *trp* regulatory region in assessing repression plus attenuation is that any unanticipated consequence of separating the operator and attenuator was avoided.

To validate this test system we measured *trpE* enzyme levels (indicative of the combined effects of repression and attenuation) and *lacZ* enzyme levels (indicative of repression) in strains that are repressor deficient or attenuation defective. The results of these analyses (Table I) established that the regulatory indicators responded properly. Thus, *trpR* and *trp*[+] strains containing the attenuation-defective alleles *trpL82* or *trpL75*, exhibited the same ratio of *trpE* to *lacZ* enzymatic activities (see attenuation multiplier). In addition, the *trpE* and *lacZ* enzymatic activities were both 25-fold lower in *trp*[+] *trpL82* than in *trpR trpL82*. This result indicates that when regulation by attenuation has been eliminated, *trpE* and *lacZ* activities vary in parallel as a consequence of repression. Similarly, the *trpE* and *lacZ* enzyme levels of *trp*[+] *trpL75* were both ca. 3-fold lower than those of the *trpR trpL75* strain, reinforcing this conclusion. On the basis of these controls we expect that in strains that are not regulatory-defective, *trpE* and *lacZ* enzymatic activities will vary depending on whether repression and/or attenuation is being relieved. One disturbing and unexplained observation in these analyses was that in the wildtype strain grown in minimal medium the *trpE* and *lacZ* enzyme levels were 10% and 20%, respectively, of the levels in the *trpR* strain. However, mRNA measurements with the same *trp*[+] strain indicated that *trpE* and *lacZ* mRNA levels were both ca. 20% those of the *trpR* strain. Despite this one unexplained observation we believe that our experimental system can be used to measure the relative regulatory contributions of repression and attenuation under different environmental conditions.

TABLE 1   Test of repression indicator

Strain	Growth medium	% of *trpR*(λTL) control Enzyme levels		Attenuation multiplier
		trpE	*lacZ*	
trpR(λTL)	Trp	100	100	1
trpR(λTL)	Min	9	15	0.6*
trpRtrpL82 (λTL)	Trp	380	105	3.8
trpRtrpL82 (λTL)	Min	15	4	3.8
trpR trpL75 (λTL)	Trp	21	110	0.2
trpR⁺trpL75 (λTL)	Min	8	33	0.24

In strains with the *trpL82* and *trpL75* alleles regulation by attenuation does not occur, and termination at the *trp* attenuator is reduced and increased, respectively. Data taken from Yanofsky, Kelley, and Horn (1984). The minimal medium used throughout was that of Vogel and Bonner (1956). β-Galactosidase was measured as described by Miller (1972), and the TrpE polypeptide was assayed as described in Kelley and Yanofsky (1982). The attenuation multiplier is assumed to be 1 in *trpR⁺* and *trpR* cultures growing with excess tryptophan. In other strains or other conditions the multiplier is obtained by dividing the TrpE level by the LacZ level.
*See text.

Attenuation and repression relief as a function of the severity of tryptophan starvation

We used a variety of mutant strains and growth in a medium containing acid-hydrolyzed casein (a tryptophan-free supplement that mildly starves the bacterium of tryptophan) to determine whether repression and attenuation were relieved in parallel as the severity of tryptophan starvation was increased (Yanofsky, Kelley and Horn, 1984). As can be seen in Table 2, there was no relaxation of attenuation-mediated termination until repression was relieved about 50 percent. Thus, over an appreciable range of tryptophan deprivation, repression was reduced, but attenuation was not. Only when repression was nearly totally relieved was there substantial relief of termination at the attenuator. The interpretation of these and similar experiments is diagrammatically presented in Figure 1. The figure shows that transcription of the structural genes of the *trp* operon is regulated by repression during mild tryptophan starvation, and by attenuation during severe tryptophan starvation. Interestingly, this conclusion was anticipated in the outstanding book by Ingraham, Maaløe and Neidhardt (1983). As an independent verification of this conclusion,

TABLE 2　Test of repression indicator

Strain	Growth medium	% of $trpR$(λTL) control Enzyme levels		Attenuation multiplier
		$trpE$	$lacZ$	
$trpR^+ trpE^{FBR}$(λTL)	Min	3	3	1
$trpR^+$(λTL)	Min	8	17	0.5*
$trpR^+ trpE^{FBR}$ (λTL)	ACH	32	24	1.3
$trpR^+$ (λTL)	ACH	45	44	1
$trpR^+ trpA234$-Ala (λTL)	Min	71	54	1.3
$trpR^+ trpA234$-Ala (λTL)	ACH	238	89	2.7
$trp^+ trpA211$-Val (λTL)	Min	270	100	2.7

$TrpE^{FBR}$ is a $trpE$ allele that specifies a feedback-resistant poly-peptide. $trp234$-Ala is a $trpA$ allele that specifies a partially defec-tive TrpA protein, resulting in mild tryptophan starvation and pro-duction of elevated levels of the Trp polypeptides. $TrpA211$-Val is a $trpA$ allele that specifies a functional, highly defective TrpA protein. Data taken from Yanofsky, Kelley, and Horn.
*See text.

we have shown that the tRNA^{Trp} of the Trp bradytroph trp^+ $trpA234$-Ala is moderately charged, ca. 65%, under conditions in which repression is 50% relieved (Yanofsky, Kelley and Horn, 1984). These findings suggest that the intracellular concentration of tryptophan can drop, thereby relieving repression, without appreciably affecting the extent of charging of tRNA^{Trp}, and the extent of attenuation. This conclusion is consistent with the existence of only two Trp codons in the coding region for the trp leader peptide. Only severe tryptophan starvation would appear to be sufficient to stall a translating ribosome at one of these Trp codons and cause termination relief.

Regulatory adjustment to the temperature of growth

As the temperature of a growing bacterial culture changes, the metabolic balance of each cell is certain to be perturbed, setting into motion the existing regulatory machinery. We have used our measurement system to determine enzymatic activities of $trpE$ and $lacZ$ in E. coli cultures growing at different temperatures, and following a temperature shift-up. We wished to determine how repression and attenuation were varied as the culture adjusted to the effects of the temperature of growth. We of course appreciated that when one changes the temperature of growth of a bacterial culture, many poorly understood variables are introduced that may profoundly affect the interpretation of an experiment. Thus, as the growth temperature is changed, the rate of translation of a mRNA molecule may vary, and this variation may differ for different mRNA species. Similarly the

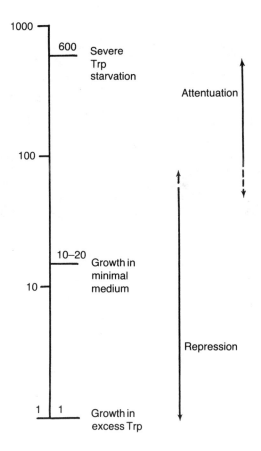

Relative *trp* operon
expression

FIGURE 1 Regulatory range of repression and attenuation. The arrows indicate the range of trp operon expression over which repression and attenuation vary, as the severity of tryptophan starvation is increased. Expression is plotted on a log scale.

functional lifetime of different messengers may change with temperature. In addition, the relative levels of the active forms of the key macromolecules involved in general metabolic processes will change, thereby imposing constraints on regulatory strategies. Despite these uncertainties measurements of expression of the *trp* operon in *trpR* strains with the attenuator either present or deleted suggested that the effect of growth temperature on termination at the attenuator could be experimentally determined (Atlung and Hansen, 1983; C. Yanofsky, unpublished).

In Table 3 we see the effect of temperature on *trpE* and *lacZ* enzyme levels in a *trp*+ strain grown in minimal medium and a *trpR* strain grown with excess tryptophan. In the latter case the beta-

TABLE 3   Enzyme levels at 30°C and 41°C

Strain	Growth medium	Relative activity of			
		trpE		*lacZ*	
		30°	41°	30°	41°
trp⁺ (λTL)(λ)	Min	1(63)	1.3	1(1840)	3.5
trpR (λTL)(λ)	Trp	1(960)	0.61	1(10,530)	2.0

Five cultures of each strain were grown in the medium indicated, at 30°C or 41°C. At a Klett value of 50 (ca. 3×10^8 cells/ml) the cells were harvested by centrifugation, washed, and assayed for TrpE and LacZ enzymatic activities. The 30° relative values were set at 1. The values in parentheses are the actual measured enzyme levels in the samples taken at 30°C.

galactosidase specific activity is twice as high at 41 C than at 30 C. This finding suggests that the *trp* promoter is twice as active at 41 C. However, the *trpE* specific enzymatic activity in the same cultures was depressed to 0.61 at 41 C. One interpretation of these results is that despite the fact that termination at the *trp* attenuator was maximal at each temperature, termination was more efficient at 41 C than at 30 C. Termination at the attenuator appeared to be approximately fourfold greater at 41 C than at 30 C. In agreement with this conclusion, measurements of tryptophan synthetase (TrpB and TrpA protein) levels in attenuator deletion and nondeletion strains by Atlung and Hansen (1983) and by ourselves (Yanofsky, unpublished) also indicated that termination at the attenuator was approximately two- to threefold greater at 41 C than at 30 C. In addition, *in vitro* transcription experiments with RNA polymerase and a restriction fragment containing the *E. coli trp* promoter/operator-leader region, indicated that transcription termination at the attenuator was markedly increased with increasing temperature (Winkler and Yanofsky, unpublished). Conceivably the termination signal is recognized more readily by polymerase at high temperatures or the antiterminator forms more readily at low temperatures.

In *trp⁺* cultures growing in minimal medium, *lacZ* enzyme levels are 3.5-fold higher at 41 C than at 30 C. This observation, combined with the results obtained with the *trpR* strain, suggest that there is about a twofold reduction in repression at 41 C relative to 30 C, resulting in *trpE* enzyme levels only slightly higher at 41 C than at 30 C. If, as we concluded, the *trp* promoter is twice as active at 41 C, and if termination at the attenuator is fourfold greater at 41 C than at 30 C, the level of *trpE* enzyme could only be maintained at a relatively constant value by reducing repression twofold. The observed changes in the activities of the *trpE* and *lacZ* enzymes therefore are internally consistent.

Temperature and nutritional shifts severely tax the regulatory versatility of the bacterial cell. In fact, one might argue that the most extreme regulatory challenge the bacterium faces occurs during the shift period. When we measured *trpE* and *lacZ* enzymatic activities during a temperature shift-up (Table 4), we found that there was an immediate change in the rate of synthesis, which persisted, and which was roughly consistent with the steady-state enzyme levels observed at the two temperatures (Table 3). Thus, in the *trpR* culture there was a slight increase in the rate of synthesis of beta-galactosidase and an insignificant change in the rate of synthesis of the TrpE protein (the rate of synthesis of the TrpE protein should eventually drop). In the *trp*[+] culture there was a substantial increase in the rate of synthesis of beta-galactosidase but only a slight increase in the rate of synthesis of the TrpE protein. These observations are in general consistent with the behavior of steady-state cultures and suggest that following the temperature shift-up, repression is partially relieved to compensate for the increased efficiency of transcription termination at the *trp* attenuator at the elevated temperature.

The simple analyses presented in this article only scratch the surface in our attempts to understand *in vivo* regulatory behavior. Nevertheless, they are beginning to reveal the dynamic features of the use of different regulatory mechanisms in the control of transcription of the *trp* operon.

TABLE 4   Enzyme production during a temperature shift from 30°C to 41°C

Strain	Growth medium	Percent change in 60 min	
		trpE	*lacZ*
trp[+] (λTL)(λ)	Min	+31	+250
trpR (λTL)(λ)	Trp	+4	+25

Duplicate cultures were grown at 30°C in the medium indicated to a Klett value of 50. A sample was taken, and then the cultures were shifted to 41°C and samples taken at 15, 30, 45, and 60 min. The four cultures grew at the same rate, and the final turbidities indicated that each culture had just about doubled by the end of the 60-min period. Each sample was assayed for TrpE and LacZ enzymatic activities. In each culture enzyme levels increased approximately linearly with time over the 60-min period.

ACKNOWLEDGEMENTS

The author is indebted to Virginia Horn for constructing most of the strains employed, and for growing the cultures and performing the beta-galactosidase assays. The author is also indebted to Valley Stewart for helpful comments. The studies described in this paper were supported by grants from the National Science Foundation (PCM 8208866), and the American Heart Association (69-015). The author is a Career Investigator of the American Heart Association.

REFERENCES

Atlung, T. and Hansen, F. G. (1983). *J. Bacteriol.,* 156, 985.

Ingraham, J. L., Maaløe, O., and Neidhardt, F. C. (1983). *Growth of the Bacterial Cell,* Sinauer Associates.

Kelley, R. L. and Yanofsky, C. (1982). *Proc. Natl. Acad. Sci. USA,* 79, 3120.

Maaløe, O., (1979). In *Biological Regulation and Development,* Vol. 1, R.F. Goldberger (ed.) Plenum Press.

Maaløe, O. and Kjeldgaard, N. O. (1966). *Control of Macromolecular Synthesis,* W. A. Benjamin.

Miller, J. (1972). *Experiments in Molecular Genetics.* Cold Spring Harbor Laboratory.

Vogel, H. J. and Bonner, D. M. (1956). *J. Biol. Chem.,* 218, 97.

Yanofsky, C. (1981). *Nature,* 289, 751.

Yanofsky, C., Kelley, R. L., and Horn, V. 1984. *J. Bacteriol.,* 158, 1018.

Regulation of Pyrimidine Nucleotide Biosynthesis in *Escherichia coli* and *Salmonella typhimurium*

Jan Neuhard
University Institute of Biological Chemistry B;
Copenhagen, Denmark

INTRODUCTION

In 1962 when I started as a post-doctoral fellow at the Institute of Biological Chemistry in Copenhagen, I was going to work with Dr. Agnete Munch-Petersen on nucleotide metabolism in bacteria. Since we were in the same building as the Institute of Microbiology and we had no experience in working with bacteria, we approached Ole Maaløe and his group for help and inspiration. As a result, we were talked into developing a system for quantitating the intracellular concentration of the nucleic acid precursors (Neuhard et al., 1965), and we used this technique to determine nucleoside triphosphate pools under numerous different growth conditions. These early studies taught me at least three things: (1) the ratios between the pool sizes of the eight nucleoside triphosphates are held constant under most growth conditions, whereas the absolute concentrations may vary greatly; (2) the overall rate of stable RNA synthesis *in vivo* is quite insensitive to large fluctuations in the intracellular concentrations of the nucleoside triphosphate substrates, as long as the ratio between them remains constant (Beck et al., 1973); (3) nucleotide synthesis is highly regulated.

In the following I will summarize our present knowledge of the regulation of pyrimidine nucleoside triphosphate synthesis, especially with respect to the transcriptional regulation of expression of the genes encoding the six enzymes responsible for the *de novo* synthesis of UMP.

BIOSYNTHESIS OF dTTP FROM UMP

The pathway by which the four pyrimidine nucleoside triphosphates–
UTP , CTP, dCTP, and dTTP—are synthesized is shown in Figure
1. It may be regarded as an unbranched pathway with dTTP as the
ultimate end product. Thus, UTP, CTP, and dCTP are intermediates
in the biosynthesis of dTTP. Some of the intermediates of the pathway
may be supplied from exogenous sources via the salvage pathways,
as indicated on Figure 1. All the enzymes involved have been
characterized, and mutants defective in each step have been isolated.
The genetic loci encoding these enzymes are scattered on the
chromosome (Møllgaard and Neuhard, 1983).

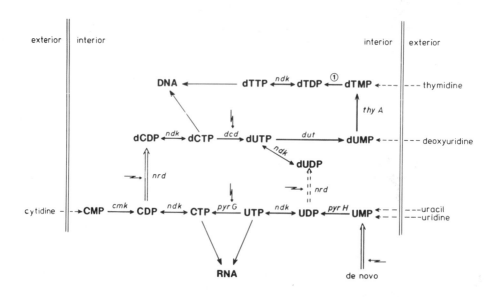

FIGURE 1   Pyrimidine nucleotide biosynthesis in E. coli and S.
typhimurium. The enzymes involved are identified by their gene
symbols as follows: cmk, CMP (dCMP) kinase; dcd, dCTP deaminase;
dut, dUTPase; nuk, nucleoside diphosphokinase; nrd, ribonucleoside
diphosphate reductase; pyrG, CTP synthetase; pyrH, UMP kinase;
thyA, thymidylate synthase. dTMP kinase is identified by 1. Double
arrows indicate enzymes whose synthesis is regulated. Broken arrows
indicate allosterically regulated enzymes.

The conversion of UMP to dTTP is regulated at three points by
allosteric enzymes: CTP synthetase (Long and Pardee, 1967),
ribonucleoside diphosphate reductase (Thelander and Reichard, 1979),
and dCTP deaminase (Beck et al., 1975). In addition, the synthesis
of ribonucleoside diphosphate reductase is regulated at the level of
transcription of the nrdAB operon (Hanke and Fuchs, 1983). It
appears that any condition that tends to decrease the DNA/mass ratio

of the cell, i.e., causes initiation of new DNA replication forks, result in derepression of the rate of ribonucleotide reductase synthesis.

DE NOVO SYNTHESIS OF UMP

The *de novo* pathway consists of six enzymes which convert bicarbonate, the amide group of glutamine, and aspartate into UMP (Figure 2). Carbamoyl phosphate, the first intermediate in the pathway, is also an intermediate of the arginine biosynthetic pathway. Thus, carbamoylphosphate synthetase may as well be regarded as an arginine biosynthetic enzyme. The six enzymes are coded for by the six unlinked loci, *pyrA* (*carAB* in *E. coli*) through *pyrF*. The two first enzymes of the pathway, i.e., carbamoylphosphate synthetase and aspartate transcarbamoylase, are allosteric enzymes consisting of two nonidentical subunits each. In both cases, the genes encoding the two subunits are adjacent on the chromosome and constitute operons: the *pyrA* or *carAB* operon (Gigot et al., 1980) and the *pyrB* operon (Panza et al., 1982).

REGULATION OF THE DE NOVO PATHWAY

The carbamoylphosphate synthetase reaction is regulated in accordance with its dual metabolic role. Its activity is inhibited by UMP and activated by ornithine, an intermediate in the arginine pathway (Pierard, 1966; Abdelal and Ingraham, 1975). In addition, the

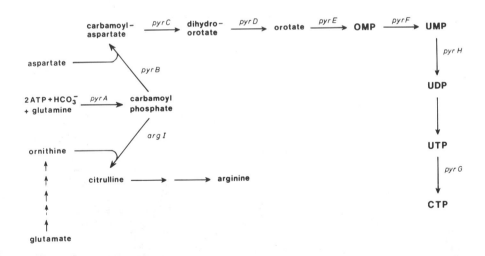

FIGURE 2 De novo synthesis of UPT and CTP. The individual enzymes are identified by their gene symbols as follows: ndk, nucleoside diphosphokinase; pyrA, carbamoylphosphate synthase; pyrB, aspartate carbamoyl transferase; pyrC, dihydroorotase; pyrD, dihydroorotate oxidase; pyrE, orotate phosphoribosyltransferase; pyrF, orotidine 5'-phosphate decarboxylase; pyrG, CTP synthetase; pyrH, UMP kinase. Taken from Neuhard, 1983.

synthesis of the enzyme is subject to cumulative repression by both arginine and pyrimidines (Pierard et al., 1976). The common regulatory molecule of the arginine regulon, the *argR* gene product, participates in this cumulative repression (Pierard et al., 1972). Transcription of *carAB* is initiated from two adjacent promoters 70 base pairs apart. The Pribnow box of the proximal promoter is part of an 18-bp DNA sequence, which appears to constitute the basic arginine operator sequence (Cunin et al., 1983). Transcription from the distal promoter appears to be pyrimidine controlled by a mechanism which still has to be disclosed (Piette et al., 1984; Bouvier et al., 1984).

The flow through the remaining five steps of the pathway is also highly regulated. The first enzyme, aspartate transcarbamoylase is feedback-inhibited by CTP and activated by ATP (Gerhart, 1970), and the synthesis of all five enzymes is controlled in a complicated, noncoordinated manner by the intracellular nucleotide pools. By using mutant strains that allow the independent manipulation of individual nucleotide pools, it was shown (Table 1) that expression of *pyrB*, *pyrE*, and *pyrF* is repressed by UTP (or UDP), whereas expression of *pyrC* and *pyrD* is controlled predominantly by CTP (or CDP) (Schwartz and Neuhard, 1975). In addition, starvation for guanine nucleotides results in derepression of *pyrB*, *pyrE*, and *pyrF* and repression of *pyrC* and *pyrD* expression (Jensen, 1979).

TABLE 1   Repressibility of the *pyr* genes in *Salmonella typhimurium*

Genetic locus	Derepression/ repression*	Negative effectors
pyrA	35▽	Arginine; PyrTP (PyrDP)
pyrB	200	UTP (UDP); guanine nucleotide
pyrC	15	CTP (CDP)
pyrD	20	CTP (CDP)
pyrE	30	UTP (UDP); guanine nucleotide
pyrF	10	UTP (UDP); guanine nucleotide

*Ratio between specific activities in conditions of depression (obtained by growing pyrimidine auxotrophic strains with UMP as sole pyrimidine source), and of repression (grown with uracil and cytidine added to the growth medium).

▽Fully repressed levels were determined in cultures grown in the presence of uracil, cytidine, and arginine.

MUTANT STUDIES

Many attempts have been made over the years to select mutants showing constitutive synthesis of the six pyrimidine biosynthetic enzymes. Invariably, such selections yield mutants that are constitutive owing to low endogenous pools of the "corepressors"

(Justensen and Neuhard, 1975). No true regulatory mutants have been obtained. The presence of multicopy recombinant plasmids, containing the individual *pyr* genes, does not affect the expression of the other five *pyr* genes present on the chromosome in single copies (Jensen et al., 1984). Thus, no evidence for a titratable common regulatory protein was obtained. These results combined with the results of the effector studies (Table 1) indicate that whereas the *pyr* genes all belong to the pyrimidine-starvation *stimulon,* they do not constitute a *regulon.* Several genes encoding proteins involved in the pyrimidine salvage pathway (*upp, udk, codAB,* (Neuhard, 1983)) are also part of the pyrimidine-starvation stimulon.

Two types of mutants that show altered regulation of the expression of certain *pyr* gene and that are not "pool mutants" have recently been characterized (Jensen et al., 1982; Neuhard et al., 1982). In both types the mutation maps within the *rpoBC* gene cluster. The first type consists of mutants that become auxotrophic for arginine when uracil is present in the growth medium. This uracil-sensitivity phenotype is caused by hyperrepression of *pyrA* expression by exogenous uracil (Neuhard et al., 1982). The expression of other *pyr* genes is not affected in these mutants. The second type shows high constitutive expression of *pyrB, pyrE,* and probably also *pyrF,* despite a high endogenous pool of the effector nucleotide UTP (Jensen et al., 1982). RNA polymerase from a mutant of the latter type has been purified and found to display a four-fold increased value of K_M towards UTP in the elongation reaction, both on synthetic and on T7 DNA templates (K. F. Jensen, pers. commun.). Therefore, it appears that RNA polymerase is directly involved in the regulation of expression of the *pyr* gene and that the RNA polymerase itself may be the receptor that senses the intracellular concentrations of the nucleotide effectors.

REGULATION OF *pyrB* AND *pyrE* EXPRESSION

Studies of recombinant plasmids containing the *pyrE* gene from *E. coli* or from *S. typhimurium* have shown that the *pyrE* promoter is located about 800 base pairs upstream from the start of the structural part of the gene (Poulsen et al., 1984; Neuhard, unpublished results). The entire *pyrE* gene, including the leader region but not the promoter, was cloned into the *bla* or *tet* genes of pBR322 in such a way that transcription of *pyrE* originates from the antibiotic-resistance promoters. Expression of *pyrE* from these plasmids is regulated normally by pyrimidines (unpublished results). From a *pyrE::lac* fusion strain of *S. typhimurium* we have isolated a mutant which shows constitutive expression of beta-galactosidase from the *pyrE* promoter. This mutation maps within the leader region, several hundred base pairs separated from the *pyrE* promoter (unpublished observations). Thus, it may be concluded that *pyrE* expression is not regulated at the level of transcription initiation.

STRUCTURAL STUDIES

The primary structure of the *pyrBI* operon (Hoover et al., 1983; Navre et al., 1983; Roof et al., 1982; Turnbough et al., 1983) and the *pyrE* gene (Poulsen et al., 1983) of *E. coli* has recently been determined. In addition, the leader regions of S. typhimurium pyrB and pyrE have been sequenced (Kelln and Neuhard, unpublished results). Figure 3 compares some of the essential features of these genes. 25-50 base pairs before the start of the structural genes the structures indicate a Rho-independent transcriptional terminator. Another region of dyad symmetry capable of encoding a rather stable hairpin, followed by a thymidine-rich cluster, precedes the attenuator. Finally, the leader RNAs of all four genes contain open reading frames that may encode leader peptides.

Based on these structural features the following model for transcriptional regulation of the expression of these genes has been proposed (Turnbough et al., 1983). When the supply of UTP is low, the region of dyad symmetry followed by a uridine-rich region will act as a strong pause site for RNA polymerase, allowing the leading ribosome, translating the leader peptide, to catch up with the RNA polymerase. When the transcribing RNA polymerase reaches the attenuator, the highly coupled leading ribosome precludes the formation of the attenuator hairpin necessary for transcriptional termination, allowing the RNA polymerase to continue transcription into the structural gene. When the supply of UTP is large, the RNA polymerase does not pause in the leader region. This will prevent the leading ribosome from catching up with the RNA polymerase before the attenuator hairpin is formed and transcription is terminated.

It is interesting to note that the position of the stop codons of the putative leader peptides are located differently relative to the attenuator. Since pyrBI expression in both E. coli and S. typhimurium responds to pyrimidine starvation by a factor of 200 (Table 1), it would imply that whether the leader peptide stops within the attenuator, as in S. typhimurium, or distal to it as in E. coli, is of minor importance. However, in the pyrE gene the UGA stop codon of the leader peptide is located five base pairs upstream from the attenuator. This could imply that the leading ribosome translating the leader peptide may offer less protection against the formation of the attenuator hairpin, even under conditions of tight coupling to the transcribing RNA polymerase. This could explain why pyrE expression responds less (30x) to pyrimidine starvation than pyrBI (Table 1).

S1 mapping experiments have shown that transcription of the E coli pyrB gene in vivo starts predominantly from the P_2 promoter, though small amounts of longer transcripts were detected (Navre and Schachman, 1983). Using an in vitro transcription system Turnbough et al. (1983) showed, that the P_1 and P_2 promoters were equally effective, and that 98 percent of the transcripts from either P_1 or P_2 were terminated at the attenuator. Furthermore they observed a clearcut transcriptional pause site within the T-rich sequence preceding the attenuator. However, this pause site was only evident when a suboptimal concentration of UTP was employed.

Recently Poulsen et al. (1984) sequenced the region upstream from the pyrE gene of E. coli. They found an open reading frame of approximately 750 base pairs, which may code for a polypeptide of about 240 amino acids (Figure 3). It is in phase with the putative 52-amino acid leader peptide, and ends at the same UGA stop codon. The existence of a 26-kdal protein of unknown function, coded for by the ORF region, was subsequently confirmed by minicell experiments. S1 mappings of pyrE mRNA synthesized in vivo showed that 95 percent of the transcriptional starts originate at the P_{ORF}

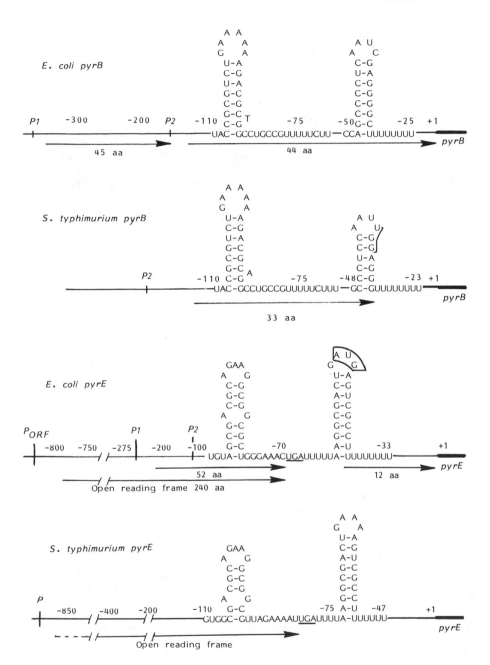

FIGURE 3 Structure of the regulatory regions of pyrB and pyrE from E. coli and S. typhimurium. The sequences shown are those of the corresponding RNA transcripts. Possible translation products are indicated by arrows. Numbers are base pairs, starting from the first base pair of the structural part of the gene.

promoter, and subcloning of the entire *ORF-pyrE* region lacking only P_{ORF} confirmed that the two genes constitute an operon with *pyrE* as the distal gene. The S1 mappings further indicated that whereas the amount of *pyrE*-specific mRNA is dependent on the pyrimidine supply of the cell, this is not the case with the amount of *ORF*-specific mRNA. Thus, transcription and translation of the first gene of an operon is significantly decoupled under normal growth condition, resulting in a high degree of transcription termination at the attenuator located immediately in front of *pyrE*. Only under conditions of low intracellular UTP concentration is this polarity relieved, probably due to pausing of the RNA polymerase at the "pause site" located in front of the attenuator, allowing the leading ribosome to catch up with the RNA polymerase.

The model predicts that the coupling between transcription and translation of the region preceding the attenuator determines the extent of transcriptional readthrough into the structural gene. Indirect evidence for this was obtained by Turnbough (1983), who showed that *pyrA* and *pyrB* expression (*pyrE* expression was not measured) is preferentially reduced in a *hisT* mutant of *S. typhimurium*. (A *hisT* mutation causes a reduced rate of translational elongation.) This repression is reversed by addition of DNA-binding agents such as proflavin, which presumably inhibit the rate of transcriptional elongation. In a *hisT*+ strain of *S. typhimurium*, sublethal concentrations of streptolydigin, which inhibits transcriptional elongation, preferentially stimulate *pyrB* expression. Finally, it should be recalled that the *S. typhimurium rpoBC* mutants, which contain an RNA polymerase with increased K_M for UTP in the elongation reaction, show high constitutive expression of *pyrB* and *pyrE* (Jensen et al., 1982).

REGULATION OF *pyrC* AND *pyrD* EXPRESSION

The expression of *pyrC* and *pyrD* are predominantly regulated by the intracellular concentration of a cytidine nucleotide (Table 1). From measurements of the levels of dihydroorotase (*pyrC*) and dihydroorotate oxidase (*pyrD*) in different *S. typhimurium* mutants, which contain varying intracellular concentrations of CTP, it appears (Figure 4) that the synthesis of these two enzymes is regulated in a coordinated manner (Schwartz and Neuhard, 1975). The primary structure of the regulatory region of the *S. typhimurium pyrC* gene is given in Figure 5 as the corresponding transcript. The region allows for a secondary structure very similar to that of *pyrE*, consisting of two hairpin structures of which the distal looks very much like a Rho-independent transcriptional terminator. Furthermore, the putative transcript contain an open reading frame from the cloning site to a UGA stop codon located between the two hairpins. A cluster of pyrimidines follows immediately after the UGA codon. This structural organization is very similar to that of the leader region of *pyrE*, suggesting a similar regulation. However, it is not apparent from the structures why *pyrC* expression is regulated predominantly by CTP, whereas the expression of *pyrE* is controlled by UTP (Table 1).

It should therefore be considered that the putative attenuator is the terminator of another gene located upstream from *pyrC*, and that the *pyrC* promoter and regulatory region are located between position -74 and +1. Three lines of evidence suggest that this is the

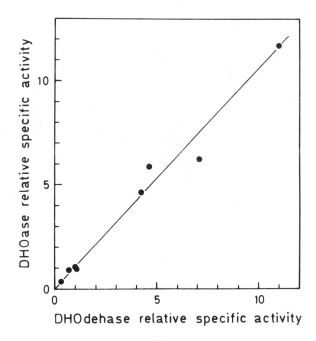

FIGURE 4 Coordinate expression of the pyrC and pyrD genes. The relative specific activities of DHOase are plotted against the relative specific activity of DHOdehase in different mutants of S. typhimurium growing exponentially with altered UTP and CTP pools. (From Schwartz and Neuhard, 1975.)

case. First, plasmid pJRC21, in which the first 260 base pairs up to the *NruI* site are deleted (Figure 5), shows normal expression of *pyrC*, indicating that translation of the open reading frames in front of *pyrC* is not required for normal regulation. Furthermore, the nucleotide sequence of the remaining DNA up to the attenuator does not show homology with the consensus sequence of a prokaryotic promoter (Hawley and McClure, 1983). The second line of evidence was obtained by opening the DNA at the *BssHII* site located in the attenuator (Figure 5), digesting the 5' overhangs away, and religating. The derived plasmid, which lacks the *BssHII* site, shows normal *pyrC* expression and regulation. Thirdly, a sequence showing considerable homology with the consensus Pribnow box (Hawley and McClure, 1983) is found around position −45, suggesting a transcriptional start at position −40 (Figure 6). Such a transcript would be able to form a rather stable hairpin due to the region of dyad symmetry centered around position −2. It is of interest that part of the Shine–Dalgarno sequence for *pyrC* would be included in the stem of such a hairpin.

By using *pyrC::Mud*(ApR,*lac*) fusion strains of *S. typhimurium*, we have isolated mutants that show high-level constitutive expression of beta-galactosidase from the *pyrC* promoter. Analyses of such mutants will hopefully reveal the molecular basis of the cytidine

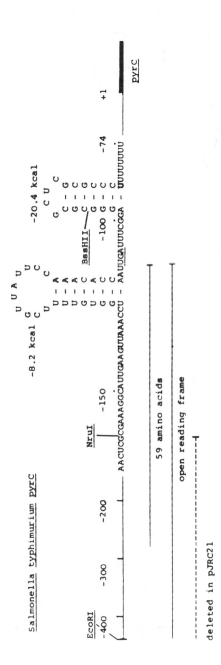

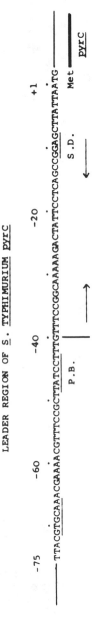

FIGURE 5 Structure of the regulatory region of pyrC from S. typhimurium. See legend to Figure 3 for explanation.

FIGURE 6 Nucleotide sequence of the first 75 bases of the leader region of pyrC from S. typhimurium. P.B., Pribnow box; S.D. Shine-Dalgarno sequence; arrows, regions of dyad symmetry; vertical line, possible transcription start site.

nucleotide promoted repression of *pyrC* (and *pyrD*) expression. Most certainly, it will turn out to differ both from the cumulative repression that controls *pyrA* expression and from the attenuation mechanism that regulates the expression of *pyrBI* and *pyrE*.

ACKNOWLEDGEMENTS

I thank R. A. Kelln and K. F. Jensen for making some of their data available to me before publication. This work was supported by grants from the Danish Natural Science Research Council.

REFERENCES

Abdelal, A. T. H. and Ingraham, J. L. (1975). *J. Biol. Chem.*, 250, 4410.

Beck, C. F., Eisenhardt, A. R., and Neuhard, J. (1975). *J. Biol. Chem.*, 250, 609.

Beck, C. F., Ingraham, J., Maaløe, O., and Neuhard, J. (1973). *J. Mol. Biol.*, 78, 117.

Bouvier, J., Patte, J-C., and Stragier, P. (1984). *Proc. Nat. Acad. Sci. USA*, 81, 4139.

Cunin, R., Eckhardt, T., Piette, J., Boyer, A., Pierard, A., and Glansdorff, N. (1983). *Nuc. Acid. Res.*, 11, 5007.

Gerhart, J. C. (1970). In *Current Topics in Cellular Regulation*, B. L. Horecker and E. R. Stadtman (eds.). Academic Press. p. 276.

Gigot, D., Crabeel, M., Feller, A., Charlier, D., Lissens, W., Glansdorff, N., and Pierard, A. (1980). *J. Bacteriol.*, 143, 914.

Hanke, P. D. and Fuchs, J. A. (1983). *J. Bacteriol.*, 154, 1040.

Hawley, D. K. and McCure, W. R. (1983). *Nucl. Acid. Res.*, 11, 2237.

Hoover, T. A., Roof, W. D., Foltermann, K. F., O'Donovan, G. A., Bencini, D. A., and Wild, J. R. (1983). *Proc. Nat. Acad. Sci. USA*, 80, 2462.

Jensen, K. F. (1979). *J. Bacteriol.*, 138, 731.

Jensen, K. F., Larsen, J. N., Schack, L., and Sivertsen, A. (1984). *Eur. J. Biochem.*, 140, 343.

Jensen, K. F., Larsen, J. N., and Schack, L. (1982). *EMBO J.*, 1, 69.

Justesen, J. and Neuhard, J. (1975). *J. Bacteriol.*, 123, 851.

Long, C. W. and Pardee, A. B. (1967). *J. Biol. Chem.*, 242, 4715.

Møllgaard, H. and Neuhard, J. (1983). In *Metabolism of Nucleotides, Nucleosides, and Nucleobases in Microorganisms,* A. Munch-Petersen (ed.). Academic Press. p. 149.

Navre, M. and Schachman, H. K. (1983). *Proc. Nat. Acad. Sci. USA,* 80, 1207.

Neuhard, J. (1983). In *Metabolism of Nucelotides, Nucleosides and Nucleobases in Microorganisms.* A. Munch-Petersen (ed.), Academic Press. p. 95.

Neuhard, J., Jensen, K. F., and Stauning, E. (1982). *EMBO J.,* 1, 1141.

Neuhard, J., Randerath, E., and Randerath, K. (1965). *Anal. Biochem.,* 13, 211.

Panza, C. D., Karels, M. J., Navre, M., and Schachmann, H. K. (1982). *Proc. Nat. Acad. Sci. USA,* 79, 4020.

Pierard, A. (1966). *Science,* 154, 1572.

Pierard, A. N., Glansdorff, N., Gigot, D., Crabeel, M., Halleux, P., and Thiry, L. (1976). *J. Bacteriol.,* 127, 291.

Pierard, A. N., Glansdorff, N., and Yashphe, J. (1972). *Mol. Gen. Genet.,* 118, 235.

Piette, J., Nyonoya, H., Lusty, C. J., Cunin, R., Weyens, G., Crabeel, M., Charlier, D., Glansdorff, N., and Pierard, A. (1984). *Proc. Nat. Acad. Sci. USA,* 81, 4134.

Poulsen, P., Bonekamp, F., and Jensen, K. F. (1984). *EMBO J.,* 3, 1783.

Poulsen, P., Jensen, K. F., Valentin-Hansen, P., Carlsson, P., and Lundberg, L. G. (1983). *Eur. J. Biochem.,* 135, 223.

Roof, W. D., Foltermann, K. F., and Wild, J. (1982). *Mol. Gen. Genet.,* 187, 391.

Schwartz, M. and Neuhard, J. (1975). *J. Bacteriol.,* 121, 814.

Thelander, L. and Reichard, P. (1979). *Ann. Rev. Biochem.,* 48, 133.

Turnbough Jr., C. L. (1983). *J. Bacteriol.,* 153, 998.

Turnbough Jr., C. L., Hicks, K. L., and Donahue, J. P. (1983). *Proc. Nat. Acad. Sci. USA,* 80, 368.

Intercistronic Sites in the *lac* Operon

Donald P. Nierlich, Catherine Kwan*, George J. Murakawa, Paul A. Mahoney, Alvin W. Ung, and Daniel Caprioglio

*The University of California, Los Angeles and *Mount St. Mary's College, Los Angeles*

One of the implications of the work of Ole Maaløe and his collaborators—and appropriately of the papers presented here—is that underlying the growth of bacteria are many regulatory mechanisms, of which a substantial proportion function in seemingly subtle ways. For example, the *lac* genes of *Escherichia coli* are among the most studied of bacterial genes, and yet there are still important features of their organization and regulation that remain to be understood. One of these clearly is the function of the intercistronic spaces, that is, the 54 nucleotides that separate the end of the *lacZ* (*Z*) and the beginning of the *lacY* (*Y*) coding regions, and the provisionally 66 nucleotides that separate the *Y* and the *lacA* (*A*) coding regions (Figure 1a,b) (Büchel et al., (1980). These regions encode known functions, those being translational stop codons, three tandem TAA's in the case of *lacZ* and one TAA in the case of *Y*, and the presumptive Shine–Dalgarno sequences, which specify the sites of initiation of translation. In addition, the sequences display features that suggest unknown functions. Büchel et al. (1980) identified one prominent example. In the RNA transcript extending from the end of the *Z* gene and into the beginning of *Y*, a discontinuous stem–loop structure can be seen that encompasses a total of 109 nucleotides and is stabilized by 28 base pairs. The 30–nucleotide loop contains the start of the *lacY* gene and its Shine–Dalgarno sequence—similar loops have been proposed for a number of bacterial intercistronic spaces (Selker and Yanofsky, (1979).

THE *lac* Z-Y INTERCISTRONIC REGION

Our interest in the regulatory functions of the *lacZY* region grew from a study, still ongoing in the laboratory, of the decay of the *lac* messenger RNA. Blot hybridization was performed on bacterial RNA using as probes DNA from the *Z* gene or from the *Y* and *A* genes. The RNAs were purified from induced and uninduced (as control) cultures of *E. coli* CR63, separated on formaldehyde-agarose gels, and blotted onto DBM paper so that the same filter paper could be repetitively hybridized. These experiments show that the predominant species of *lac* messenger in strain CR63 is the 3 kb *Z* messenger, and that there is several-fold less *Y-A* messenger in the cell than *Z* messenger (unpublished observations).

These results in part could be explained if there was an RNA processing site near the *lac Z-Y* boundary. Such a site has been proposed by Lim and Kennell (1979), and its existence is still a possibility, but the reduction observed in the amount of distal messenger prompted us to examine the DNA sequence for potential termination sites. Figure 1a shows that there is a region of dyad symmetry that immediately follows the translational stop codons of the *Z* gene (3121-3146). This sequence has many of the features generally associated with Rho-independent termination sites: a GC-rich dyad, followed by a sequence rich in T residues (Holmes et al., 1983). Figure 2 shows a comparison of the transcript of this sequence with that of *trpAt,* which follows the tryptophan operon (Wu and Platt, (1978). Other sequences in this region may also function as terminators. Although the features distinguishing the Rho-dependent terminators are not known (Richardson, 1983), the sequence immediately following the one just discussed and extending into the second codon of the *Y* gene (Figure 1a, 3147-3169) has homology with sequences associated with a number of Rho-dependent terminators (Platt et al., 1983). It also contains an AT-rich dyad symmetry. A number of AT-rich dyads are also present further into the *Y* sequence. The first of these is shown in Figure 1a beginning at 3172 in a region very rich in T; such sequences may also be associated with termination. It should be noted that two sequences distal to *trpA* have been associated with termination of the operon. There is a sequence, *trpAt',* downstream from *trpAt,* whose deletion abolishes termination *in vivo* (Wu et al., 1981). The two sequences function independently *in vitro* as terminators (*trpAt'* requires Rho factor), but their role and interrelationship *in vivo* are not yet understood.

To determine the endpoint of the *lacZ* messenger, we have carried out S1 nuclease mapping (Kwan et al, 1984). Total bacterial RNA was hybridized to an end-labeled DNA probe that spans the *Z-Y* region, and unprotected DNA was digested with nuclease (Berk and Sharp, 1977). The results of this experiment are shown in Figure 3. RNA from induced cells protects a specific DNA fragment terminating in the *lac Z-Y* intercistronic region, consistent with a termination site being located following the U's associated with the GC-rich stem-loop (Figure 2). It is also possible that this RNA endpoint has arisen as the result of messenger processing.

THE lac Y-A INTERCISTRONIC REGION

We have also searched the *lac Y-A* intercistronic space for recognizable features (Figure 1b). Although there are no sequences readily associated with termination (see below), there is another recognizable

(a) *lac* Z-Y intercistronic space

```
              3120                    3140        _____ _ 3160_ __
CAAAAATAATAATAACCGGGCAGGCCATGTCTGCCCGTATTTCGCGTAAGGAAATCCATT
                                                      \__/
GlnLys  *   *   *                                    S.D.

              3180                    3200
ATGTACTATTTAAAAAACACAAACTTTTGGATGTTCGGTTTATTCTTTTTCTTT

MetTyrTyrLeuLysAsnThrAsnPheTrpMetPheGlyLeuPhePhePhePhe
```

(b) *lac* Y-A intercistronic region

```
      4420              4440                  4460
GTCGCTTAAGCAATCAATGTCGGATGCGGCGCGACGCTTATCCGACCAACATATCA

ValAla  *                        rep

      4480              4500
TAACGGAGTGATCGCATTGAACATGCCAATGACCG

      S.D.        ? AsnMetProMetThr
```

(c) *lac* Y-A repetitive extragenic palindrome

lac	TGTCGGATGCGGCGC GA CG CTTATCCGACCAACA
Consensus	YGCCgGATGCGRCGXXTXXXGCGYCTTATCcGGCCTACR

FIGURE 1 Sequences of the lac intercistronic regions. Shown underlined are initiation and termination sequences, Shine and Dalgarno (S.D.) sequences, and regions of dyad symmetry discussed in the text. Stop codons are shown with *, and the ? indicates a probable start of translation for lacA (Büchel et al., 1980). The dyads indicated include potential $G-\overline{U(T)}$ pairs that would form in the corresponding transcript. The palindrome shown overlined in (a) is part of a sequence 3147-3169 in the Z-Y space that shows homology with sequences common to some Rho-dependent terminators (Platt et al., (1983):

GCGCT--AA--CGCAATTAAT--TCATTA--A

The broken dyad symmetry discussed in the text as that indicated by Büchel et al. extends from 3206 to 3214 (not shown). The consensus sequence indicated in (0c) is from Higgens et al., (1982. The numbering of sites is that of the lac messenger RNA, designating as 1 the first nucleotide of the sequence AATTGTG...; the sequence was obtained from the Genbank (TM; Bolt, Beranek and Newman, Inc., Cambridge, MA) file for the lac genes (ECOLAC). Except for the features introduced here, the sites identified are from Büchel et al. (1980). Abbreviations used in (c): Y, pyrimidine; R, purine; g, G or T; c, C or A; X, variable.

sequence. Higgens et al. (1982) have described a conserved sequence found in intercistronic spaces or distal to the last gene of some operons of *E. coli* or *Salmonella typhimurium*. This sequence possesses a dyad symmetry that is often repeated in alternating opposite orientations, thus contributing to a larger symmetry of the region. The short palindromes are highly conserved among the different intercistronic spaces that have this feature, and it is a single copy that we have found in the *Y–A* intergenic space. This is shown in Figure 1c; shown below the *lac* sequence is the consensus sequence given by Higgens et al. (1982). Interestingly, two deviations from the consensus sequence occur in positions that preserve the integrity of the palindrome.

The short conserved sequences have been designated repetitive extragenic palindromes, or *rep* (M. J. Stern, *Cell,* in press). When originally found, these sequences were hypothesized to be associated with intercistronic regulation of transcription, specifically to reduce expression of the distal genes of an operon. However, transcription does not seem to be affected by the *rep* sites, as direct manipulation of the site in the histidine transport operon has shown, and currently their function is unknown (Giovanna Ames, personal communication). Approximately 20 *rep*-containing intercistronic spaces have now been identified and it is estimated that there are more than 100 in the *E. coli* genome.

DISCUSSION

The intercistronic spaces within operons differ in size. The *trpE* stop codon overlaps the *trpD* gene by 1 base pair, whereas the space separating the *rplL* and the *rpoB* genes is 324 base pairs. The spaces between the *lac Z* and *Y* genes, and *Y* and *A* genes, 54 and about 66 bp, respectively, are intermediate in this range. The spaces encode the expected stop and start sequences, but in addition, they contain

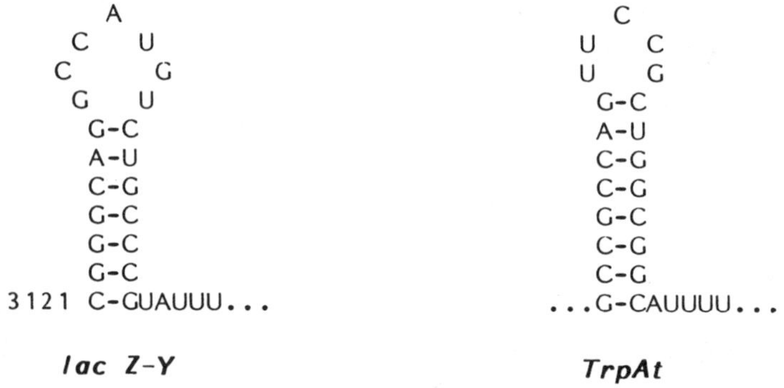

```
            A                              C
        C       U                      U       C
        C           G                  U           G
            G       U                      G-C
            G-C                            A-U
            A-U                            C-G
            C-G                            C-G
            G-C                            G-C
            G-C                            C-G
            G-C                            C-G
        3121 C-GUAUUU...              ...G-CAUUUU...

            lac Z-Y                        TrpAt
```

FIGURE 2 Comparison of the terminator-like sequence of the lac Z-Y region with that of trpAt (Wu and Platt, (1978). The drawings show the presumed secondary structure of the transcripts.

a number of features highly suggestive of sites involved in termination of transcription or in RNA processing. Associated with the *lac Z-Y* intercistronic space, there is a long region of broken dyad symmetry that when transcribed would possess features common to many sites of RNase III processing (Robertson, (1982). There is a GC-rich stem-loop structure that strongly resembles a Rho-independent termination sequence, and also several other sequences near the beginning of the *Y* gene, particularly one at 3147-3169 (Figure 1a) with homologies to

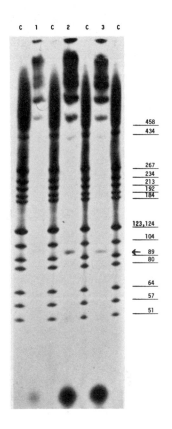

FIGURE 3   Messenger RNA endpoints in the lac Z-Y region. Nuclease S1 analysis according to Berk and Sharp (1977). The probe was a fragment of lac DNA cut at the EcoR1 site (prior to 3058) of the Z gene and extending distally into Y and A; it was end-filled with TTP and dATP. Lane 1, RNA from uninduced bacterial cells digested 60 min. with S1 nuclease; lanes 2 and 3, RNA from induced cells digested for 30 and 60 min.; C, size markers provided by a HaeIII digest of nick-translated pBR322 DNA. The protected band is indicated by an arrow.

other terminators and a weak internal AT-rich dyad; each of these sequences could be Rho-dependent terminators. Another AT-rich site is shown in Figure 1a at 3172-3196; it is part of a particularly T-rich region (coding for the Y gene signal polypeptide). Others begin downstream at nucleotides 3286 and 3508. That termination does occur in this region is argued by our evidence that the abundance of lacYA messenger is several-fold less than that of the lacZ messenger and by the demonstration that lac messenger protects a DNA fragment extending through the T-rich sequence adjacent to the G-C hairpin in the intercistronic space (Figure 2). In the lac Y-A intercistronic region, there is also a short region of dyad symmetry and a homology to the rep sequences described by Higgens et al. (1982). These sequences were also considered at one time to be involved in transcription termination.

Our experimental observations of the Z-Y intergenic space can be interpreted in different, but not necessarily exclusive, ways. The first would be that the intergenic space is the site of a processing reaction that initiates decay of the Y messenger (Lim and Kennell, (1979); the 3' terminus of the cut would then be subject to exonuclease degradation back to a point determined by the GC-rich hairpin. The lacZ messenger would thus be protected from degradation from tnat site. A second possibility, which we favor, is that the intergenic sequence is the site of attenuation and the basis of polarity in lac.

Polarity of gene expression, a reduction in the expression of promoter distal genes, was first discovered in the lac operon and was an element in recognizing its polycistronic organization. There are, however, several forms of polarity recognizable in the lac operon. Natural polarity refers to the fact that the A-gene product, thiogalactoside transacetylase is synthesized at a lower rate than that of the Z gene, beta-galactosidase. The polypeptide molar ratio is generally about 8 to 9, depending somewhat on the bacterial strain (Zabin and Fowler, 1970; Kennell and Riezman, (1977). This ratio is not affected by rho mutations, or replacement of the lac promoter region with the trp promoter; although some strains have a much lower ratio, this is apparently related to an underproduction of beta-galactosidase (Silverstone and Magasanik, 1972; Reznikoff et al., 1974; Guarente et al., 1977). The ratio is also dependent on growth temperature, decreasing by about a factor of two between 40 C and 30 C (Nishi and Zabin, (1963). Various explanations for the natural polarity have been offered, most significantly that loading of ribosomes at the sites of the distal genes is less efficient than that at Z, and that the distal mRNA decays somewhat faster than the Z gene messenger (Kennell and Riezman, 1977; Lim and Kennell, 1979).

Aside from the temperature effects noted, the ratio of beta-galactosidase to transacetylase activity is fairly constant over most steady-state growth conditions and during transients associated with either induction of the operon or change of growth conditions (Silverstone and Magasanik, 1972). However, there are reports of a relative decrease in transacetylase expression during specific severe conditions, for example, the polarity induced with the use of antibiotics that inhibit protein synthesis (Alpers and Tomkins, 1966; Varmus et al., 1971; Ullmann et al., 1979). Some of these observations may stem from an underestimation of transacetylase production at very low rates of synthesis (Silverstone and Magasanik, (1972). There is an indication that some induced-polarity effects are mediated by Rho, and may involve cAMP and receptor protein acting at a site other than the promoter (Ullmann et al., 1979).

The most extensively studied polarity effects are those induced by the presence of nonsense mutations in the *Z* gene. The strength of the polar effect is strongly dependent on the location of the mutation in *Z,* there is a gradient, punctuated with peaks and troughs, of decreasing strength from the 5' end to the 3' end of the gene (Zipser et al., 1970). This observation and the fact that the polarity of these mutations is suppressed by mutations in *rho* (*supA*) suggests that there are Rho-dependent sites of transcription termination at several points within the *Z* gene that function when the messenger is exposed by disassociation of the ribosomes at the nonsense sites (Zipser et al., 1970; Franklin and Yanofsky, 1976). One such site is presumably near the *Z-Y* intercistronic space as indicated by the gradient of polarity.

We believe that many of these polarity effects can be understood based on termination events taking place at the *Z-Y* boundary. Our model initially postulates that there are at least two types of termination events occurring here: Rho-independent and Rho-dependent. This view is supported by the observation that the major component of the natural polarity, the approximately 8-fold reduction in expression of the *lacA* gene relative to *Z,* is unaffected by *rho* mutation or cAMP effects (Guarente et al., 1977; Figure 1, Table 2 of Ullmann et al., 1979). On the other hand, the effects of temperature and antibiotic treatment are largely reduced by *rho* mutation (Ullmann et al., 1979). Indeed, the presence of a *rho* mutation in cells treated with chloramphenicol restores the synthesis of *lacYA* messenger species (Graham et al., 1982). We postulate that the GC-rich termination sequence functions in the Rho-independent mode, and a nearby Rho-dependent site, presumably within the 5'-proximal region of *Y,* provides for that component of the operon response that is affected by growth conditions. Alternative models can be made incorporating the fact that some Rho-independent terminators are stimulated *in vitro* by Rho (Holmes et al., (1983). Finally, if the long stem-loop structure that extends through this region does prove to be an RNase III site, it could provide the basis for some polarity events. Since its 5'-stem overlaps the carboxyl terminus of beta-galactosidase and its 3'-stem overlaps the amino terminus (fifth codon onward) of the lactose permease (*Y*) gene, the availability of ribosomes might well modulate its exposure to processing. Cleavage in this region would inactivate the *Y* messenger and might correspond to the intercistronic cleavage hypothesized by Lim and Kennell (1979) to initiate messenger decay. The various factors that would be involved in these functions were they correct, Rho, NusA and RNase III, are known to influence *lac* expression, but in no case is their precise role known (De Crombrugghe et al., (1973; Greenblatt et al., (1980; Gitelman and Apirion, (1980). We are currently investigating the functions of the *lac Z-Y* sequences directly by their *in vitro* manipulation.

The intercistronic spaces of the *lac* operon display a number of features that suggest complex functions. The *Y-A* sequence is associated with an unknown function that appears widely distributed within, and unique to, extracistronic regions. The sequences of the *Z-Y* space potentially can provide an explanation for the polarity observed in that region and possibly for the processing of the messenger. Currently, mechanisms that regulate bacterial gene expression by termination are primarily associated with attenuation mechanisms that occur in leader sequences. These controls also could be regarded as one type of intercistronic regulation. Recently the controls in the large intercistronic space of the *rplJL-rpoBC* transcription unit have been investigated. The controls in this region include both attenuation and RNase III processing (Barry et al., (1980). We suspect that such intercistronic regulatory modes will be more common than previously recognized.

REFERENCES

Alpers, D., and Tomkins, G. (1966). *J. Biol. Chem.*, 241, 4434.

Barry, G., Squires, C., and Squires, C. L. (1980). *Proc. Natl. Acad. Sci. USA*, 77, 3331.

Berk, A. J. and Sharp, P. A. (1977). *Cell*, 12, 721.

Büchel, D. E., Gronenborn, B., and Muller-Hill, B. (1980). *Nature*, 283, 541.

De Crombrugghe, B., Adhya, S., Gottesman, M., and Pastan, I. (1973). *Nature New Biol.*, 241, 260.

Franklin, N. C. and Yanofsky, C. (1976). In *RNA Polymerase*. R. Losick and M. Chamberlin (eds.) p. 693. Cold Spring Harbor Lab.

Gitelman, D. R. and Apirion, D. (1980). *Biochem. Biophys. Res. Comm.*, 96, 1063.

Graham, M. Y., Tal, M., and Schlessinger, D. (1982). *J. Bacteriol.*, 151, 251.

Greenblatt, J., Li, J., Adhya, S., Friedman, D. I., Baron, L. S., Redfield, B., Kung, H.-F., and Weissbach, H. (1980). *Proc. Natl. Acad. Sci. USA*, 77, 1991.

Guarente, L. P., Mitchell, D. H., and Beckwith, J. (1977). *J. Mol. Biol.*, 112, 423.

Higgens, C. F., Ames, G. F.-L., Barnes, W. M., Clement, J. M., and Hofnung, M. (1982). *Nature*, 298, 760.

Holmes, W. M., Platt, T., and Rosenberg, M. (1983). *Cell*, 32, 1029.

Kennell, D. and Riezman, H. (1977). *J. Mol. Biol.*, 114, 1.

Kwan, C., Murakawa, G. J., Mahoney, P. A., and Nierlich, D. P. (1984). *Ann. Am. Soc. Microbiol.*, 84, 115.

Lim, L. W. and Kennell, D. (1979). *J. Mol. Biol.*, 135, 369.

Nishi, A., and Zabin, I. (1963). *Biochem. Biophys. Res. Comm.*, 13, 320.

Platt, T., Horowitz, H., and Farnham, P. J.. (1983). In *Microbiology—1983*. Schlessinger, D. (ed.), American Society for Microbiology, p. 21.

Reznikoff, W. S., Michels, C. A., Cooper, T. G., Silverstone, A. E., and Magasanik, B.. (1974). *J. Bacteriol.*, 117, 1231.

Richardson, J. P. (1983). In *Microbiology--1983*. Schlessinger, D. (ed.) p. 31. American Society for Microbiology.

Robertson, H. D. (1982). *Cell*, 30, 669.

Selker, E. and Yanofsky, C. (1979). *J. Mol. Biol.*, 130, 135.

Silverstone, A. E., and Magasanik, B. (1972). *J. Bacteriol.*, 112, 1184.

Ullmann, A., Joseph, E., and Danchin, A. (1979). *Proc. Natl. Acad. Sci. USA*, 76, 3194.

Varmus, H. E., Perlman, R. L., and Pastan, I. (1971). *Nature New Biol.*, 230, 41.

Wu, A. M., Christie, G. E., and Platt, T. (1981). *Proc. Natl. Acad. Sci. USA*, 78, 2913.

Wu, A. M. and Platt, T. (1978). *Proc. Natl. Acad. Sci. USA*, 75, 5442.

Zabin, I. and Fowler, A. V. (1970). In *The Lactose Operon*. J. R. Beckwith and D. Zipser (eds.) p. 27. Cold Spring Harbor Laboratory.

Zipser, D., Zabell, S. Rothman, J. Grodzicker, T., Wenk, H., and Novitski, M. (1970). *J. Mol. Biol.*, 49, 251.

Regulation of the Arabinose Operon

Robert Schleif
Brandeis University; Waltham, Massachusetts

The L-arabinose operon was discovered by Ellis Englesberg to possess an unusual type of regulation (Englesberg et al., 1965). At the time of his pioneering genetic and physiological work, only the *lac* system was well characterized. Unfortunately Englesberg's initial findings of a positive regulatory mechanism in the *ara* system were not well received, in part owing to the mechanistic simplicity of a negative regulatory system in which a repressor blocks access of RNA polymerase to the DNA. Why should any other mechanism exist since a negative system could explain all the phenomena? Now, twenty years later, we are a little more open-minded. As we are seeing in this symposium, nature seems loathe to use the same regulatory mechanism more than once. The problem of a positive regulatory system is not so much that it is different from the *lac* system, but that of how it works.

This small review will mention the fundamental properties of the *ara* system, describe an evolutionary conjecture, and then turn to firmer data and explain the data suggesting the presence of loops in DNA that are involved in gene regulation.

THE BASIC REGULATORY PHENOMENA OF THE *ara* OPERON

In order for arabinose to be catabolized by *Escherichia coli* cells, it must be transported from the growth medium to the cytoplasm. This is accomplished by two independent active-transport systems. Once within the cell, three arabinose-induced enzymes convert arabinose to D-xylulose-5-phosphate that then enters the pentose phosphate shunt (Figure 1). The three arabinose enzymes and the transport systems are induced by the presence of arabinose (Table 1). In the absence of arabinose, these proteins are synthesized at low levels, and during growth in the presence of arabinose, their rate of synthesis is increased from 100 to 400 times. Under these conditions several of the *ara* enzymes can constitute as much as 2% of the cell protein.

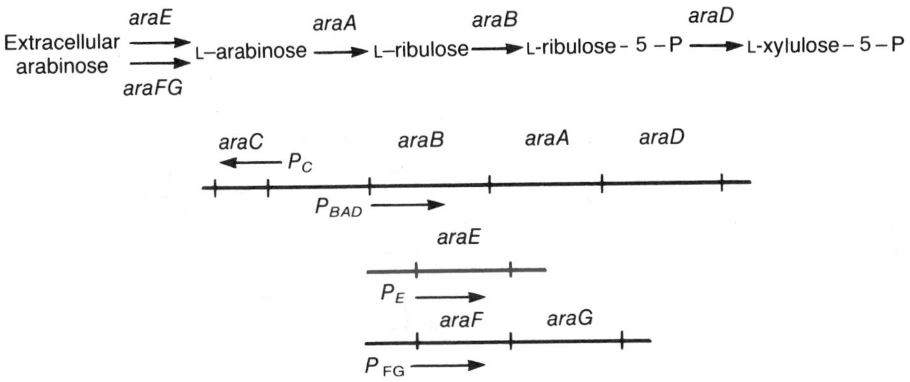

FIGURE 1 The metabolic pathway for catabolism of arabinose and the genetic structure of the arabinose operons.

TABLE 1   Activity of promoter responsive to the AraC protein for various genotypes

Genotype	Promoter activity
C^+	1
C^+ + arabinose	300
C^- + arabinose	1
C^+, crp^- + arabinose	10

Induction of the arabinose enzymes (the *araBAD* enzymes), as well as the active-transport proteins *araE* and *araFG*, requires the presence of the product of the *araC* gene. From experiments performed *in vitro* with purified components, AraC protein has been shown to activate transcription of the *araBAD* and *araE* genes after binding adjacent to RNA polymerase. In addition to the positive control by AraC protein, the *ara* operons are subject to catabolite repression. Therefore, their transcription requires the presence of the cyclic AMP receptor protein CRP, which binds adjacent to AraC protein (Figure 2) (Ogden et al., 1980).

A third basic property of the *ara* system is that transcription of the *araBAD* genes is also negatively regulated by AraC protein. That is, AraC protein acts as a repressor in addition to acting as an inducer. Paradoxically, repression is exerted from a site that lies upstream from all of the sites that are required for induction. Englesberg found that a deletion, the infamous Δ719, left the operon fully inducible but repression-negative (Englesberg et al., 1969). With the determination of the regulatory-protein binding sites in the *ara* regulatory region and the sequencing that located the endpoint of Δ719 in the upstream *araC* binding site (Miyada et al., 1980), it appeared that the upstream repression site in *ara* was found (Figure 2). Below I show that the site required for repression lies still further upstream than the *araO₁* site and that the DNA likely loops back from this site to the *araI* site to generate repression.

The fourth basic property of the *ara* system is that AraC protein represses its own synthesis (Casadaban, 1978; Table 2). While this provides a simple mechanism for maintaining a low and steady synthesis of this important protein, a variation of the repression may be even more important. If arabinose is added to growing cells, the synthesis of AraC protein is derepressed about fivefold for about 20 minutes (Ogden et al., 1980; Hahn and Schleif, 1983; Stoner and Schleif, 1983). During this time the absolute level of AraC protein doubles or triples in the cell. Then, its rate of synthesis falls back to normal, and the excess protein decays and is diluted by cell growth back to its normal level.

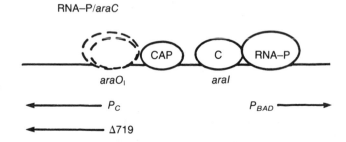

FIGURE 2   Regulatory protein binding sites in the regulatory region of the P<u>C</u>-P<u>BAD</u> promoters.

TABLE 2   Dependence of steady-state activity of P_C on *araC* genotype

Genotype	Steady-state P_C activity
C^+	1
C^-	6

THE USE OF REGULATING AraC PROTEIN SYNTHESIS

Why transiently induce synthesis of AraC protein when arabinose appears? One habitat of *E. coli* is the human gut. In this location arabinose is a free lunch because our intestinal cells cannot take up arabinose or metabolize it. The bacteria are subjected to pulses of arabinose when we have eaten vegetables, since arabinose is a common component in the walls of plant cells. Given an environment subject to such sporadic feasts of arabinose, it seems sensible for bacteria not to squander their limited resources in synthesizing *ara* enzymes constitutively, but instead to devise a system permitting rapid synthesis of the *ara* enzymes only on demand.

One way to respond to the presence of arabinose would be to induce the *ara* promoters and to leave them on at that rate thereafter. The consequences of this mode of regulation are that one doubling time after induction of the enzymes, their cellular concentration would be only half the value that they would have, if they had instead been induced for many generations. This is easily seen by "Copenhagen thinking." If the enzyme were induced many generations ago, then as the number of cells doubles from C to $2C$ to $4C$, the concentration of enzyme in the culture would increase from E to $2E$ to $4E$, and so forth (Table 3). Now consider the situation in which induction of the enzyme started when the cell concentration is C. Previously, when the cells grew from C to $2C$, they synthesized an additional E of enzyme. The newly induced cells can still synthesize this amount of enzyme in this doubling time. Therefore, induction at a concentration of C yields a level of E enzyme when the cells reach $2C$. That is, the level of enzyme is 50% the value obtained by induction for many

TABLE 3   Enzyme levels in various conditions

Steady-State Enzyme Levels

	Cell doublings			
	0	1	2	3
Cell number	C	$2C$	$4C$	$8C$
Total enzyme in culture	E	$2E$	$4E$	$8E$
Enzyme synthesized per doubling		$1E$	$2E$	$4E$
Enzyme/cell	1	1	1	1

Enzyme Levels if Induced at Doubling Zero

	Cell doublings			
	0	1	2	3
Cell number	C	$2C$	$4C$	$8C$
Enzyme synthesized per doubling		$1E$	$2E$	$4E$
Total enzyme in culture	0	$1E$	$3E$	$7E$
Enzyme/cell	0	$1/2$	$3/4$	$7/8$

generations. Analogously, the enzyme concentration after 2, 3, and 4 doublings, is 3/4, 7/8, and 15/16, the value it would have in the steady state.

Now that we have established the consequences of inducing a promoter at a fixed rate, let us see the consequences for the cells. If the *ara* enzymes are only 50% induced after one generation, their level could be rate limiting. Alternatively, if this level is sufficient to support the most rapid growth possible in this medium, then additional enzyme that they synthesize is wasted.

A way around this conundrum is to synthesize the *ara* enzymes at a high rate until they have reached a value near the desired final steady-state value, and then to shift and synthesize them at a lower rate, one just sufficient to maintain the enzyme level (Figure 3). Such a program for rapid induction is possible in the *ara* operon. If AraC protein is normally present at levels insufficient to induce the *ara* promoters to their maximal rates, then the transient increase in the level of *ara* protein following addition of arabinose would permit the desired transient hypersynthesis of the *araBAD* enzymes. Even if this effect is relatively minor in the *ara* operon, the consequences of being able to induce the *ara* enzymes a few seconds faster than one's competitors would be of significant value to the organism.

UPSTREAM REPRESSION AND DNA LOOPING

The properties of upstream repression as revealed by Englesberg's deletion Δ719 have greatly bothered me over the years. We repeated the basic observations of repression from upstream (Schleif and Lis, 1975), but made no real progress until genetic engineering was possible. Recently we cloned the *ara* regulatory region on a plasmid with the *araBAD* promoter P_{BAD} driving the relatively easily assayed *galK* gene. Then we opened the plasmid well upstream of the *ara* regulatory region and made a series of deletions entering the *ara*

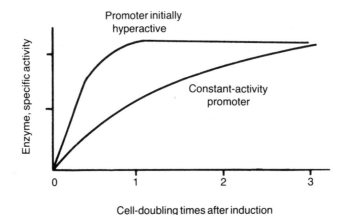

FIGURE 3 Induction kinetics for a promoter that is induced at a fixed rate and for a promoter whose initial activity is greater than its activity at later times.

regulatory region various distances. As expected, these reproduced the phenomenon of repression from upstream (Dunn et al., 1984).

A surprising observation about the deletions was that the upstream repression site was not the expected $araO_1$ site, but instead a site located 280 base pairs upstream from the start of P_{BAD} transcription! This new site is called $araO_2$. The finding of this site is, of course, consistent with the properties of Δ719, since this deletion removes the upstream site in addition to deleting part of $araO_1$.

By repeating DNAse footprinting experiments under conditions appropriate for detecting AraC protein binding to this region, we indeed found that $araC$ can specifically bind to a site located between 270 and 290 base pairs upstream from the transcription start site of P_{BAD}. The sequence of this region possesses partial homology to the AraC protein binding sites at $araI$ and $araO_1$, but the homology is not easily seen without knowing the pattern of bases contacted by AraC as it binds DNA. This aspect to the story cannot be told here, though it should be mentioned that AraC binds *three* adjacent major grooves rather than the two seen by phage repressors, *lac* repressor, and the CRP protein.

How can repression from $araO_2$ work? Three semiplausible schemes can be envisioned (Figure 4). The conformation of the DNA could be altered by the binding of $araC$ and this change could be propagated down the DNA, much like skewing a ladder. Alternatively, AraC protein could polymerize on the DNA beginning from the $araO_2$ site. Then, repression could be a simple steric hindrance but require

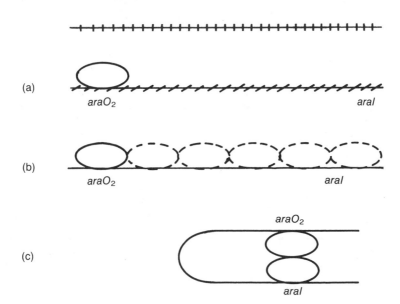

(a) $araO_2$ $araI$

(b) $araO_2$ $araI$

(c) $araO_2$ $araI$

FIGURE 4 Three possible modes for repression from a site well upstream of a promoter: (a) tilt, (b) polymerization, and (c) DNA looping.

the $araO_2$ site. The third possibility is that the DNA could loop back, and either the DNA or AraC protein bound to it could contact a protein in the induction region to generate repression.

The third possibility appears correct. Inserting and deleting small numbers of nucleotides between $araO_2$ and the remainder of the regulatory region changes both the distance between the sites as well as their angular orientation. Insertion of five base pairs rotates one site with respect to the other by half a helical turn of the DNA (assuming it to be right helical with a pitch of about 10.5 base pairs per turn) (Figure 5). Repression was lost when the insertions added nonintegral numbers of turns and was possible when 11 and 31 base pairs were inserted. DNA looping seems the most simple explanation for these findings. The other end of the loop appears to be AraC protein bound at the $araI$ site, since mutation of the $araI$ site eliminates repression.

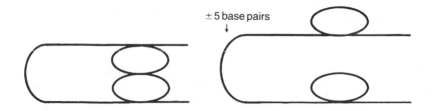

± 5 base pairs

FIGURE 5 The effects of inserting or deleting five base pairs between <u>$araO_2$</u> and the remainder of the <u>P_{BAD}</u> promoter.

In the vein of conjecture established by the earlier part of this review, let us examine one possible reason that nature may have had for evolving such a bizarre repression scheme. With AraC protein acting as a positive activator, which functions by binding adjacent to RNA polymerase, how can catabolite repression be generated? Conceivably it could work the way we first thought, by CRP assisting the binding of AraC, which in turn assists the binding of RNA polymerase at P_{BAD}. However, *in vitro* there seems to be no cooperativity in the binding of $araC$ and CRP.

Nature could have invented upstream repression in order to provide the system with a CRP dependence. That is, the function of CRP could be to assist the breaking of repression, and that is just what is found. In the absence of $araO_2$, $araBAD$ can be induced without CRP, a feat impossible if $araO_2$ is present. How CRP manages to stimulate P_{BAD} by breaking repression, to provide a little stimulation (about fourfold) to P_{BAD} in the absence of repression, and to stimulate P_C remains the object of future research.

ACKNOWLEDGEMENTS

The research reported here was supported by National Institutes of Health Grant 18277 and was mainly carried out by Teresa Dunn, Steve Hahn, William Hendrickson, and Robert Schleif.

REFERENCES

Casadaban, M. (1976). *J. Mol. Biol.,* 104, 557.

Englesberg, E., Power, J., and Lee, N. (1965). *J. Bacteriol.,* 90, 946.

Englesberg, E., Squires, C., and Meronk, F. (1969). *Proc. Natl. Acad. Sci. USA,* 62, 1100.

Hahn, S. and Schleif, R. (1983). *J. Bact.,* 155, 593.

Miyada, C., Horwitz, A., Cass, L., Timko, J., and Wilcox, G. (1980). *Nucleic Acids Res.,* 8, 5267.

Ogden, S., Haggerty, D., Stoner, C., Kolodrubetz, D., and Schleif, R. (1980). *Proc. Natl. Acad. Sci. USA,* 77, 3346.

Schleif, R. and Lis, J. (1975). *J. Mol. Biol.,* 95, 417.

Stoner, C. and Schleif, R. (1983). *J. Mol. Biol.,* 170, 1049.

Growth-Rate-Dependent Regulation of a Central Metabolism Gene

Richard E. Wolf
University of Maryland, Catonsville

Enzymes of glycolysis and the hexose monophosphate shunt in enteric bacteria were once called "constitutive". Studies by D. G. Fraenkel and others had shown that the level of several of the *Escherichia coli* enzymes (e.g., phosphoglucose isomerase, glucose 6-phosphate dehydrogenase) did not change much during growth on different sugars or under conditions of metabolic stress, such as when metabolism of glucose was restricted to the shunt by a mutation that blocked glycolysis (reviewed by Fraenkel and Vinopal, 1973). In this instance "constitutive" was used to suggest only that specific exogenous inducers or repressors were probably not involved in regulating expression of the central metabolism genes. However, some undoubtedly interpreted the use of the word as meaning "unregulated." Perhaps for this reason, there has been a general lack of interest in studying the genetic regulation of these pathways.

In any event, as a postdoctoral fellow in Fraenkel's laboratory, I became interested in understanding how the level of central metabolism enzymes was kept fairly constant and whether any unique or fundamental genetic regulatory mechanisms were involved. I isolated, as a potentially useful reagent, a ϕ80 specialized transducing phage carrying one such gene, *gnd,* which encodes 6-phosphogluconate dehydrogenase (6PGD), an enzyme of the shunt (Wolf and Fraenkel, 1974). T. Isturiz and I used the phage DNA to direct the *in vitro* synthesis of 6PGD (Wolf and Isturiz, 1975). We were interested in the possibility that fractionating the cell-free system might reveal the components essential for *gnd* expression. To determine whether the phage was in fact a suitable template for such studies, i.e., whether the entire *gnd* control region was intact, we measured the *in vivo* expression of 6PGD from the ϕ80 *gnd* prophage. We found that the amount of 6PGD activity in the lysogen during growth on glucose was three- to fourfold higher than the amount in succinate-grown cells.

The result surprised us because the dogma concerning the central metabolism enzymes was that their levels were fairly invariant. The observation in turn led me to consider whether the genes that produce proteins whose levels are proportional to growth rate were regulated, while genes for proteins whose levels are independent of growth rate were unregulated. A personal discovery of the work and theoretical considerations of O. Maaløe (Maaløe and Kjeldgaard, 1966; Maaløe, 1979) suggested that the distinction was irrelevant: both types of genes must be regulated, either actively or passively, since their respective rates of expression increase as a function of growth rate. Three general classes of genes were implicit in Maaløe's analysis of the effect of growth rate on ribosome biosynthesis (Table 1). "Constitutive" best applies to proteins in class A; they are synthesized at constant absolute rates, and hence each represents a decreasing fraction of total protein as growth rate increases. The level of a class B protein is independent of growth rate, the result of a synthesis rate that is directly proportional to growth rate. The level of class C proteins increases in proportion to growth rate, and hence their rate of synthesis is proportional to the square of the growth rate. That these are the basic categories in *E. coli* was demonstrated unequivocally by Pedersen et al. (1978).

TABLE 1   Relations between growth rate and the relative amount and absolute rate of synthesis of specific proteins

Class	$\dfrac{\text{Specific protein}}{\text{Total protein}}$	Absolute synthesis rate of specific protein
A	Proportional to $1/\mu$	Constant
B	Constant	Proportional to μ
C	Proportional to μ	Proportional to μ^2

μ is the growth rate expressed in doublings per hour.

Since the level of all proteins is in fact coordinated with the cellular growth rate (Pedersen et al., 1978), growth rate-dependent regulation must be the fundamental genetic regulatory mechanism in the bacterial cell. In this formulation other types of regulation of increasing specificity (ranging from global circuits, like heat shock, SOS induction, and anaerobiosis, to highly specific units, like a biosynthetic operon) are viewed as being superimposed on the basic system. Work described elsewhere in this volume promises that we will soon understand the mechanism of growth-rate control of ribosome biosynthesis. However, understanding the overall process of growth rate-dependent regulation requires study of the way the level of other *E. coli* proteins (e.g., the class B proteins and the proteins of class C that are not members of the protein synthesizing system) is coordinated with growth rate. I next summarize briefly our current knowledge of the growth rate-dependent regulation of 6PGD level, work that was stimulated by my encounter with Maaløe's ideas.

PHYSIOLOGY OF *gnd* EXPRESSION

The specific activity of 6PGD in *E. coli* K12 increases three- to fourfold with increasing growth rate during steady state growth on different carbon sources, in the range of growth rates between $k = 0.1$ (acetate, serine) and $k = 0.6$ (glucose) (k, the specific growth rate constant, is defined as ln2/(mass doubling time in hours)). The level does not increase further in fully supplemented glucose medium, for which k = 1.6 (Wolf et al., 1979). (A similar effect of growth rate has also observed by Wolf et al., (1979) for glucose 6-phosphate dehydrogenase, G6PD, encoded by *zwf*.) The variation in specific activity of 6PGD reflects variation in the amount of the enzyme (Wolf et al., 1979). The level of 6PGD varies as a function of growth rate per se, not carbon source: when growth rate on glucose is reduced by an increasing ratio of alpha-methylglucoside to glucose, by anaerobic growth, or by introducing a mutation in *pgi* (phosphoglucose isomerase), there is a proportionate decrease in the level of the enzyme (Wolf et al., 1979; Farrish et al., 1982). Thus, 6PGD level remains coordinated with growth rate when the shunt is nonfunctional and during metabolic stress. This contrasts with the enzymes of glycolysis, which are induced twofold by anaerobiosis (Smith and Neidhardt, 1983a), and with the Krebs cycle enzymes, which are repressed during anaerobic growth (Smith and Neidhardt, 1983b).

The effect of growth rate on the level of 6PGD (and G6PD) is similar to that observed for protein components of the translational machinery, the "core." However, three lines of evidence have led to the conclusion that *gnd* (and *zwf*) are not part of the same regulatory network as the core genes. (1) During a nutritional shift-up the rate of RNA accumulation increases immediately, while there is a substantial lag before accumulation of 6PGD, G6PD, and total protein attains the post-shift rate (Farrish et al., 1982). Since the differential rates of synthesis of r-proteins (Schleif, 1967) and other core components increase immediately, proteins whose levels are directly proportional to growth rate can be subdivided into two classes, those that respond immediately and those that display a delayed response. (Interestingly, after a shift-down, accumulation of 6PGD resumes concomitant with total RNA, significantly after total protein.) (2) Synthesis of 6PGD is not subject to stringent control (Farrish et al., 1982). (3) The level of 6PGD depends on gene dosage: growth-rate control of 6PGD level is intact at high *gnd* gene dosages, and expression of the chromosomal gene is unaffected by the presence in *trans* of a high copy number of *gnd* control regions (Farrish et al., 1982). Thus, growth-rate control of 6PGD level is not subject to autogenous control like r-proteins (Nomura, this volume), nor is it dependent on positive or negative regulators that are present in limiting amounts.

MOLECULAR BIOLOGY

The DNA sequence of *gnd* has been determined (Nasoff et al., 1984). The beginning of the 6PGD coding sequence was located by determining the N-terminal amino acid sequence of purified 6PGD and comparing it to the decoded DNA sequence; the MW of the polypeptide encoded by the single long open reading frame, 51,600, agreed perfectly with the MW of the 6PGD subunit estimated by SDS polyacrylamide gel electrophoresis (Wolf and Shea, 1979). The 5' end of *in vivo gnd* mRNA was positioned on the DNA sequence by S1 mapping (Nasoff et al., 1984). In addition, *in vitro* transcription of small restriction fragments presumed to contain the *gnd* promoter yielded run–off RNAs of a size that accorded with the S1 mapping. Analysis of the DNA sequence yielded the following information, some of which served to rule out several previously proposed models (Farrish et al., 1982) for regulation of *gnd* expression. (1) The region upstream from the transcription startpoint is typical of *E. coli* promoters. (2) *Gnd* mRNA has a 56–nucleotide leader that contains a strong ribosome binding signal; a Shine–Dalgarno sequence, AGGAG, is located an optimal 7 nucleotides before the AUG start codon. (3) The codon composition of the 6PGD–coding sequence resembles that of r–proteins in that it shows a distinct avoidance of codons that are recognized by tRNAs present in cells in low relative molar amounts, and a preference for codons that optimize the codon–anticodon interaction energy. (4) The leader does not contain a Rho–independent transcription termination signal. Thus, the growth–rate control of *gnd* expression is not carried out by an attenuation mechanism like the one proposed to regulate the *E. coli ampC* gene (Jaurin et al., 1981), which shows the same physiology as *gnd* (Jaurin and Normark, 1979). (5) The leader contains regions of dyad symmetry that have the potential to form secondary structures able to sequester the Shine–Dalgarno sequence and the start codon. Formation of these structures *in vivo* might limit the translation initiation frequency, and hence the yield of 6PGD per *gnd* mRNA.

PROPERTIES OF *gnd–lac* OPERON AND GENE FUSIONS

Recent results obtained with *gnd–lac* operon and gene (protein) fusions have suggested that growth rate–dependent regulation of *gnd* expression is carried out by a translational control mechanism that requires a site within the 6PGD coding sequence. The data supporting this hypothesis are depicted in Figure 1 and summarized below.

    Gnd–lac operon fusions were prepared and characterized (Baker and Wolf, 1983) as a means to determine whether the growth rate–dependent regulation is carried out by a transcriptional or posttranscriptional control mechanism. In a *gnd–lac* operon fusion expression of beta–galactosidase depends upon transcription that initiates at the *gnd* promoter and is subject to any transcriptional regulation that takes place at the *gnd* promoter or any other site of transcriptional regulation lying between the promoter and the fusion joint. Thus, if beta–galactosidase levels in the *gnd–lac* operon fusions were to show the same dependence on growth rate as 6PGD itself, then a transcriptional control mechanism would probably be involved in the regulation; on the other hand, if the level of beta–galactosidase did not vary with growth rate, like 6PGD, then the regulation would likely derive from a translational control mechanism. (We point out that using operon fusions to assess possible transcriptional control avoids the problem inherent with measuring directly, by DNA–RNA hybridization, the effect of growth rate on the amount and rate of synthesis of an mRNA: namely, the necessity of normalizing to total RNA, when the fraction of total transcript that is mRNA varies with growth rate (Gausing, 1977).)

    Four operon fusions were prepared with phage Mud1 and characterized genetically. The fusion joints mapped at different sites

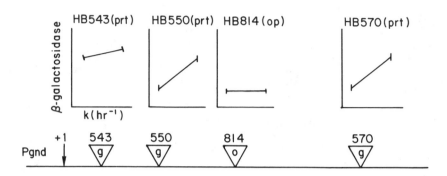

FIGURE 1 Growth-rate-dependent regulation of β-galactosidase level in gnd-lac operon and gene fusions. Cultures were grown in acetate and glucose minimal medium and β-galactosidase levels assayed in sonic extracts. The level of β-galactosidase in the operon (op) and gene (prt) fusion strains is plotted against the growth rate of the culture; k = ln2/doubling in hr. An inverted triangle represents the position in gnd of a fusion joint. The gnd map is drawn to scale, except the precise location of the HB814 fusion joint is not known. Data is taken from Baker and Wolf (1983, 1984).

across the structural gene, and in each case *lac* was shown by a genetic test to be fused to the *gnd* promoter. The level of beta-galactosidase was invariant with growth rate in the four fusions, and was unaffected by the presence of a *gnd⁺* gene in *trans*. These results suggest that the fraction of total mRNA is constant and independent of growth rate and thus that the growth rate-dependent regulation of 6PGD level is carried out by a translational control mechanism that increases the yield of 6PGD per *gnd* mRNA molecule in direct proportion to growth rate. However, the data do not rule out the possibility that the regulation of 6PGD level is transcriptional and requires a site downstream from the most promoter-distal operon fusion joint. Another possibility not ruled out by the data and discussed later in this article is that the regulation is due to a growth rate dependence of the functional half life of *gnd* mRNA.

        To examine the possibility that growth rate-dependent regulation of *gnd* expression is transcriptional and requires a site in the promoter-distal portion of the gene, and to determine more precisely the location of regulatory sites, we prepared eight independent *gnd-lac* gene (protein) fusions (Baker and Wolf, 1984) with phage Mud11 (Casadaban and Chou, 1984). In a *gnd-lac* gene fusion strain expression of *lacZ* will depend on transcription initiation at the *gnd* promoter and on translation initiation at the *gnd* ribosome-binding site; however, if such gene fusions are to be useful in studying regulation, the fusion joint must be in the reading frame that allows synthesis of a hybrid 6PGD-beta-galactosidase protein, and the hybrid protein should be enzymatically active. The rationale for preparing *gnd-lac* gene fusions was as follows. First, if the level of beta-galactosidase in any gene fusion were to vary with growth rate like 6PGD, and if its fusion joint were to lie upstream from an operon fusion joint, then the possibility that the growth rate-dependent regulation is transcriptional and requires a distal regulatory site could be eliminated. Second, the position in *gnd* of the fusion joints of growth rate-regulated gene fusions can be used to map the sites necessary for the growth rate-dependent regulation. Thus, if the level of beta-galactosidase in all the gene fusions were to vary with growth rate like 6PGD, then the necessary sites for regulation must lie upstream from the most promoter-proximal gene fusion joint. On the other hand if beta-galactosidase levels were invariant in a set of gene fusions with promoter-proximal fusion joints, but were proportional to growth rate in gene fusions with more promoter-distal fusion joints, then an essential regulatory site could be assigned to the region bracketed by the two classes of gene fusions. Finally, if no gene fusion were to show growth-rate control, then the regulation would require a site downstream from the 6PGD coding region.

The DNA sequence of each gene fusion joint was determined and compared to that of *gnd*. In each case the fusion was between the first and second base pair of a codon in the *gnd* structural gene, the reading frame necessary for production of a hybrid 6PGD-beta-galactosidase polypeptide. Moreover, each fusion produced a polypeptide whose mobility in an SDS-polyacrylamide gel correlated perfectly with the molecular weight predicted from the DNA sequence for the respective hybrid protein.

The level of beta-galactosidase was growth rate-dependent in fusion HB550, whose fusion joint is within codon 118, and in the six other fusions (e.g., HB570) with more promoter-distal fusion joints (see Figure 1). In addition, the level was unaffected by the presence of an $F'gnd^+$ plasmid in *trans* to the fusions. Thus, all of the sites necessary for the growth rate-dependent regulation are upstream from the site defined by the HB550 fusion joint, and the regulation is not autogenous. The fact that the fusion joint of a growth rate-regulated fusion, viz., HB550, lies upstream from an operon fusion joint (e.g., HB814; see Figure 1) eliminates the above-mentioned possibility that the regulation requires a site in the distal end of *gnd*.

The level of beta-galactosidase produced from the HB543 fusion (fusion joint at codon 48) was threefold higher than that of the other gene fusions when the strain carrying it was growing slowly, with acetate as carbon source; moreover, the level was increased only slightly by growth on glucose (see Figure 1). In other words, compared to the growth rate-"inducible" phenotype of the other gene fusions, fusion HB543 appears to be "derepressed." These data suggest that a site essential for growth rate-dependent regulation of *gnd* expression includes the sequence between codon 48 and codon 118 of the structural gene, and that this is a site for negative control.

It is possible, of course, that the growth rate-derepressed phenotype of strain HB543, and hence the hypothesis that the regulation requires a site in the 6PGD coding sequence, is due to an artifact of the fusion joint. However, we think this explanation is unlikely for the following reason. Coincidentally, the DNA sequences surrounding the fusion joints of strains HB543 and HB570 are identical, except for 1 bp, starting at a position 13 bp upstream from the fusion joints, though the growth rate dependence of beta-galactosidase level is different in the two strains (Baker and Wolf, 1984).

The properties of the operon fusions just described suggested that the growth rate-dependent regulation of *gnd* expression is carried out by a translational control mechanism. Accordingly, the internal regulatory region would be a site for translational control.

The genetic structure of phage Mud11 (Casadaban and Chou, 1984) provided an opportunity to test further the hypothesis that the growth rate-dependent regulation of *gnd* expression is carried out at the translational level. In the *gnd::Mud11* gene-fusion strains expression of galactoside transacetylase, encoded by *lacA*, is subject to regulation by the transcriptional, but not the translational, control signals that lie upstream from the fusion joints. In other words, the *gnd-lacZ* gene (protein) fusion strains also carry *gnd-lacA* operon fusions. Accordingly, assay of galactoside transacetylase activity in the *gnd::Mud11* fusion strains can be used to assess the growth-rate dependence of *gnd* transcription.

Strains carrying the HB543 and HB550 fusions were grown in acetate and glucose minimal media under standard physiological conditions, and assayed for galactoside transacetylase activity by the method of Miller (Miller, 1972). With the HB543 fusion strain the specific activity of the enzyme was 4.4 and 4.8 U/mg for the acetate and glucose cultures, respectively. With the HB550 fusion strain, the respective values were 3.7 and 4.9 U/mg. Two other experiments gave similar results. Thus, expression of the transacetylase in the phage Mud11-induced *gnd-lac* gene fusion strains, like expression of beta-galactosidase in the phage Mud1-induced *gnd-lac* operon fusion strains is growth rate-independent. We conclude that the growth-rate control of *gnd* expression is carried out at the translational level.

We know of no other example of translational control in bacteria that requires a site deep within the protein-coding sequence of the regulated gene. However, several types of negative control at the translational level that involve the ribosome binding region of mRNA have been recognized, and any of them could theoretically be responsible for the growth rate-dependent regulation of *gnd* expression. For example, both protein (Campbell, Stormo, and Gold, 1983; Nomura, this volume) and RNA (Simons and Kleckner, 1983; Mizuno, Chou, and Inouye, 1984) molecules have been shown to be able to act as translational repressors by binding to the leader region of mRNAs. Thus, the internal regulatory region of *gnd* might be such an "operator," with the amount of translational repressor being inversely proportional to growth rate.

The internal regulatory region of *gnd* contains a sequence that is highly complementary ($\Delta G = -13$ kcal/mole) to the ribosome binding site of the mRNA. This interaction could be important to *gnd* regulation, since it is known that the formation of mRNA secondary structures that sequester the ribosome binding site can reduce translation initiation frequency (Iserentant and Fiers, 1980; Queen and Rosenberg, 1981; Ghysen et al., 1982; Hall et al., 1982; Mamelak et al., this volume). Unknown structural features of *gnd* mRNA might also be important for control. For example, the rate of polypeptide chain elongation has been shown recently to differ between certain mRNAs and to vary with the cellular growth rate (Pedersen, this volume).

Another model for growth rate-dependent regulation of 6PGD level can be put forth that accounts for the properties of the operon and protein-fusion strains without proposing a growth rate-dependent variation in the yield of 6PGD per *gnd* mRNA molecule. The essential elements of the model are the following. (1) The functional half-life of *gnd* mRNA varies in direct proportion to growth rate. (2) The site for the rate-limiting step in the inactivation of the mRNA lies in the region of *gnd* delimited by the HB543 and HB550 fusion joints. (3) The functional half-life of the *lac* portion of the *gnd-lac* fusion mRNAs is independent of growth rate and is unaffected by events occurring upstream, for example, because degradation of mRNA proceeds in the 3'-to-5' direction following an initial endonucleolytic cleavage event. According to this model, the relative amount of *gnd* mRNA in a normal strain would increase in direct proportion to growth rate as a result of both the rate of synthesis and the functional half-life of the mRNA varying directly with growth rate; thus, a constant yield of 6PGD per mRNA molecule would produce a growth rate-dependent increase in the level of 6PGD. The same argument applies to the production of beta-galactosidase from *gnd-lacZ* mRNA in all the protein fusions but HB543. Similarly, the level of beta-galactosidase in the operon fusions and the level of transacetylase in all the protein-fusion strains but HB543 would be independent of growth rate, because the rates of synthesis of the mRNAs, initiating at the *gnd* promoter, are proportional to growth rate, and the functional half-lives of the *lac* portion of the fused mRNAs are invariant. The model predicts that the ratio of *gnd*-specific mRNA sequences upstream from codon 48 in total RNA prepared from *gnd-lac* operon fusions or from protein-fusion strains other than HB543 would increase in direct proportion to growth rate, whereas the ratio would be invariant in RNA prepared from strain 543.

It is not at all clear what the regulatory effector for growth-rate control of *gnd* expression might be. However, the central role of ribosome biosynthesis in the control of growth and the fact that upon a nutritional shift-up the differential synthesis rate of ribosomal components increases before that of 6PGD suggests that the concentration of ribosomes or ribosomal components might be involved. Clearly elucidating the molecular mechanism for growth rate-dependent regulation of *gnd* expression will require the combined approaches of microbial physiology, genetics and molecular biology.

It will also be interesting to determine whether the particular mechanism is involved in growth-rate control of other central metabolism genes, especially *zwf,,* since the physiology of its expression and the enzymatic reaction carried out by its gene product are similar to those of *gnd*. In this context one wonders whether the regulatory region in the *gnd* structural gene is embedded within the coding sequence for NADP binding.

ACKNOWLEDGEMENTS

I thank my collaborators and colleagues for their help and thoughtful interest in this work. H.V. Baker II carried out the experiments on the growth-rate independence of galactoside transacetylase activity. The research was supported by USPHS Grant GM27113.

REFERENCES

Baker II, H. V. and Wolf Jr., R. E. (1983). *J. Bacteriol.*, 153, 771.

Baker II, H. V. and Wolf Jr., R. E. (1984). *Proc. Nat. Acad. Sci. USA,* 81, 7669.

Campbell, K. M., Stormo, G. D., and Gold, L. (1983) In *Gene Function in Prokaryotes,* J. Beckwith, J. Davies,, and J. Gallant (eds.), Cold Spring Harbor Laboratory, p. 185.

Casadaban, M. J. and Chou, J. (1984). *Proc. Natl. Acad. Sci. USA,* 81, 535.

Farrish, E. E., Baker II, H. V., and Wolf Jr., R. E. (1982). *J. Bacteriol.,* 152, 584.

Fraenkel, D. G. and Vinopal, R. T. (1973). *Ann. Rev. Microbiol.,* 27, 69.

Gausing, K. (1977). *J. Mol. Biol.,* 115, 335.

Ghysen, D., Iserentant, D., Derom, C., and Fiers, W. (1982). *Gene,* 17, 55.

Hall, M. N., Gabay, J., Debarbouille, M., and Schwartz, M. (1982). *Nature,* 295, 616.

Iserentant, D. and Fiers, W. (1980). *Gene,* 9, 1.

Jaurin, B., Grundstrom, T., Edlund, T., and Normark, S. (1981). *Nature,* 290, 221.

Jaurin, B. and Normark, S. (1979). *J. Bacteriol.,* 138, 896.

Maaløe, O. (1969). *Develop. Biol.,* Suppl., 3, 33.

Maaløe, O. and Kjeldgaard, N. O. (1966). *Control of Macromolecular Synthesis,* Benjamin.

Mamelak, L., Christensen, T., Fiil, N., Friesen, J. D., and Johnsen, M. (1984). This volume.

Miller, J. H. (1972). *Experiments in Molecular Genetics,* Cold Spring Harbor Laboratory, New York.

Mizuno, T., Chou, M-Y., and Inouye, M. (1984). *Proc. Natl. Acad. Sci. USA,* 81, 1966.

Nasoff, M. S., Baker II, H. V., and Wolf Jr., R. E. (1984). *Gene,* 27, 253.

Nomura, M. (1984). This volume.

Pedersen, S., Bloch, P. L., Reeh, S., and Neidhardt, F. C. (1978). *Cell,* 14, 179.

Pedersen, S. (1984). This volume.

Queen, C. and Rosenberg, M. (1981). *Cell,* 25, 241.

Schleif, R. (1967). *J. Mol. Biol.,* 27, 41.

Simons, R. W. and Kleckner, N. (1983). *Cell,* 34, 683.

Smith, M. W. and Neidhardt, F. C. (1983a). *J. Bacteriol.,* 154, 336.

Smith, M. W. and Neidhardt, F. C. (1983b). *J. Bacteriol.,* 154, 344.

Wolf Jr., R. E. and Fraenkel, D. G. (1974). *J. Bacteriol.,* 117, 468.

Wolf Jr., R. E. and Isturiz, T. (1975). *Proc. Natl. Acad. Sci. USA,* 72, 4381.

Wolf Jr., R. E., Prather, D. M., and Shea, F. M. (1979). *J. Bacteriol.,* 139, 1093.

Wolf Jr., R. E. and Shea, F. M. (1979). *J. Bacteriol.,* 138, 171.

From Prokaryotic to Eukaryotic Cells: Similarities in Transport Regulation

Herman M. Kalckar and Donna B. Ullrey
Boston University, Massachusetts

The symposium in genetics and molecular biology, honoring Ole Maaløe, addressed itself mainly to basic biological concepts as disclosed from advanced studies on prokaryotes. There were also a few reports on new observations on eukaryotic cells, including a transition case, classified between prokaryotes and eukaryotes (the Archaebacteria as studied by P. Dennis).

In discussing some interesting features regarding mediated regulation of hexose transport, it might not be too much out of place to report, as a start, some observations on transport regulations in conidia of *Neurospora crassa*. This so much the more since these studies represent, as far as we know, the first reports on a type of transport regulation that was later observed in cultured avian and mammalian fibroblasts.

During some studies of transport of hexoses into conidia of *Neurospora*, it was found that glucose starvation of the conidia greatly enhanced the rate of uptake of labeled glucose or some of the analogues (Scarborough, 1970; Neville et al., 1971). Replacement of glucose by fructose gave the same effect as glucose starvation, i.e., the rate of uptake of labeled glucose or glucose analogues was greatly increased (ibid., 1970, 1971). In fructose media, the transport of the glucose analogues, even that of 3-o-methyl-glucose (3OMG), was found to be active, i.e., collecting the ligand against a concentration gradient (Scarborough, 1970; Neville et al., 1971).

Interest in this type of transport regulation was greatly stimulated when it was found that cultured avian and mammalian fibroblasts also respond to glucose deprivation by a marked enhancement of their hexose transport (Martineau et al., 1972; Kalckar and Ullrey, 1973). The uptake or transport rates were monitored by ^{14}C-labeled glucose, 2-deoxyglucose or galactose (5 or 10 min uptake at 37 C), or by ^{14}C-labeled 3OMG transport tests over 10-20 seconds at 23 C. The rapid test with radioactive 3OMG was used in studies on chick embryo fibroblasts as well as on various hamster fibroblast cultures, especially in kinetics studies (Christopher et al., 1976a; Christopher, 1977; Ullrey et al., 1982).

A number of metabolizable D-aldohexoses (glucose, mannose, and D-glucosamine) are able, after 12 to 20 hrs incubation with cultured chick embryo or hamster fibroblasts, to down-regulate the hexose transport or uptake system very markedly (Ullrey et al., 1975; Christopher et al., 1976b,c). The transport step is essentially that of facilitated diffusion (Christopher et al., 1976a; Christopher, 1977); the uptake is coupled to phosphorylation and subsequent steps (ibid). The mediated down-regulation by glucose is perhaps best illustrated after a period of glucose starvation (using fructose or D-xylose as replacements for glucose) by refeeding of the cultures with glucose. The refeeding brings about a marked down-regulation of the hexose transport system. However, interference of this down-regulation can be brought about by inhibiting either protein synthesis (using cycloheximide) or oxidative energy metabolism (using 2,4-dinitrophenol, DNP). This is illustrated in Table 1.

The abolishment of the glucose-mediated down-regulation of hexose uptake during refeeding (Christopher et al., 1976b,c) is a striking phenomenon that we are trying to explore. The release of the down-regulation by addition of DNP, oligomycin, or malonate is a phenomenon of particular importance, since it indicates that the glucose-mediated down-regulation by the aldohexoses is an active process (Kalckar et al., 1979). In order to underline this, we have replaced the long word "down-regulation" with the short word "CURB."

At the symposium, Dr. Munch-Petersen described several interesting features regarding the regulation of the purine nucleoside transport system in some *E. coli* mutants. These mutants were of the so-called *unc* type (i.e., uncoupled with regard to aerobic glucose metabolism). The transport rates for purine nucleosides were found to be 2.5- to 3-fold higher in glucose-starved cells, whereas the cellular ATP levels were only 1/3 of those found in glucose-fed cells (Munch-Petersen and Pihl, 1980). This reverse relationship between the size of the ATP pool and regulation of a transport system seems similar to some of our recent observations on hamster fibroblast cultures, as summarized in Table 1 (modified from Rapaport et al., 1983).

Related examples of such reverse relationships can perhaps be found in earlier studies, both in prokaryotic cells (Reider et al., 1979) and in eukaryotic cells (cf. review by Gould, 1979).

We were able to obtain further clues as to the nature of the mediated down-regulation, or transport curb, from studies on a metabolic fibroblast mutant in which glucose is unable to contribute to energy metabolism, be it anaerobic or aerobic (Ullrey et al., 1982). The energy metabolism in this mutant is probably derived from L-glutamine, a major component in Dulbecco-modified Eagle (DME) medium. A comparison with the parental strain, which, unlike the mutant, is able to convert glucose-6-P to fructose-6-P (cf. Pouyssegur et al., 1980), showed the following differences in the patterns of the mediated transport curb (Ullrey et al., 1982).

In the parental line, D-glucosamine mannose, as well as glucose, was able to contribute to aerobic energy metabolism and the

TABLE 1   Hexose uptake rates and the size of cellular ATP pools

Medium	Uptake of [^{14}C]gal, nmol/mg protein/10 min	Total cell ATP pool, nmol/mg protein
Glucose	0.9	25
Glucose + NaF*	1.9	13
Glucose + DNP*	1.7	18
0	3.2	16

After rinsing, near-confluent NIL cultures were fed DME* containing 4 mM L-glutamine, 10% calf serum, and 22 mM glucose. The cultures were maintained for 20 hr at 37°C in the presence of 6.2 mM DNP or 0.5 mM NaF.

*NaF, sodium fluoride; DNP, 2,4-dinitrophenol; DME, medium without pyruvate.

concomitant mediated curb of the hexose transport system (Ullrey et al., 1982; see also Ullrey and Kalckar, 1981). In the mutant, called the PGI mutant (i.e., the mutant with the defective phosphoglucose isomerase, Pouyssegur et al., 1980; Ullrey et al., 1982), only glucose elicited a transport curb (see Table 2). Mannose was inactive in the mutant, and fructose was inactive in both lines (Kalckar and Ullrey, 1982).

It is evident from Table 2 that the mutant, unlike the parental 023 line, was unable to use D-glucosamine for the mediation of the transport curb. L-glutamine, being present in excess (4 mM), contributed presumably mainly to energy metabolism.

In this PGI mutant, glucose is unable to contribute to energy metabolism, but is able to contribute to the glucose-6-P dehydrogenase pentose shunt. It is therefore of interest that Gay and Amos (1983) have found that purine nucleotide precursors, and especially PRPP, seem to play an important role in the mediation of the hexose transport curb. Apparently the pentose shunt is crucial for the regulation, at least in chick embryo fibroblasts.

Table 2 also shows that when glucose brought about a marked transport curb in the mutant, malonate was able to release the curb. This type of release was also accomplished by addition of DNP (Ullrey et al., 1982). Since the PGI mutant is unable to generate oxidative metabolism from glucose, the energy was derived largely from L-glutamine (Ullrey and Kalckar, 1982). In the absence of L-glutamine, glucose was only able to "generate" a transport curb in the PGI mutant if supplemented with mannose, a hexose that upon phosphorylation can be converted to fructose-6-P by a specific phosphomannose isomerase (Ullrey and Kalckar, 1982), and this ester contributes to energy metabolism, along the Embden-Meyerhof-Krebs pathway (cf. Kalckar and Ullrey, 1984a,b). In fact, the PGI mutant,

TABLE 2 Catabolic down-regulation of 3-o-MeGlc transport by glucose and glucosamine in the *PGI⁻* mutant and its parental strain

Conditioning medium	3-o-[^{14}C]MeGlc transported, pmol/mg cell protein/20 sec*	
	DS-7(*PGI⁻*)	023(parental)
No hexose	13.0	17.8
No hexose + malonate	23.9	20.3
Glucose	3.6	4.6
Glucose + malonate	15.3	12.0
Glucosamine	10.7	5.7
Glucosamine + malonate	10.7	17.5
Fructose	15.2	18.2
Fructose + malonate	21.7	20.8

Concentrations in medium: glucose, 22 mM; glucosamine, 5 mM; fructose, 22 mM; malonate, 25 mM; DME medium without pyruvate with 45 mM L-glutamine; dialyzed fetal calf serum, 10%.
*Transport was measured at 23°C.

maintained in media devoid of L-glutamine, was unable to preserve its ATP pool in media with glucose, whereas mannose was highly effective (Plesner et al., 1985).

All the hexoses mentioned, with the exception of fructose (which is not an aldohexose), are ligands of the hexose transport system. In order to elicit a curb of this transport system, it seems that at least three requirements have to be fulfilled: (1) a hexose that is a ligand of the transport system must be present; (2) the glucose-6-P dehydrogenase pentose shunt must be operating; (3) oxidative energy metabolism must be generated. The development of the hexose transport curb was not found to be correlated with the capacity to generate lactic acid (Ullrey and Kalckar, 1982; Kalckar and Ullrey, 1984a).

Correlations between hexose metabolism and the transport curb still remain a puzzling question. Thus, it was recently reported that the nonmetabolizable analogue 3OMG, if incubated with human fibroblast cultures over 18 hr, provoked a marked transport curb (Germinario et al., 1985). We have found that another supposedly nonmetabolizable transport ligand, D-allose, also seems able to provoke a transport curb (Ullrey and Kalckar, 1982, unpublished). This was also independently found by Germinario et al. We have observed this resposne in the PGI mutant as well and checked that the response to allose is not an inhibition but an energy-depedent curb, since it is counteracted by addition of malonate (much like the glucose-mediated curb illustrated inb Table 2. Finally, high glycolysis, claimed to be one of the characteristic of tumorigenic animal cell lines, has been shown to be a nonessential factor in this respect. As shown by Franchi et al. (1981), the nonglycolytic PGI mutant has preserved a significant part of the tumorigenic capacity of the parental strain. This seemed particularly evident when the mutant was reclaimed in cell culture after an animal passage. The reclaimed cell cultures showed higher tumorigenic capacity, yet they still remained unable to generate any lactic acid from glucose (Franchi et al., 1981). Thus, the old hypothesis about the linkage between glycolysis and tumor growth seems to have lost most of its importance.

AFTERWORD

I am indeed thankful to the organizing committee for inviting me as a participant in the symposium celebrating Ole Maaløe's 70th birthday. My friendship with Ole dates back to 1946, when I returned to liberated Denmark. Ole and a couple of other very special Danish scholars helped me in their active and creative way to get over the cultural shock that inevitably hits you after 7 years' absence. Let me confine myself at this place to express my gratitude to Ole.

This work was supported by National Science Foundation Grants PCM 8021552 and 8302034.

REFERENCES

Christopher, C. W. (1977). *J. Supramol. Struct.*, 6, 485.

Christopher, C. W., Colby, W. W., and Ullrey, D. (1976c). *J. Cell. Physiol.*, 89, 683.

Christopher, C. W., Johnson, W. C., and Ullrey, D. (1975). *J. Gen. Physiol.*, 66, 20a.

Christopher, C. W., Kohlbacher, M. S., and Amos, H. (1976a). *Biochem. J.*, 158, 439.

Christopher, C. W., Ullrey, D., Colby, W. W., and Kalckar, H. M. (1976b). *Proc. Nat. Acad. Sci. USA*, 73, 2429.

Franchi, A., Silvestre, P, and Pouyssegur, J. (1981). *Inter. J. Cancer*, 27, 819.

Gay, R. J. and Amos, H. (1983). *Biochem. J.*, 214, 133.

Germinario, R. J., Change, Z., Manuel, S., and Olveira, M. (1985). *Biochem. Biophys. Res. Commun.*, 128, 1418.

Gould, M. K. (1979). *Trends Biochem. Sci.*, 4, 10.

Kalckar, H. M., Christopher, C. W., and Ullrey, D. (1979). *Proc. Nat. Acad. Sci. USA*, 76, 6453.

Kalckar, H. M. and Ullrey, D. (1973). *Proc. Nat. Acad. Sci. USA,* 70, 2502.

Kalckar, H. M. and Ullrey, D. (1984a). *Proc. Nat. Acad. Sci. USA,* 81, 1126.

Kalckar, H. M. and Ullrey, D. (1984b). *Fed. Proc.,* 43, 2242.

Martineau, R. M., Kohlbacher, M., Shaw, S. N., and Amos, H. (1972). *Prov. Nat. Acad. Sci. USA,* 69, 3407.

Munch-Petersen, A. and Pihl,J. (1980). *Proc. Nat. Acad. Sci. USA,* 77, 2518.

Neville, M. M., Suskind, S. B., and Roseman, S. (1971). *J. Biol. Chem.,* 246, 1294.

Plesner, P., Ullrey, D., and Kalckar, H.M. (1985). *Proc. Nat. Acad. Sci. USA,* 82, 2761.

Pouyssegur, J., Franchi, A., Salmon, M. C., and Silvestre, P. (1980). *Proc. Nat. Acad. Sci. USA,* 77, 2698.

Rapaport, E., Christopher, C. W., Svihovec, S., Ullrey, D., and Kalckar, H. M. (1979). *J. Cell. Physiol.,* 104, 229.

Reider, E., Wagner, E. F., and Schweiger, M. (1979). *Proc. Nat. Acad. Sci. USA,* 76, 5529.

Scarborough, G. A. (1970). *J. Biol. Chem.,* 245, 1694.

Ullrey, D. B., Franchi, A., Pouyssegur, J., and Kalckar, H. M. (1982). *Proc. Nat. Acad. Sci. USA,* 79, 3777.

Ullrey, D. B., Gammon, M. T., and Kalckar, H. M. (1975). *Arch. Biochem. Biophys.,* 167, 410.

Ullrey, D. B. and Kalckar, H. M. (1981). *Arch. Biochem. Biophys.,* 209, 168.

Ullrey, D. B. and Kalckar, H. M. (1982). *Biochem. Biophys. Res. Commun.,* 107, 1537.

Multigene Regulation and the Growth of the Bacterial Cell

Frederick C. Neidhardt
University of Michigan Medical School, Ann Arbor

INTRODUCTION

I left *Drosophila* genetics for good in the autumn of 1952, on the day I first constructed the growth curve of a bacterial culture. The precision with which bacterial growth could be measured and its beautiful simplicity fascinated me, and I have been studying it ever since.

That the growth curve was a resultant of complex factors, and could be used to extract useful information about the metabolism and physiology of cells was quickly demonstrated to me by my mentor, Boris Magasanik, who introduced me to the elegant work of Jacques Monod on diauxie (Monod, 1942; Monod, 1947; Monod, 1949). Indeed the central phenomenon of diauxie, the glucose effect, became the subject of my Ph. D. thesis, and my approach to this long-standing problem in bacterial and yeast physiology was rooted deeply in an analysis of growth response (Neidhardt and Magasanik, 1957).

This work brought me immediately to an interest in multigene regulation, i.e., to those aspects of gene expression that transcend the controls unique to individual genes or operons. That such controls existed seemed self-evident to me. In the glucose effect (catabolite repression, as we came later to call this complex control network) the presence of the inducing agent was clearly necessary to turn on many catabolic operons, but it was not sufficient to ensure their expression; each operon seemed to have a separate control, but also a control shared among them. The stringent response is clearly another instance of the response of otherwise independently controlled genes to a common, overriding regulator (Gallant, 1979). Many other examples have been subsequently discovered, including some within the protein-synthesizing system. Regulation of the 20 aminoacyl-tRNA synthetases, for example, was shown to include both individual and group controls (Neidhardt, Parker, and McKeever, 1975).

Overriding controls, not well defined biochemically, but of great potency and of great potential for integrating cell activities, clearly existed. How to identify and study them systematically was not at all certain until it became possible to monitor the expression of the total cellular genome during changes in growth conditions (by means of two-dimensional gel electrophoresis to resolve total cell protein) and to probe the genome for units that share behavioral patterns (by means, for example, of Mud*lac* fusions).

Thanks to these and related technical advances, it has been possible to learn that the bacterial cell possesses multigene networks ("global control systems" of unlinked, separately controlled genes that share some common regulatory signal and, in many instances, common regulatory genes) that it uses, for example, in coping with sudden stress. Nutrient depletion, or chemical or physical assault (temperature shifts, irradiation, osmotic shock, toxic solutes and solvents), is now known to activate one or another such network. Bacterial growth physiology has entered a new phase, one concerned explicitly with discovery and analysis of multigene systems (Gottesman and Neidhardt, 1983), such as the cyclic-AMP/CRP network, the SOS response, the phosphate starvation-induced response, and the heat-shock response.

Recent work on the reaction of *E. coli* to a shift-up in temperature illustrates this new direction. [A few key references are given; for a complete list the reader should consult the recent review by Neidhardt, VanBogelen, and Vaughn (1984)].

THE HTP REGULON

Upon a shift from a temperature in the midrange for *E. coli* growth, 30 C, to 40 C or higher, many proteins are abruptly induced. Seventeen of them are called heat-shock proteins by analogy with the response to heat of other organisms. Their synthesis accelerates manyfold within seconds, the magnitude depending on the magnitude of the temperature shift. Peak synthesis rates are reached in 5-10 minutes, and by 20 minutes the induction has largely subsided, except upon a shift to very high temperature. At 46 C, which permits balanced growth in rich medium, the 17 proteins constitute nearly 15% of the protein mass of the cell. At 50 C, where net growth is not possible, *E. coli* exhibits a nearly exclusive synthesis of the heat-shock proteins.

The genetic unit of regulation in multigene control systems is the *regulon*—two or more operons that share a common regulatory molecule (e.g., a protein activator or repressor)—and the 17 heat-induced genes of *E. coli* have been shown to constitute a regulon, which is called the HTP (high temperature production) regulon. Their shared regulatory molecule is protein F33.2, the product of a gene, *htpR*, that maps at 76 minutes. This protein is required as a positive effector of the heat-shock response.

The *htpR* gene has been cloned (Neidhardt, VanBogelen, and Lau, 1983) and sequenced (Landick, Vaughn, Lau, VanBogelen, Erickson, and Neidhardt, 1984). The predicted amino acid sequence of its protein product is homologous to the predicted sequence of the C-terminal half of the *E. coli* RNA polymerase sigma factor (43% exact matches plus conservative replacements). This high degree of structural similarity has functional significance, because mutations in *htpR* can be suppressed by some mutations in *rpoD*, the gene for sigma factor, and vice versa (A. G. Grossman, Y. N. Zhu, C. A. Gross, G. E. Christie, J. S. Heilig, R. Calendar; and A. G. Grossman, Y. N. Zhu, T. Baker, and C. A. Gross, cited in Neidhardt, VanBogelen, and Vaughn, 1984). Also, the cellular levels of sigma inversely affect the magnitude of the heat-shock response (Yamamori, Osawa, Tobe, Ito, and Yura, 1982). Recent information from *in vitro* systems indicates that F33.4, in the absence of sigma factor, can direct transcription initiation at heat-shock promoters (Grossman, Erickson, and Gross, 1984). The five of these that have been sequenced share a number of features that may contribute to their recognition by HtpR (Neidhardt, VanBogelen, and Vaughn, 1984). It appears, therefore, that the heat-shock response in *E. coli* involves a molecular mechanism—re-programming RNA polymerase by a substitute sigma factor—that has been implicated in the heat-shock response of *Drosophila* (Parker and Topol, 1984) and in the sporulation of *Bacillus subtilis* (Losick and Pero, 1981), but which has been hitherto unknown in *E. coli*. Many important details remain to be worked out, such as how heat turns on the heat-shock sigma factor, and how one of the heat-shock genes, *dnaK*, is involved as a modulator of the response (Tilly, McKittrick, Zylicz, and Georgopoulos, 1983).

The 17 heat-shock proteins are readily visualized on two-dimensional gels. Aided by the extensive work underway to identify all the gene products of *E. coli* on these gels (cf. Neidhardt, Vaughn, Phillips, and Bloch, 1983), seven heat-shock polypeptide spots have been identified as known proteins or the products of previously characterized genes. They are a heterogeneous lot, consisting of GroEL and GroES (products of the *groE* or *mop* operon), DnaK and DnaJ (products of the *dnaK* operon), sigma factor itself (product of the *rpoD* gene of a complex operon), Lon (product of the *lon* or *capR* gene), and LysU (an isozyme of lysyl-tRNA synthetase and product of the *lysU* gene).

These heat-shock genes are scattered on the genetic map, but with some clustering in the northwest quarter. Their protein products are functionally diverse: DnaK and DnaJ are implicated in DNA replication; sigma factor is involved in RNA synthesis; LysU is involved in protein synthesis; Lon is implicated in protein degradation; and GroES and GroEL are implicated in protein processing and assembly (reviewed in Neidhardt, VanBogelen, and Vaughn, 1984). In short, virtually all of the major macromolecular processes of the cell have ties with the heat-shock response.

It is a striking fact that the biological function of the heat-shock response is only a point for conjecture at this moment in any organism, in spite of the astonishing finding that its elements appear, with unprecedented genetic conservation, in all three of the major divisions of living things, prokaryotes, eukaryotes, and archaebacteria (cf. Bardwell and Craig, 1984). Perhaps tracking down the remaining heat-shock proteins in *E. coli* will uncover the meaning of the HTP regulon. If so, *E. coli* will once again have proved its utility as a subject for biological inquiry. Even if the answer in *E. coli* proves unique to the biology of this organism (which seems highly unlikely), its solution will be important to growth physiology, because the heat-shock response is absolutely necessary for the growth and survival of *E. coli*.

THE ROAD TO COPENHAGEN

My work as a graduate student uncovered the fact that the growth-supporting ability of a substrate could be used to predict its ability to mimic glucose in preventing "adaptation" to alternative carbon sources (Neidhardt and Magasanik, 1957). A whole host of questions were raised by this observation: What determines the growth rate of bacterial cells in a given medium? How is the efficiency and high rate of growth on diverse substrates and under diverse chemical and physical conditions achieved? How is gene expression coordinated with overall growth rate? My fascination with this outwardly simple behavior of cells led me deep into the forest of growth physiology, along a path that led through Paris to Copenhagen.

A postdoctoral year (1956-1957) with Jacques Monod and Francois Gros at the Pasteur Institute (Neidhardt and Gros, 1957) led me directly, back at Harvard with Boris Magasanik, to analyze the variation of RNA level with growth rate in *Klebsiella aerogenes* (Neidhardt and Magasanik, 1960), a study that paralleled the independent work on *Salmonella typhimurium* by Elio Schaechter, Niels Ole Kjeldgaard, and Ole Maaløe in Copenhagen (Schaechter, Maaløe, and Kjeldgaard, 1958; Kjeldgaard, Maaløe, and Schaechter, 1958).

The work and ideas on the variation of the Protein Synthesizing System (PSS) with growth rate developed in Copenhagen reinforced our own ideas and had a large influence on my subsequent work. In Elio Schaechter's words, we were all destined to do "a lot of shifty experiments" in the decade of the 60s, learning about the "irritability" of RNA synthesis. A sabbatical leave with Ole in Copenhagen (1968-1969) capped off this research. This visit was for me a scientific high point. It led, then and later, to friendships and shared ideas with Ole, Niels Ole, Elio, and the many disciples of Ole, including Martin Pato, Lasse Lindahl, Kirsten Gausing, Nils Fiil, Kaspar von Meyenburg, Jim Friesen, Solvejg Reeh, and Steen Pedersen. The latter two played a major role in Ann Arbor in initiating current studies of multigene regulation (regulon analysis) by adapting the O'Farrell two-dimensional gel system to quantitative work with *E. coli* (cf. Reeh, Pedersen, and Neidhardt, 1977; Pedersen, Bloch, Reeh, and Neidhardt, 1978).

The central contribution of the Copenhagen school to bacterial growth physiology is sometimes thought of in terms of individual pieces of work, such as the economy of balanced growth brought about by the adjustment of the amount of the PSS to match growth conditions. For me it has always been more — a way of thinking about global coordination of cellular processes, and a style of experimentation, which have been an inspiration to my own studies and which have led to much of the pleasure I have found in exploring cell growth.

ACKNOWLEDGEMENTS

Many past and current studies, postdoctoral fellows, and co-workers played major roles in the work described in this chapter. The work on the heat-shock regulon was supported by a grant from the National Institutes of Health (GM-17892). The project on identification and mapping of gene products was supported by a grant from the National Science Foundation (PCM 8207190).

REFERENCES

Bardwell, J. C. A. and Craig, E. A. (1984). *Proc. Natl. Acad. Sci. USA*, 81, 848.

Gallant, J. (1979). *Ann. Rev. Genetics*, 13, 393.

Gottesman, S. and Neidhardt, F. C. (1983) In *Gene Function in Prokaryotes*, J. Beckwith, J. Davies, and J. A. Gallant (eds.), Cold Spring Harbor Laboratory, p. 163

Grossman, A., Erickson, J. W., and Gross, C. A. (1984). *Cell*, 38, 383.

Kjeldgaard, N. O., Maaløe, O., and Schaechter, M. (1958). *J. Gen. Microbiol.*, 19, 607.

Landick, R., Vaughn, V., Lau, E. T., VanBogelen, R. A., Erickson, J. W., and Neidhardt, F. C. (1984). *Cell*, 38, 175.

Losick, R. and Pero, J. (1981). *Cell*, 25, 582.

Monod, J. (1942). *Recherches sur la Croissance des Cultures Bactériennes*, Hermann and Cie. Paris.

Monod, J. (1947). *Growth*, 11, 223.

Monod, J. (1949). *Ann. Rev. Microbiol.*, 3, 371.

Neidhardt, F. C. and Gros, F. (1957). *Biochim. Biophys. Acta*, 25, 513.

Neidhardt, F. C. and Magasanik, B. (1957). *J. Bacteriol.*, 73, 260.

Neidhardt, F. C. and Magasanik, B. (1960). *Biochim. Biophys. Acta*, 42, 99.

Neidhardt, F. C., Parker, J., and McKeever, W. G. (1975). *Ann. Rev. Microbiol.*, 29, 215.

Neidhardt, F. C., VanBogelen, R. A., and Lau, E. T. (1983). *J. Bacteriol.*, 153, 597.

Neidhardt, F. C., VanBogelen, R. A., and Vaughn, V. (1984). *Ann. Rev. Genetics*, 18, 295.

Neidhardt, F. C., Vaughn, V., Phillips, T. A., and Bloch, P. L. (1983). *Microbiol. Revs.*, 47, 231.

Parker, C. S. and Topol, J. A. (1984). *Cell*, 37, 273.

Pedersen, S., Bloch, P. L., Reeh, S., and Neidhardt, F. C. (1978). *Cell*, 14, 179.

Reeh, S., Pedersen, S., and Neidhardt, F. C. (1977). *J. Bacteriol.,* 129, 702.

Schaechter, M., Maaløe, O., and Kjeldgaard, N. O. (1958). *J. Gen. Microbiol.,* 19, 592.

Tilly, K., McKittrick, N., Zylicz, M., and Georgopoulos, C. (1983). *Cell,* 34, 641.

Yamamori, T., Osawa, T., Tobe, T., Ito, K., and Yura, T. (1982) In *Heat Shock from Bacteria to Man,* M. J. Schlesinger, M. Ashburner, and A. Tissières (eds.), Cold Spring Harbor Laboratory.

The Pho Regulon of *Escherichia coli*

Annamaria Torriani and Douglas N. Ludtke
Massachusetts Institute of Technology; Cambridge, Massachusetts

INTRODUCTION

Elucidation of the mechanisms "which cooperate to produce the relationship between growth conditions and the development of the Protein Synthetic System" has been the goal of Maaløe (1979). His work dealt with cells which have reached an equilibrium with the medium and are growing at the maximum rate allowed by the established equilibrium. However, in exponential phase of growth the cell expresses only 30% of its chromosomal capacity. Seventy per cent of the chromosome is in storage for special needs, such as changes of medium composition, phage infection, changes of temperature or of oxygen concentration, SOS system, and survival systems at starvation (amino acids, phosphates). During these conditions bacteria are in "Overdrive." This appropriate metaphore was introduced by Gorini to explain the extra power a cell can mobilize to overcome conditions of stress and to describe the dramatic increase of enzyme synthesis due to starvation for a specific amino acid (L. Gorini and W. K. Maas, 1957). Similarly, when *E. coli* senses an inorganic phosphate (Pi) deficiency, it shifts to "overdrive," and a number of proteins and enzymes are rapidly synthesized. This provides the cell with the capability to survive by utilizing other sources of Pi and by increasing the efficiency of Pi uptake.

An overall view of metabolism leading to chemical synthesis of *E. coli* from glucose, shown in Figure 1, suggests that the very first requirement in order to obtain the fueling reactions (ATP) leading to the formation of the components of *E. coli* is Pi. Since Pi is essential and ubiquitous, we can expect a very complex regulatory mechanism to control the entry of Pi from the medium and the internal level of Pi in the cell. In this short review we will consider only one system: the pho regulon of *E. coli*.

THE PHO REGULON

The term regulon was first proposed by Maas and Clark (1964) for the arginine biosynthetic pathway to describe "a system in which the production of the enzymes (of metabolic pathways) can be controlled by a single repressor substance." In the pho regulon the single repressor substance is Pi. The genes of the pho regulon are those genes that are negatively regulated by the concentration of Pi in the medium and activated by a positive effector PhoB, the product of *phoB* gene.

A simple schematic cross-section of an *E. coli* cell (Figure 2) shows that the genes of the pho regulon are scattered around the genome, and their products are found in all fractions of the cell: PhoE porin in the outer membrane, the binding proteins for Pi and glycerol-3-phosphate (G3P) and hydrolases (alkaline phosphatase) in the periplasmic space, and components of active transport systems for Pi and G3P in the cytoplasmic membrane, as well as regulatory proteins in the cytoplasm.

Components in the outer membrane

The phosphate available in the medium as Pi, G3P, or linear polyphosphates (poly(Pi)) can enter the periplasm through pores in the outer membrane called matrix proteins or porins. These funnellike structures have a very wide specificity and allow diffusion of low-molecular-weight (800-900) hydrophilic nutrients (amino acids, carbohydrates, and salts) across the outer membrane into the periplasmic space (Lugtenberg and Van Alphen, 1983).

In *E. coli* K12 when Pi is abundant in the medium, it diffuses through the porin proteins OmpC and OmpF. These two proteins are antigenically related and have very similar molecular weights and amino acid sequences. These porins are not regulated by Pi but by changes in the osmolarity of the medium, by cAMP, and by the control elements encoded by the genes *envZ* and *ompR* (Garrett et al., 1983). Because Pi can enter the cell via OmpC and OmpF, it was astonishing to find that during Pi-limiting conditions these cells produce another very similar porin protein: PhoE (Van Alpern et al., 1978; Foulds and Chai, 1978; Argast and Boos, 1980; Overbeeke et al., 1983). Korteland et al., (1982) and Overbeeke and Lugtenberg (1982) found that a cell deprived of inorganic phosphate can utilize large linear polyphosphates (up to 200 units long), because they can enter the periplasm through the induced PhoE porin. This porin differs from OmpC and OmpF in that it is particularly efficient for negatively charged solutes like cefsulodin (a beta-lactam antibiotic) and polyphosphates, but is rather inefficient for Pi.

Components in the periplasmic space

The region between the outer membrane and the cytoplasmic membrane has a unique solute composition different from the cytosol and from the medium. This region is observed only in Gram-negative bacteria and is called periplasmic space. Sequestered within the periplasmic space are proteins with three known functions: hydrolytic enzymes, which serve obvious roles in nutrient acquisition; binding proteins for amino acids, carbohydrates, sulfur, and Pi, which are part of

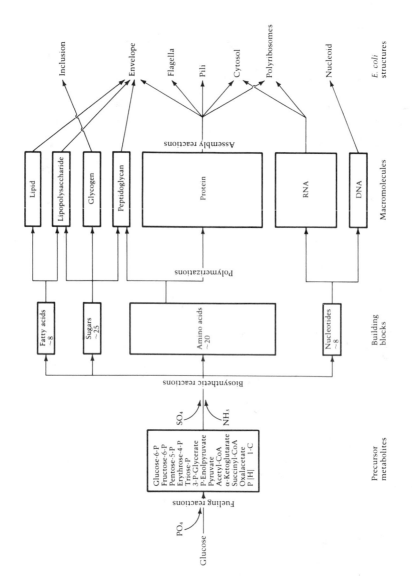

FIGURE 1 Overall view of metabolism leading to production of an E. coli cell from glucose. The size of each box is proportional to need. (From Growth of the Bacterial Cell, Ingraham, Maaløe, and Neidhardt, 1983, p. 51.)

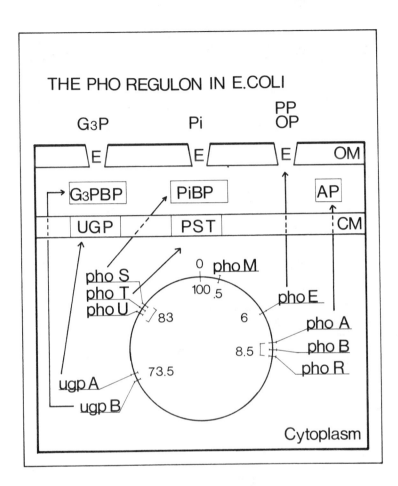

FIGURE 2 Schematic cross-section of an E. coli cell. Abbreviations: PP, polyphosphates; OP, orthophosphate; G3P, sn-glycerol-3-phosphate; OM, outer membrane; CM, cytoplasmic membrane; E, porin E; AP, alkaline phosphatase; BP, binding proteins (Pi or G3P); PST, phosphate-specific transport system; UGP, G3P uptake.

the active transport systems of these ligands; and chemoreceptors, which enable the cell to sense nutrients in the environment and move toward them. All the periplasmic proteins have the common characteristic of being exported through the cytoplasmic membrane. The mechanism of protein secretion has been the subject of many studies that have been reviewed by Michaelis and Beckwith (1982) and by Silhavy et al. (1983). Among the 30-50 proteins secreted into the periplasmic space there are at least 4 whose synthesis is known to be regulated also by Pi. These periplasmic proteins were first observed by Morris et al. (1974) and were called P1, P2, P3 and P4;

later they were also observed by Argast and Boos (1980) and called GP1, GP2, GP4, and GP3, respectively. The P1 protein was identified as alkaline phosphatase, the product of the *phoA* gene (Morris et al., 1974). P2 was found to be the G3P-binding protein (Argast and Boos, 1979) encoded by the *ugpB* gene (Schweizer et al., 1982b). P4, also known as protein R2a (Garen and Otsuji, 1964), was identified as the Pi-binding protein encoded by *phoS* (Gerdes and Rosenberg, 1974; Yagil et al., 1976). The P3 protein has not yet been identified.

One of the hydrolytic enzymes in the periplasmic space is alkaline phosphatase, which is synthesized when the cell is starved of Pi (Torriani, 1958, 1960; Horiuchi T. et al., 1959; Horiuchi, S., 1959). The protein was shown to be exported (Malamy and Horecker, 1961) as an inactive precursor with subunits of molecular weight 55,000 and with a hydrophobic N-terminal signal sequence (Inouye and Beckwith, 1977). These subunits polymerize in the periplasm and bind zinc to form an active protein of MW 110,000 (Schlesinger et al., 1969). The active precursor is processed to a protein of MW 96,000 by a proteolytic enzyme that copurifies with the outer membrane fraction (Inouye and Beckwith, 1977). This location for the active phosphatase may be useful to the cell to protect the cytoplasmic phosphomonoesters from hydrolysis and to provide Pi to the cell by hydrolysis of phosphorylated monoesters present in the medium.

The Pi liberated in the periplasmic space is captured by another exported protein, the Pi-binding protein (Garen and Otsuji, 1964; Medveczy and Rosenberg, 1970; Gerdes and Rosenberg, 1974; Morita et al., 1983), whose structural gene, *phoS*, is induced when Pi becomes exhausted in the medium. This gene maps at 83.5 min and has a negative effect on the pho regulon. It has recently been sequenced and shown to have its own promoter and to constitute a single-cistron operon (Magota et al., 1984; Surin et al., 1984). It was shown by Amemura et al. (1982) that overproduction of the Pi-binding protein represses the expression of *phoA* in cells grown in low-Pi medium as well as in *phoR*-constitutive mutants.

A novel transport system for G3P was described (Argast and Boos, 1980) as being co-regulated with *phoE* and other genes of the pho regulon. Two genes, *ugpA* and *ugpB*, which code for components of this G3P transport system, have been mapped at 75.3 minutes (Schweizer et al., 1982a). Mutants in *ugpA* synthesize the periplasmic G3P-binding protein but cannot transport G3P, and *ugpB* mutants do not make the binding protein. Therefore, the binding protein encoded by *ugpB* was proven necessary but not sufficient for the transport of G3P by this system. Recently, a regulatory region needed for the expression of the G3P-binding protein has been cloned (Schweiser and Boos, 1983) The *ugp* system does not transport Pi.

Other periplasmic proteins have been described that are regulated in a different manner from those of the pho regulon. A poly(Pi) depolymerase (phosphoanhydride phosphohydrolase) with a pH optimum of 2.5 cleaves the terminal phosphate specifically from GTP and ppGpp as well as from poly(Pi) (Dassa and Boquet, 1981). This enzyme is induced by Pi starvation during aerobic growth but is induced even in the presence of excess Pi during anaerobic growth. Induction of this enzyme is not controlled by the *phoB* gene product but is negatively controlled by cAMP (Dassa et al., 1982). Three polyphosphatases localized in the outer leaf of the cytoplasmic membrane and released in the periplasmic space have been described by Nesmeyanova et al. (1975a). These enzymes are induced by Pi starvation and positively controlled by *phoR* and *phoU* (Nesmeyanova et al., 1975b). Another periplasmic phosphatase, an acid phosphatase with pH optimum of 4.5, is synthesized constitutively and is not induced by Pi deprivation (Dvorak et al., 1967; Hofstein and von Porath, 1962; Torriani, 1960). Another periplasmic enzyme induced by Pi limitation is the aminopeptidase N but its synthesis is not dependent on *phoB* (Bally et al., 1983).

Components in the cytoplasmic membrane

Pi is actively transported across the cytoplasmic membrane by two major systems: phosphate inorganic transport (Pit) and phosphate specific transport (Pst) (Willsky et al., 1973, Willsky and Malamy, 1980a,b; Zuckier and Torriani, 1981). The Pit system is encoded by only one gene, *pit*, mapping at 77 minutes, and energy required to drive transport is derived from the proton motive force. This system is expressed constitutively and is not controlled by Pi. Strain K10 and its derivatives are *pit* and depend on the Pst system for Pi transport (Sprague et al., 1975).

The Pst system uses phosphate-bound energy such as ATP to transport Pi across the membrane (Rosenberg et al., 1979). The gene encoding the membrane component of the active transport systems appear to be *pstA*. The mutant *pst-2* of Sprague et al. 1975 is probably *pstA*, as are *phoT9* (Amemura et al., 1982), *pho32*, and *phoT34* (Surin, Rosenberg, and Cox 1984). It was demonstrated that a *pit pstA* (*pit pst2*) mutant is incapable of utilizing Pi, and this mutant requires organic phosphate such as G3P which is transported into the cell by the *glpT*-encoded system (Zuckier et al., 1980). Other genes of the Pst system like *phoS* and *phoU* (formerly called *phoT35*) have functions other than Pi active transport, since double mutant strains, *pit phoS* or *pit phoU* are able to utilize Pi (Zuckier and Torriani, 1981). These mutant strains were indeed *pit*, because they could not accumulate arsenate, a better substrate for the Pit system than of the Pst system, (Rae and Strickland, 1975), and were arsenate resistant. Pit[+] cells are arsenate sensitive when the ratio of arsenate to Pi is 10 to 1.

Surin, Rosenberg and Cox (1985) have recently completed the nucleotide sequence of the Pst region. They found that this DNA encodes four open reading frames. These genes map at 83.5 min, and the order established by the DNA sequencing is *phoU pstB pstA pstC* with all genes transcribed counterclockwise. Mutations are available in these genes, except *pstC*, and these provoke constitutive synthesis of the pho regulon. The amino acid composition, derived from the DNA sequence of the polypeptides encoded by these genes, indicated that *pstC, pstB*, and *phoU* are peripheral membrane proteins. The results of Surin, Rosenberg and Cox also show that the *pstA* gene product is an integral membrane protein. Levits et al. (1984) reported that a mutant (*phoS64*) of the Pst system, unable to synthesize the Pi-binding protein, is located distal to other *phoS* mutations (*phoS64 phoU pstB*) and belongs to a different complementation group. It has been suggested to call this gene *phoV*.

Other components of the regulon

Looking at Figure 1, we can expect that many genes will fall directly or indirectly under Pi control. To investigate this question Wanner and McSharry (1982) used a Mud1 (*bla*⁺*lacZ*) phage to direct the fusion of *lacZ* to phosphate-regulated promoters (Casadaban and Cohen, 1979). The insertion of this phage into a gene renders it nonfunctional. The phage insertion in the chromosome is detected by ampicillin resistance in the recipient cell. The expression of *lacZ*, devoid of its promoter, depends on the promoter of the gene in which the phage is inserted. By using a chromogenic substrate of beta-galactosidase it was easy to detect fusion mutants in which the enzyme synthesis was controlled by Pi. Some 18 promoters inducible by Pi starvation (*psiA* to *psiR*) were fused to *lacZ* and roughly mapped on the *E. coli* chromosome. They are located randomly throughout the chromosome. It is quite obvious that 18 is a minimum number, since the insertion of Mu into an essential gene would be a lethal event. Furthermore, insertion into Pi regulatory genes may abolish the regulation and would not be identified. Several of the *psi* genes were also found to be induced by carbon or nitrogen starvation, as well as by some of the *pho* regulatory genes. Two genes *psiF* and *psiG* are closely linked to *proC*, and the gene order is *phoA psiF proC psiG phoB phoR* (Wanner et al., 1981). However, *psiF* is not regulated by *phoB*, and hence does not constitute an operon with *phoA*. The *psi0-lacZ* fusion is induced by Pi, carbon or nitrogen starvation and during anaerobiosis, but is not controlled by the *pho* regulatory-gene products. This analysis elucidates the complexities of global control of Pi metabolism and the fact that only some of the Pi-regulated genes belong to the *pho* regulon. Neidhardt (personal communication) observed that *E. coli* K12 produces 86 new proteins upon Pi starvation.

EFFECTORS OF THE PHO REGULON

Some 26 years ago in the laboratory of J. Monod in Paris the senior author of this paper investigated the effect of depletion of various nutrients on the production of amylomaltase and lactase (now beta-galactosidase). The levels of these enzymes were compared to the levels of other enzymes produced constitutively, such as acid phosphatase. *E. coli* was grown in a chemostat (Monod, 1950; Novick and Szilard, 1950) in a medium with a low concentration of Pi. The use of the chemostat allowed the growth rate to be controlled by changing the rate of input of fresh medium. At very low growth rates Pi was the growth-limiting factor. The level of acid phosphatase with a pH optimum of 4.5 was not affected by the level of Pi, but unexpectedly another enzyme, alkaline phosphatase with a pH optimum at 8.2, was produced when Pi was growth limiting (Torriani, 1958, 1960). The same enzyme was observed by S. Horiuchi (1959) and T. Horiuchi (1959) in Japan while studying the amount of RNA in *E. coli* cells depleted of Pi. These original findings generated a vast array of studies.

Soon after the discovery of alkaline phosphatase and the mapping of its structural gene, *phoA* (Levinthal, 1959), a genetic study of its regulation was begun. Mutant strains were isolated in which alkaline phosphatase synthesis were either uninducible (Garen and Echols, 1962a) or constitutive (Torriani and Rothman, 1961; Echols et al.,

1961; Garen and Echols, 1962a,b), which suggested positive and negative regulation. These mutations mapped in two very distinct regions of the *E. coli* chromosome (9 and 83.5 min.) (Echols et al., 1961; Garen and Echols, 1962b; Garen and Garen, 1963). These regions were originally called R_1 (Garen and Echols, 1962a, b) and R_2 (Garen and Otsuji, 1964), respectively.

PhoB mutations

Mutations at the R_1 site display three different levels of *phoA* expression: R_{1a} (*phoRa*) mutants are constitutive but not fully induced (Echols et al., 1961; Garen and Echols, 1962); R_{1b} (*phoRb*) mutants are constitutive and fully induced (Echols et al., 1961); and R_{1c} (*phoRc*) mutants are uninducible (Garen and Echols, 1962). Bracha and Yagil (1973) isolated mutants similar to *phoRc*, which they called *phoB*. Fine-structure mapping of this region (Kreuzer et al., 1975) indicated that *phoRa* and *phoRb* represent a single cistron, called *phoR*, which is linked but different from *phoRc*, called *phoB*. Brickman and Beckwith (1975) proved genetically that *phoB* encodes an activator protein that is necessary for the induction of the *phoA* gene, as proposed by Morris et al. (1974). Experiments by Inouye et al. (1977) and Pratt (1980) demonstrated the requirement of the *phoB*-gene product for the induction of *phoA* expression *in vitro*.

Although mutant *phoRc* (such as *phoB63*) and *phoB* (*phoB23*) belong to the same cistron (Pratt and Torriani, 1976), they have different phenotypes. The *phoRc* mutants are uninducible for alkaline phosphatase and for the G3P-binding protein (Willsky and Malamy, 1976). The *phoB* mutants (Yagil, 1975) cannot induce these proteins nor the Pi-binding protein (Yagil et al., 1976) or PhoE (Tommassen and Lugtenberg, 1980). Recently, a mutant that maps near *phoB* was selected as a fluorouracil-plus-adenosine-resistant derivative of a *upp* strain (Heyde and Portalier, 1983). Five periplasmic proteins, including alkaline phosphatase and Pi-binding protein, were missing from the periplasm of these mutants. A more complete genetic analysis is required to confirm that this mutant is an allele of the *phoB* gene.

A mutant of *phoA* (*pho1003* (Bin)) was isolated (Wanner et al., 1979) that is independent of the *phoB* gene product and synthesizes alkaline phosphatase constitutively. This mutation was later identified as a change in a single base pair in the promoter region of *phoA* (Inouye et al., 1982). Thus, the product of *phoB* may act by binding to the promoter of *phoA,* but direct evidence substantiating this hypothesis has not been reported. The *phoB* gene product has been identified in maxicells and recognized by gel electrophoresis as a protein with a molecular weight of 31,000 (Tommassen et al., 1982a; Makino et al., 1982). The product of the *phoB* gene induces not only *phoA* but all the other genes of the pho regulon (Figure 2).

phoM and phoR mutations

The *pho* regulon is also under the positive and negative control of the *phoR* gene product and positively controlled by the *phoM* gene product. The products of the *phoR* and *phoM* genes are proteins with apparent molecular weights of 47,000 (Tommassen et al., 1982) and between 55,000 and 60,000 (Ludtke et al., 1984; Makino et al., 1984; Tommassen et al., 1984), respectively. Genetic results suggest that the *phoR* gene product is a multimeric protein with three or four subunits (Pratt and Gallant, 1972). In the majority of the *phoR* mutant strains (like *phoR$_2$*, now *phoR68*), alkaline phosphatase is expressed constitutively but at a low level. However, strains carrying the *phoR$_3$* (*phoR69*) allele express alkaline phosphatase constitutively but at a high level (Echols et al., 1961). This finding suggested the double role of *phoR* as a positive and a negative effector. Thus, *phoR69* may have lost the negative regulatory function. More recent experiments (Wanner and Letterell, 1980; Wanner, 1983) confirmed the double effect of *phoR* product. Clarification came from the identification of a new positive effector of *phoA*, the *phoM* gene product (Wanner et al., 1979).

A *phoR$^+$phoM* strain is inducible like the wild type strains, but the *phoR68 phoM* strain is uninducible like *phoB* strains. This suggests that *phoR68* has lost both the positive and negative functions of the *phoR* gene. The positive function can be replaced by *phoM$^+$*, as in the *phoR68* strain. However, a *phoR69* strain, which synthesizes alkaline phosphatase constitutively at a high level, does not require the *phoM* gene product: the *phoR69 phoM* strain has the same level of alkaline phosphatase as the *phoR69 phoM$^+$* strain. Thus, *phoR69* has lost the negative function and retained the positive one. Studies in our laboratory that determined the levels of alkaline phosphatase produced by isogenic strains carrying the *phoR* and *phoM* mutant alleles summarize these findings (Table I). Wanner and Letterell (1980) and Tommassen et al. (1982) presented a scheme in which *phoR* affects

TABLE 1   Regulation of alkaline phosphatase

Relevant genotype	Specific activity of alkaline phosphatase	
	High Pi	Low Pi
Wildtype	0.1	38
phoB	0.1	0.1
phoM	0.1	36
phoR68(C$_2$)	15	12
phoR68 phoM	0.1	0.1
phoR69(C$_3$)	36	37
phoR69 phoM	35	39

Strains are isogenic except for mutations in the *pho* genes. Wildtype is W3110 F^-. The cultures were grown overnight in MOPS medium with 4 mg/ml glucose and 2 mM (high) or 0.1 mM (low) inorganic phosphate (Pi). Levels of activity are expressed as nmol of *p*-nitrophenol produced by 1 ml of toluenized cells (4×10^8) in 1 min.

phoB expression as a repressor if bound to Pi or as an activator when free of Pi. In these schemes the regulatory effect of Pi is two steps removed from the genes of the regulon. We will discuss in the next section some results suggesting that Pi is not the cofactor that acts with the product of *phoR* gene to repress *phoB* expression.

The positive function of *phoM* is not yet clearly defined. *PhoM* is a poor inducer of *phoB* in comparison to *phoR69* and may act indirectly on *phoB* expression. This is supported by the observation (Ludtke, unpublished data) that the level of alkaline phosphatase remains low even when PhoM is produced from a $phoM^+$ multicopy plasmid.

CONTROL OF THE EFFECTORS OF THE REGULON

The technique of gene fusion (Casadaban, 1976) facilitated the study of the control of operons with unknown or nonselectable functions. The control sequences of a gene of interest is fused to the structural gene of an enzyme which is easy to assay. In the majority of cases the *lacZ* gene, encoding beta-galactosidase, is fused to the promoter of interest, for instance *phoB* (Makino et al., 1982). To study the regulation of *phoB* by Pi, *phoR, phoS, phoT,* and itself, the amount of beta-galactosidase produced by a *phoB-lacZ* fusion was measured in high- and low-Pi medium and in these mutant-strain backgrounds by Shinagawa et al. (1983). The results showed that *phoB* is in fact controlled by *phoR,* and by Pi, as suggested by Wanner and Letterell, 1980) and by Tomassen and Lugtenberg (1982). Guan et al. (1983) fused *phoB* to *cat* (chloramphenicol transacetylase) and confirmed these results, as well as providing evidence that *phoB* is autoregulated. Thus, *phoB* may be indirectly controlled by Pi via *phoR.*

There is reason to believe that *phoR* may not bind to Pi but binds to some small molecules to form the inducer of *phoB.* Evidence for this inference is that the intracellular level of Pi does not change for at least four hrs after Pi is depleted in the medium whereas *phoA* induction is seen within 15 minutes (Wilkins, 1972). *In vitro* experiments have shown that $phoA^+$ is induced only by the *phoB* gene product and is insensitive to Pi and to the products of the other control genes tested (*phoR, phoS, phoU*). Some difficulties encountered in the reproducibility of alkaline phosphatase synthesis *in vitro*, under conditions in which β-galactosidase was always synthesized, lead us to believe that a dialyzable cofactor molecule was required for *phoA* induction (Pratt, 1980). Reconstruction experiments with fractionated S30 ribosomal extracts could regenerate the capacity to synthesize β-galactosidase but not alkaline phosphatase, again suggesting a special need of a cofactor molecule. The dialyzable cofactor may be a nucleotide (Pratt, unpublished data). Experiments have been reported in which the accumulation of an adenine derivative (Wilkins, 1972) caused the synthesis of alkaline phosphatase in high Pi media.

Variations in the nucleotide pools may very well influence the entire pho regulon. The nucleotide pool may vary for many reasons, such as amino acid starvation, anaerobiosis, the growth cycle (exponential or stationary due to Pi starvation), and changes of the levels of phosphate reserve molecules (like polyphosphates) used for the synthesis of ATP during Pi starvation. Thin layer chromatographic

(TLC) analysis (Bochner and Ames, 1982) of the pool of nucleotides in *E. coli* during exponential phase and during phosphate starvation may reveal some interesting changes that could be related to the inducibility of the pho regulon. A comparative analysis of wildtype and various mutants strains of the pho regulon may better correlate the results. The TLC patterns of ^{32}P-labeled nucleotides from the stationary phase of wildtype Pi-starved cells (PhoA induced) and from *phoU* mutants in the exponential or stationary phase of growth (PhoA constitutive) are remarkably similar. The patterns differ by seven nucleotides from the pattern of the wildtype cells growing exponentially (PhoA repressed). The seven specific nucleotides can induce the synthesis of alkaline phosphatase in wildtype plasmolyzed cells, in spite of the high concentration of Pi in the *in vitro* mix (Pratt, 1980). We are in the process of identifying these nucleotides (Rao, Wang, Yashphe, and Torriani, 1985, in preparation).

Villarejo et al (1983) have observed an additional form of regulation of synthesis of alkaline phosphatase that becomes evident in *phoR* mutant strains. As described previously, the expression of the pho regulon in *phoR68* mutant strains is uncoupled from phosphate control and depends on *phoM*-gene-product activation of the *phoB* gene, which encodes the positive regulator for the system. Growth of such *phoT* mutant strains in M63-glycerol medium in which the 100 mM K-PO$_4$ buffer component has been reduced to 50 mM or 25 mM leads to lower intracellular levels of alkaline phosphatase or of beta-galactosidase if a *phoA-lacZ* protein fusion is used. The conclusion of these workers was that *phoA* expression is dependent on the osmolarity of the growth medium.

We have reproduced these results but find that the reduction in *phoA* expression correlates wtih the reduced buffering capacity of the medium rather than the osmolarity (Pahel and Torriani, in preparation). This interesting phenomenon deserves further attention since the growth conditions employed also affect the expression of the non-phosphate-regulated outer membrane porin genes, *ompC* and *ompF*.

Using a *phoM-lacZ* protein fusion, it was found that *phoM* expression was not affected by changes in Pi levels in the media or by mutations in either the *phoB* or *phoR* genes (Ludtke et al., 1984) (A. Nakata, personal communication). The expression of *phoM* is increased threefold in a *phoU* mutant background. A possible function of *phoM* was suggested by the results of M. Villarejo (Villarejo et al., 1983) showing the effect of osmolarity of the media on the level of alkaline phosphatase in a constitutive strain (*phoR68 phoM+*). We have investigated the possibility that this difference was due to changes in the expression of *phoM*. The level of *phoM* expression, was reduced 50% in media of low buffering capacity (NB) in comparison to the level found in media of high buffering capacity (M63) (Ludtke, unpublished results). Therefore, the effect of reduced buffering capacity on *phoA* expression may be the result of a change in *phoM* expression.

As seen in Table 2, strains with a mutation in any one of the genes encoding the PST system (*phoS, phoT* (*pstA*) or *pstB*) or in the nearby *phoU* gene, express *phoA* constitutively at very high levels. Therefore, the products of these genes must in some manner repress expression of the pho regulon. Of these genes it is known that *phoS* expression is induced by Pi limitation (Willsky and Malamy, 1976) and *phoU* expression is slightly higher in cells grown in excess Pi and in *phoB* or *phoR phoM* mutant strains (Nakata, personal

TABLE 2   Effect of concentration of inorganic phosphate
on the activity of alkaline phosphatase synthesized by various *E. coli* mutants

Relevant genotype▼	Activity of alkaline phosphatase*	
	High Pi	Low Pi
Wildtype	0.09	38.5
phoS	73.1	74.6
phoT (pstA)	131.8	112.8
phoU	82.2	70.4

*Levels of activity are expressed as nmol p-nitrophenol
produced per ml of 4×10^8 toluenized cells per min.
▼Strains are isogenic except for mutations in the pho
regulon. The wildtype strain is MC4100. Cultures were
grown overnight in MOPS medium, as described in Table 1.

communication). Control of *phoT* (*pstA*) and *pstB* expression has not
yet been established; however, recent results of Surin et al. (personal
communication) suggest that they are repressed by Pi and regulated
by *phoB*.

The regulatory pathways controlling the pho regulon are still
very sketchy. In agreement with existing data (Table 2), we suggest
that a reaction modifies a dialyzable factor, possibly a nucleotide,
from a coinducer "X" to a corepressor "Y." These factors could act
together with the product of the *phoR* gene to induce or repress the
expression of *phoB* (Figure 3). During conditions in which an excess
Pi is present in the medium, an enzyme (perhaps the product of the
phoU gene) changes the coinducer "X" to a corepressor "Y." When
Pi levels become growth limiting or when any one of the gene encoding
the Pst systems (*phoS, phoT* (*pstA*), *pstB,*) or the *phoU* genes are
mutated, the level of coinducer "X" increases, because it is no longer
converted to the corepressor "Y." Induction of *phoB* expression by
the proposed PhoR "X" complex would result in PhoB induction of the
pho regulon.

POLYPHOSPHATES AS ENERGY RESERVES

A number of years ago we observed that *phoU* mutants, constitutive
for synthesis of alkaline phosphatase, gave rise to uninducible mutants
at a high frequency when stored in stabs. It was established that
the *phoU* population was replaced by a *phoU phoB* population. This
phenomenon was reproduced under anaerobic conditions in rich medium
(Penassay broth, but not in LB or synthetic medium). Zuckier and
Torriani (1981) showed that the *phoU* mutants accumulated poly(Pi)
and that *phoU phoB* mutants did not. They argued that the high
synthesis of poly(Pi) in the *phoU* strain was deleterious to the cells.
Thus, two genes *phoU* and *phoB* of the regulon appear to regulate
the synthesis of poly(Pi).

A MODEL OF THE PHO REGULON

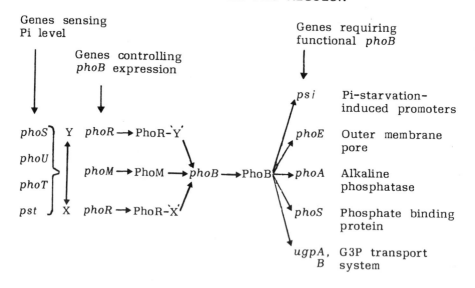

FIGURE 3   A model for the control of the pho regulon.

E. coli strain K10 is known to be a poor producer of poly(Pi) during aerobic growth. However, the addition of Pi to Pi-starved cells or a rapid shift to either anaerobiosis, or into media deprived of sulfur causes these cells to accumulate poly(Pi) (Kulaev and Vagabov, 1983). Recent studies showed that a *phoU* mutant strain accumulates up to five times as much poly(Pi) as a wildtype strain if suddenly shifted to anaerobiosis and confirmed that *phoB* or *phoU phoB* mutant strains do not. The accumulation reaches a peak after 6 hrs of anaerobiosis; then, the poly(Pi) molecules are gradually degraded (Rao et al., 1985). Nuclear magnetic resonance analysis of poly(Pi) in extracts from cells grown anaerobically showed the presence of both short-chain (15-25 phosphate units) and long-chain (more than 200 phosphate units) polymers. Short-chain polymers predominated in the cell extracts of cultures grown for six hours anaerobically (6-hour culture), whereas long-chain polymers appeared in cultures grown anaerobically for 24 hours (24-hour cultures). ^{31}P NMR spectra of intact cells indicated that neither aerobically grown wildtype *E. coli* K10 nor a *phoU* mutant strain had any appreciable amount of polyphosphates. On the contrary, ^{31}P NMR signals for polyphosphates were strong for an anaerobically grown *phoU* strain and were stronger in the 6-hour culture than in the 24-hour culture. The above observations were verified on poly(Pi), extracted, purified and analyzed by gel filtration with Sepharose-4B columns (Rao et al, 1985).

The synthesis of poly(Pi) by an ATP-Mg^{2+} kinase has been described (Kornberg et al., 1956). *In vitro* this enzyme synthesizes large molecules of poly(Pi) with lengths in excess of 200 units. During preliminary studies of this kinase, we found it to be synthesized at a time when poly(Pi) was being degraded. We are investigating the hypothesis that *in vivo* this kinase works in reverse to utilize accumulated poly(Pi) to make ATP needed during anaerobiosis (Rao and Torriani, unpublished).

There is no doubt that the *phoU* and *phoB* genes play a role in the metabolism of poly(Pi) (Nesmeyanova et al., 1975a; Zuckier et al., 1980), but their function is far from being understood at this time.

The studies of polyphosphate metabolism in *E. coli*, though extensive (Kulaev and Vagabov, 1983), are rather incomplete. The available results suggest that under conditions of Pi starvation *E. coli* can obtain Pi from poly(Pi) present in the medium via the induced porin PhoE. These poly(Pi) molecules, containing up to 200 phosphate units, will be hydrolyzed to Pi in the periplasm by polyphosphatases (Nesmeyanova, 1975b). Recent studies indicate that the presence of alkaline phosphatase is vital and that porin PhoE is useful but not essential for the utilization of poly(Pi) in the medium (Mikich and Torriani, unpublished results). On the other hand, linear poly(Pi) containing as many as 1600 monomers can be synthesized in the cytoplasm by the ATP-Mg^{2+} kinase (Kornberg et al., 1956). These poly(Pi) molecules may be used in the cytoplasm in certain reactions in place of ATP (viz., reactions involving polyphosphate kinase; poly(Pi) glucokinase) or merely serve as a phosphate reserve. These hypotheses need testing. The long-chain linear poly(Pi) molecules we have isolated and purified are those synthesized in the cytoplasm and are being used as substrates for studies of a cytoplasmic polyphosphatase. Our preliminary results show that *phoB* regulates this polyphosphatase.

CONCLUSION

The pho regulon includes a minimum of 20-24 genes that share a common positive regulatory element, *phoB*. The *phoB* gene in turn is positively regulated by two genes, *phoM* and *phoR*, and is also negatively controlled by the *phoR* gene product (Shinagawa et al., 1983; Guan et al., 1983) (see Figure 3). One hypothesis to explain the double role of PhoR has been that it is a repressor if bound to Pi and an inducer when free of Pi (Garen and Garen, 1963; Morris et al., 1974; Wanner and Latterell, 1980; Tommassen et al., 1982). We have listed some reasons why we do not believe that PhoR binds Pi. Rather we hypothesize that a cofactor "X" may bind to PhoR to form the inducer, PhoR-X. The cofactor may be a phosphorylated derivative, whose level is controlled by extracellular levels of Pi, as suggested by Wilkins (1972) and by our current studies (Torriani, unpublished data). It is possible that the components of the PST system, encoded by the *phoS, phoT* and *pst* genes, function as sensors of the extracellular Pi concentration. This would be analogous to the function that components of other transport systems play as chemoreceptors (Adler and Epstein, 1974). The product of the *phoU* gene may be an enzyme that changes "X" to a "Y," or it may regulate the gene encoding this enzyme.

If *phoR* is inactivated by mutation, the product of *phoM* can act, presumably directly, upon the promoter of *phoB* to induce *phoB* expression. The level of *phoB* expression, and therefore expression of the pho regulon, resulting from PhoM induction is not as high as that resulting from PhoR "X" induction (Table 1 and Table 2). It is not clear what role PhoM plays in the cell, because levels of induction of the pho regulon in wildtype and *phoM* mutant strains are indistinguishable (see Table 1).

To study their regulation many of the genes of the regulon have been cloned (*phoA, phoB, phoR, phoS, phoU, phoE, phoM*) and some have been sequenced (*phoA, phoS, phoE*). Investigations to determine if the genes regulated by *phoB* have similar promoter sequences indicate that there is a consensus sequence that may be the site of binding of the *phoB* gene product (Surin et al., 1984; Magota et al., 1984).

Why *E. coli* needs so many proteins to overcome Pi starvation and why the regulation is so complex are matters for speculation. Obviously the global metabolism of the cell requires Pi, and it is not surprising to find that about 3% (86 genes) of the genome is regulated by Pi. About 20 genes constitute the pho regulon, but 40 to 60 additional genes controlled by Pi may be found in other regulons. Maaløe suggested (1983) that such a set of regulons should be called a *stimulon*.

To study the expression of the 70% of genome that is silent in cells growing rapidly in equilibrium with their surroundings, the chemostat is a valuable instrument. Continuing growth equilibria can be established at various levels of a single limiting factor. Maaløe's experiments dealt with cell doublings per hour between 0.5 and 2.0, while the alkaline phosphatase is induced at a doubling time of 10 hours when phosphate is the growth-limiting factor (Torriani, 1960). Analysis of cultures at equilibrium at various levels of Pi may reveal "survival signals" (or alarmones) specifically linked to this limiting factor.

Thus, Project K, the complete solution of *E. coli* (EMBO meeting on developments in Molecular Biology and the EMBO Laboratory Project, November 27-30, 1969), suggested by Crick as an exhaustive analysis of one cell (*E. coli* K12) still has a long way to go!

ACKNOWLEDGEMENTS

Work carried out in our laboratory is supported by U. S. Public Health Service Grant GM-24009.

REFERENCES

Adler, J. and Epstein, W. (1974). *Proc. Nat. Acad. Sci. USA,* 71, 2895.

Amemura, M., Shinagawa, H., Makino, K., Otsuji, N., and Nakata, A. (1982). *J. Bact.,* 152, 692.

Anraku, Y. (1964). *J. Biol. Chem.,* 239, 3413, 3420.

Argast, M. and Boos, W. (1979). *J. Biol. Chem.,* 254, 10931.

Argast, M. and Boos, W. (1980). *J. Bact.,* 143, 142.

Bally, M., Murgier, M., and Lazdunski, A. (1983). *FEMS Microbiol. Lett.,* 19, 261.

Bochner, B. R., and Ames, B. W. (1982). *J. Biol. Chem.,* 257, 9759.

Bracha, M. and Yagil, E. (1973). *Mol. Gen. Genet.,* 122, 53.

Brickman, E., and Beckwith, J. (1975). *J. Mol. Biol.*, 95, 307.

Casadaban, M. J. (1976). *J. Mol. Biol.*, 104, 541.

Casadaban, M. J. and Cohen, S. N. (1979). *Proc. Nat. Acad. Sci. USA*, 76, 4530.

Cox, G. B., Rosenberg, H., Downie, J. A., and Silver, S. (1981). *J. Bact.*, 148, 1.

Dassa, E. and Boquet, P. L. (1981). *FEBS Letters*, 135, 148.

Dassa, E., Cahu, M., Desjsyaux-Cherel, B., and Boquet, P. L. (1982). *J. Biol. Chem.*, 257, 6669.

Dvorak, H. F., Brockman, R. W., and Heppel, L. A. (1967). *Biochem.*, 6, 1743.

Echols, H., Garen, A., Garen, S., and Torriani, A. (1961). *J. Mol. Biol.*, 3, 425.

Foulds, J. and Chai, T. J. (1978). *J. Bact.*, 133, 1478.

Garen, A. and Echols, H. (1962a). *J. Bact.*, 83, 297.

Garen, A. and Echols, H. (1962b). *Proc. Nat. Acad. Sci. USA*, 48, 1398.

Garen, A. and Garen, S. (1963). *J. Mol. Biol.*, 6, 433.

Garen, A. and Otsuji, N. (1964). *J. Mol. Biol.*, 8, 841.

Garrett, S., Taylor, R. K., and Silhavy, T. J. (1983). *J. Bact.*, 156, 62.

Gerdes, R. G. and Rosenberg, H. (1974). *Biochem. Biophys. Acta*, 351, 77.

Gorini, L. and Maas, W. K. (1957). *Biochim. Biophys. Acta*, 25, 208.

Guan, C. D., Waner, B. D., and Inouye, H. (1983). *J. Bact.*, 156, 710.

Heyde, M. and Portalier, R. (1983). *Mol. Gen. Genet.*, 190, 122.

Hofsten, B. and von Porath, J. (1962). *Biochim. Biophys. Acta*, 64, 1.

Horiuchi, T., Horiuchi, S., and Mizuno, D. (1959). *Nature*, 183, 1529.

Horiuchi, S. (1959). *Japan. J. Med. Sci. Biol.*, 12, 429.

Inouye, H. and Beckwith, J. (1977). *Proc. Nat. Acad. Sci. USA*, 74, 1440.

Inouye, H., Pratt, C., Beckwith, J., and Torriani, A. (1977). *J. Mol. Biol.*, 110, 75.

Inouye, H., Barnes, W., and Beckwith, J. (1982). *J. Bact.*, 149, 434.

Kornberg, A., Kornberg, S. R., and Simms, E. (1956). *Biochim. Biophys. Acta*, 20, 215.

Korteland, J., Tommassen, J., and Lugtenberg, B. (1982). *Biochim. Biophys. Acta*, 690, 282.

Kreuzer, K., Pratt, C., and Torriani, A. (1975) *Genetics*, 81, 459.

Kulaev, I. S. and Vagabov, V. M. (1983). *Adv. Microb. Physiol.*, 24, 83.

Levinthal, C. (1959). *Brookhaven Symp. Biol.*, 12, 76.

Levitz, R., Klar, A., Sar, N., and Yagil, E. (1984). *Mol. Gen. Genet.*, 197, 98.

Ludtke, D. N., Bernstein, J., Hamilton, C., and Torriani, A. (1984). *J. Bact.*, 159, 19.

Lugtenberg, B. and van Alpen, L. (1983). *Biochim. Biophys. Acta*, 737, 51.

Maaløe, O. (1979). In *Biological Regulation and Development*, R. F. Goldberg (ed.), Plenum, Vol. 1.

Maaløe, O. (1983). In *Growth of the Bacterial Cell*, J. Ingraham, O. Maaløe, and F. C. Neidhardt. Sinauer.

Maas, W. and Clark, A. J. (1964). *J. Mol. Biol.*, 8, 365.

Makino, K., Shinagawa, H., and Nakata, A. (1982). *Mol. Gen. Genet.*, 187, 181.

Makino, K., Shinagawa, H., and Nakata, A. (1984). *Mol. Gen. Genet.*, 195, 381.

Magota, K., Otsuji, N., Miki, T., Horiuchi, T., Kondo, J., Sakiyama, F., Amemura, M., Morita, T., Shinagawa, H., and Nakata, A. (1984). *J. Bact.*, 157, 909.

Malamy, M. and Horecker, B. L. (1961). *Biochem. Biophys. Res. Commun.*, 5, 104.

Medveczy, N. and Rosenberg, H. (1970). *Biochim. Biophys. Acta*, 211, 158.

Michaelis, S. and Beckwith, J. (1982). *Ann. Rev. Microbiol.*, 36, 435.

Michaelis, S., Inouye, H., Oliver, D., and Beckwith, J. (1983). *J. Bact.*, 154, 366.

Monod, J. (1950). *Ann. Inst. Pasteur*, 79, 390.

Morita, T., Amemura, M., Makino, K., Shinagawa, H., Magota, K., Otsuji, N., and Nakata, A. (1983). *Eur. J. Biochem.*, 130, 427.

Morris, H., Schlesinger, M. J., Bracha, M., and Yagil, E. (1974). *J. Bact.*, 119, 583.

Nesmeyanova, M. A., Maraeva, O. B., Severin, A. I., and Kulaev, I. S. (1975a). *Doklady Akad. Nauk. USSR*, 223, 1266.

Nesmeyanova, M. A., Gonina, S. A., and Kulaev, I. S. (1975). *Doklady Akad. Nauk. USSR*, 224, 710.

Novick, A. and Szilard, L. (1950). *Science*, 112, 715.

Overbeeke, N. and Lugtenberg, B. (1982). *J. Biol. Chem.*, 126, 113.

Overbeeke, N., Bergman, H., van Mansfield, F., and Liugtenberg, B. (1983). *J. Mol. Biol.*, 163, 513.

Pratt, C. and Gallant, J. (1972). *Genetics*, 72, 217.

Pratt, C. and Torriani, A. (1977). *Genetics*, 85, 203.

Pratt, C. (1980). *J. Bact.*, 143, 1265.

Rae, A. S. and Strickland, K. P. (1976). *Biochim Biophys Acta*, 433, 564.

Rao, N. N., Roberts, M. F., and Torriani, A. (1985). *J Bact*, 162, 242.

Rosenberg, H., Gerdes, R. G., and Chegwidden, K. (1977). *J Bact.*, 131, 505.

Rosenberg, H., Gerdes, R. G., and Harold, F. M. (1979). *Biochem J.*, 178, 133.

Schlesinger, M. J., Reynolds, J. A., and Schlesinger, S. (1969). *Ann. NY Acad. Sci.*, 166, 368.

Shinagawa, H., Makino, K., and Nakata, A. (1983). *J. Mol. Biol.*, 168, 477.

Schweizer, H., Argast, M., and Boos, W. (1982a). *J. Bact.*, 150, 1154.

Schweizer, H., Grussenmeyer, and Boos, W. (1982b). *J. Bact.*, 150, 1164.

Schweizer, H. and Boos, W. (1983). *Mol. Gen. Genet.*, 192, 177.

Silhavy, T. J., Spencer, B., and Emr, S. D. (1983). *Microbiol. Rev.*, 47, 313.

Sprague, G. F., Bell, R. M., and Cronan Jr., J. E. (1975). *Mol. Gen. Genet.*, 143, 71.

Surin, B. P., Jans, D. A., Fimmel, A. L., Shaw, D. C., Cox, G. B., and Rosenberg, H. (1984). *J. Bact.*, 157, 772.

Surin, B. P., Rosenberg, H., and Cox, G. B. (1985). *J Bact.*, 161, 189.

Tommassen, J. and Lugtenberg, H. (1980). *J. Bact.*, 143, 151.

Tommassen, J., de Genes, P., Lugtenberg, B., Hackett, J., and Reeves, P. (1982). *J. Mol. Biol.*, 157, 265.

Tommassen, J., Himstra, P., Overduin, P., and Lugtenberg, H. (1984). *Mol. Gen. Genet.*, 195, 190.

Torriani, A. (1958). *Fed. Proc.*, 18, 33.

Torriani, A. (1960). *Biochim. Biophys. Acta*, 38, 460.

Torriani, A. and Rothman, F. (1961). *J. Bact.*, 81, 835.

van Alphen, W., van Selm, N., and Lugtenberg, B. (1978). *Mol. Gen. Genet.*, 159, 75.

Villarejo, M., Davis, J. L., and Granett, S. (1983). *J. Bact.*, 156, 975.

Wanner, B. L., Sarthy, A., and Beckwith, J. (1979). J. Bact., 140, 229.

Wanner, B. L., Wieder, S., and McSharry, R. (1981). *J. Bact.*, 146, 93.

Wanner, B. L. and Letterell, P. (1980). *Genetics*, 96, 353.

Wanner, B. L. and McSharry, R. (1982). *J. Mol. Biol.*, 158, 347.

Wanner, B. L. (1983). *J. Mol. Biol.*, 166, 283.

Wilkins, A. (1972). *J. Bact.*, 110, 616.

Willsky, G. R. and Malamy, M. (1976). *J. Bact.*, 127, 595.

Willsky, G. R. and Malamy, M. (1980a). *J. Bact.*, 144, 356.

Willsky, G. R. and Malamy, M. (1980b). *J. Bact.*, 144, 366.

Yagil, E. (1975). *FEBS Letters*, 55, 124.

Yagil, E., Silberstein, N., and Gerdes, R. G. (1976). *J. Bact.*, 127, 656.

Zuckier, G. and Torriani, A. (1981) *J. Bact.*, 145, 1249.

Zuckier, G., Ingenito, E., and Torriani, A (1980). *J. Bact.*, 143, 934.

Recovery from Near-Ultraviolet Radiation Damage in Bacteria

Abraham Eisenstark
University of Missouri, Columbia

Many of us have been influenced by a central theme of Ole Maaløe's research: when cells are stressed by an environmental shift, a switch takes place in the "off-on" activities of a battery of enzymes.

Perhaps two of the best-described stress inductions are those (1) in which a heat shock will turn on synthesis of about a dozen new enzymes while simultaneously shutting off others, and (2) in which specific DNA lesions, such as those produced by far-ultraviolet radiation (FUV; 254 nm), will evoke the synthesis of repair enzymes (SOS response). In our current studies, we are comparing the responses to these stresses to that of near-ultraviolet radiation (NUV; 300-400 nm) and/or H_2O_2.

Specifically, we would like to know the overlaps and differences vis-a-vis NUV, heat-shock and SOS responses. Although DNA of bacterial cells may also be damaged by NUV (300-400 nm), the effects of NUV stress differ from (but also partially overlap) the responses following heat shock or FUV stress. For example, NUV lesions do not signal the SOS response; indeed, such damage by NUV will block subsequent induction of the SOS response by FUV. Thus, NUV stress obviously evokes signals that differ from FUV stress.

We have also found that H_2O_2 is a photoproduct of NUV (McCormick et al 1976), and unless specifically noted, the effects of NUV and H_2O_2 may be considered interchangeable. However, it should be kept in mind that NUV has effects in addition to generating H_2O_2, and these will be discussed. Also, we have recently found overlaps between NUV and heat stress. For example, in our electrophoresis gels some (but not all) of the new bands following NUV stress are at the same positions as some new bands following heat stress. This report summarizes our knowledge of the chain of events whereby NUV photons (300-400 nm) strike photoreceptors of bacterial cells, produce hydrogen peroxide and oxygen radicals, lead to DNA and other cellular

243

damages, and ultimately evoke recovery from such damage. Also, it summarizes overlaps between metabolic shifts upon NUV, FUV, or heat stresses.

NEAR-ULTRAVIOLET RADIATION STUDIES: BACKGROUND

Soon after it was observed that NUV is lethal and mutagenic for bacteria (Eisenstark, 1971), it became evident that the biological effects of NUV irradiation are unique; i.e., they are distinctly different from effects of FUV or alkylating agents (Table 1). This uniqueness of NUV has become a focus of research concerning those events that lead to DNA damage and subsequent recovery following exposure to NUV and H_2O_2 (a photoproduct of NUV). Photoeffects on other components (e.g., membranes, certain tRNA molecules, endogenous photosensitizers) will be described only briefly, though it is recognized that these photochemical events may also influence DNA metabolism (Jagger, 1983; Lee et al., 1983). Some key points in our knowledge of the NUV effects may be summarized as follows:

1. *Ubiquity of NUV biological effects.* Many microbial, plant and animal (including human) biochemical activities may be cued, modulated, damaged, and induced to synthesize enzymes by NUV radiation. Some of the mechanisms utilize common (or related) chromophores, photochemical reactions, and/or biomolecular alterations. Thus, photoreceptor and photobiological studies in any one system could lead to an understanding of fundamental NUV photo-processes in other living systems.

2. *DNA lesions.* Some of the same DNA damages of FUV irradiation have been observed for NUV (e.g., pyrimidine dimers, DNA-protein crosslinks, and DNA breaks). Despite these qualitative similarities in photoproducts, the biological effects are strikingly different; FUV photoproducts may act as effectors to activate RecA protease, but NUV photoproducts may not (Turner and Eisenstark 1984; Eisenstark, 1983; Turner, 1983).

3. *Hydrogen peroxide is a photoproduct of NUV.* We observed that the amino acid, tryptophan, when irradiated with NUV, yields a photoproduct (H_2O_2) that is lethal for *rec* mutants (Yoakum et al., 1975; McCormick et al., 1976). We have since found that H_2O_2, plus certain wavelengths of NUV, synergistically kill bacteria and phage (Hartman et al., 1979; Hartman and Eisenstark, 1978; Ananthaswamy et al., 1979). We have evidence that this synergistic action is probably the result of NUV conversion of H_2O_2 to O_2^- (Ahmad, 1977), and this action can generate a new NUV photoreceptor (McCormick, et al. 1982). Thus, NUV not only generates H_2O_2, it also converts it to O_2^-, and may then act on a new photoreceptor created by O_2^- action. This knowledge of H_2O_2 involvement, together with our knowledge of DNA damage by direct exposure to H_2O_2 (Ananthaswamy and Eisenstark, 1977), will allow us to utilize *xthA, polA, kat, nur,* and *sod* mutants (known mutants specifically sensitive to NUV) to sort out the chemical nature of NUV lesions and their relationship to H_2O_2 damage.

TABLE 1 Differences between far and near UV damage in bacteria

Characteristic	Far UV (260 nm)	Near UV (300–405 nm)	Reference*
DNA degradation in recA⁻	Yes	No	9
Mutation enhancement	High	Limited	6,20
Oxygen demand for killing	Low	High	5
Division delay	Low	High	13
Liquid-holding recovery	Yes	No	22
Induction of SOS response	Yes	No	18
β-gal synthesis in recA::lac	Yes	No	18
Weigle reactivation	Yes	No	18
Chromophore for lethality	Pyrimidine	Unknown	14
Log cells more sensitive than stationary cells	Slight	Great	5,10
β-gal and tryptophanase inhibition	No	Yes	13
Sensitivity of nur⁻	Low	High	21
Sensitivity of xthA⁻	Low	High	17
Sensitivity of polA⁻	Low	High	9
Production of DNA-protein links	Low	High	8,12
Phage injection blocked	No	Yes	11,12
Single-strand breaks in xthA⁻	Yes	No	18
Sensitivity of M. radiodurans	Resistant	High	4,23
Sensitivity of S. typhimurium Q1	Low	High	7
Enhancement of resistance and mutation by pKM101	High	None	7
Protection by radical traps	Low	High	1,16
Inhibition of membrane transport	No	Yes	15
Synergistic action with nonlethal doses of H_2O_2	No	Yes	2,3,12
Phage produced in UV'd host	Yes	No	6,13

*(1) Ahmad, 1981; (2) Ananthaswamy et al., 1979; (3) Buzard, 1980; (4) Caimi, 1983); (5,6,7) Eisenstark, 1970, 1971, 1983; (8) Eisenstark et al., 1982; (9) Ferron et al., 1972; (10,11) Hartman and Eisenstark, 1978, 1982; (12) Hartman et al., 1979; (13) Jagger, 1983; (14) McCormick et al., 1982; (15) Moss and Smith, 1981; (16) Peak et al., 1982; (17) Sammartano and Tuveson, 1983; (18) Turner, 1983; (19) Turner and Eisenstark, 1984; (20) Turner and Webb, 1981; (21) Tuveson and Jonas, 1979; (22) Tuveson and Violante, 1982; (23) Webb et al., 1978.

H_2O_2 has an important role in NUV stress; this is evident since catalase in plating media will reduce the lethal effect of NUV (Tuveson, personal communication and our own observations). Interestingly, we have found that if cells are stressed by a number of agents that we have tried, including heat, catalase in the medium will give higher colony counts (unpublished); H_2O_2 may be a common product of generalized cellular stress (Ames, 1983; Lee et al., 1983).

4. *NUV-sensitive mutants.* NUV research was sparked by the recent finding that *xthA* mutants (coding for ExoIII enzyme) are unable to repair damage by H_2O_2, though such mutants are no different from wildtype in repair of FUV damage and in SOS response (Demple, et al., 1983). Recognizing that H_2O_2 is a photoproduct of NUV, Sammartano and Tuveson (1983) noted that such *xthA* mutants are also unable to repair NUV damage. Furthermore, the types of DNA breaks normally found in NUV damage of wildtype cells do not occur in *xthA* mutants (Turner, 1983; Eisenstark and Turner, 1984). Knowledge of the involvement of ExoIII has revived earlier interest in the roles of the genes *kat* (catalase) (Demple and Halbrook, 1983; Loewen, 1984; Eisenstark and Turner, 1984; Finn and Condon, 1975), *sod* (superoxide dismutase) (Moody and Hassan, 1982; Touati, 1983; Ahmad, 1981), and *nur* (product unknown; Tuveson and Jonas, 1979) in cellular response to stress.

Mutants in the *polA* gene demonstrate particular sensitivity to H_2O_2 and NUV (Ananthaswamy and Eisenstark, 1977; Morimyo, 1982). The *polA* mutant that is deficient in the 5'-to-3' exonuclease activity (but not 3'-5') of DNA polymerase I (PolI) is specifically sensitive to NUV, but this sensitivity can be overcome by avoiding O_2 toxicity (e.g., anaerobic growth, addition of quenchers of oxygen radicals; unpublished). In summary, mutations in the *E. coli* genes *nur, kat, sod, polA,* and *xthA* in *E. coli* produce sensitivity to H_2O_2, indicating that they are specifically involved in recovery from NUV damage. Genes controlling other aspects of DNA metabolism also have roles.

5. *Inducible DNA repair.* Gel electrophoresis of extracts of cells labeled with [^{35}S]methionine after irradiation with NUV shows a number of new protein bands (unpublished; Peters and Jagger, 1981). However, the proteins have not yet been identified or shown to be associated with repair of NUV damage. The *kat* and *sod* genes are inducible by both H_2O_2 (Demple and Halbrook, 1983) and NUV (Sammartano and Tuveson, 1983; Eisenstark and Turner, 1984). Neither *polA, nur,* nor *xthA* are inducible (Eisenstark and Turner, 1984).

6. *Physiological effects.* NUV also inhibits respiration, membrane transport, and a number of enzyme activities (Wagner and Snipes, 1982; Moss and Smith, 1981; Kelland et al., 1983). Although the fluences used may not lead to lethality, the subtle biological effects may have important implications. A number of other complex physiological effects of NUV have been described, a subject that has been critically reviewed by Jagger (1983).

7. *tRNA chromophore.* Recent studies (Jagger 1983; Thomas and Favre, 1980) concerned with growth delay in *E. coli* show that a thiolated tRNA absorbs radiation at 334 nm, resulting in the

demodulation of the stringent control of protein synthesis. Certain mutants (*rel*) lack this particular tRNA, and therefore 334 nm radiation fails to exhibit growth delay in these strains.

8. *Action spectra*. In detailed studies of action spectra biological effects on bacteria and phage occur in the range ca. 320-340 nm, quite far from the DNA absorption peak of 260 nm (Ananthaswamy et al., 1979; Mackay et al., 1976; Hartman, 1981). This observation raises the question whether DNA is actually the chromophore for monochromatic NUV, or whether energy is instead transmitted to the DNA by another molecule. Studies with transforming DNA do not answer this question unequivocally, especially since sensitivity to NUV certainly cannot be accounted for by formation of pyrimidine dimers alone (Hartman et al., 1979). The question of other (photoreceptor) molecules binding tenaciously to transforming DNA is difficult to eliminate (Jagger 1983).

9. *Oxygen dependence*. NUV lethality (as well as single-strand DNA breakage) is strongly dependent on molecular oxygen (Eisenstark, 1971; Ferron et al., 1972), as is NUV mutagenesis. The oxygen effect suggests that NUV alters macromolecules only indirectly, probably by first generating reactive radicals (H_2O_2, OH$\cdot$, and/or O_2^-). This is supported by the quenching effect of glycerol and other radical-absorbing molecules (Ahmad, 1981; Peak et al., 1982).

CURRENT STUDIES

Our activities are now based on a working model which has been kept flexible to accommodate new experimental data. This model has the following features.

1. NUV photons strike molecules that generate H_2O_2, and then OH$\cdot$, and/or O_2^- (Ahmad, 1981; McCormick et al, 1976; Peak et al, 1982) In bacteria the initial NUV photoreceptors could be in membranes, which are known to harbor molecules that readily yield oxygen radicals (Lee et al., 1983), or may be a thiolated tRNA, absorbing at ca. 335 nm. NUV may also strike H_2O_2, one of its own photoproducts, to hasten O_2^- conversion (Ananthaswamy, et al. 1979; Ahmad, 1981).

2. One or more of these radicals in turn may react with cysteine (McCormick et al.,1982) in polypeptides (or a base in DNA), yielding a new photoreceptor and perhaps a DNA-protein crosslink (Hartman et al., 1979).

3. This reaction leads (perhaps enzymatically) to a critical lesion such as an apurinic or apyrimidinic DNA site.

4. In a wildtype cell, an enzyme (i. e., ExoIII) makes an incision next to the lesion (Weiss, 1981).

5. Following excision of a DNA segment on one of the strands, this segment (or a degraded portion of it) may bind to RecA protein and block its subsequent protease activity, thereby preventing any SOS response (Turner and Eisenstark, 1984; Eisenstark, 1983, Turner, 1983). Other alternatives may be possible also.

6. The damaged DNA is repaired by PolI and constitutive quantities of SOS enzymes, including the recombinase form of RecA protein (recombinational repair), and/or other enzymes not yet identified.

7. Genes (e.g., *nur*) may influence NUV sensitivity, but the gene products are unknown.

8. The *kat* and *sod* gene products catalyze toxic oxygen radicals generated by NUV (Demple and Halbrook, 1983; Eisenstark and Turner, 1984; Finn and Condon, 1975; Hassan and Fridovich, 1978).

9. Following NUV stress "alarmones" are synthesized, as is the case for heat and ethanol stress (Lee et al., 1983). Indeed, the fact that certain thiolated tRNAs absorb at ca. 335 nm may indicate that NUV directly triggers alarmone synthesis (Ames, 1983).

10. The set of new alarmones and proteins that are synthesized following NUV and/or H_2O_2 stress probably overlap those found made during SOS or heat shock responses.

Some of the experiments that have led to this model will be described briefly.

Blockage of SOS and mutation-prone repair

Two experimental results indicate that the SOS repair system is inhibited by NUV damage. (1) Plasmid pKM101, whose *mucA* and *B* genes endow cells with enhanced mutation frequency and enhanced resistance to FUV (Walker, 1984), has no influence on these properties when cells are damaged by NUV (300-400 nm). Thus, NUV lesions do not induce SOS repair, nor subsequent expression of *mucA* and *B* genes on plasmid pKM101. Furthermore, when cells are preirradiated with NUV and subsequently irradiated with FUV, SOS repair is blocked, including the repair normally controlled by genes on pKM101 (Eisenstark, 1983). (2) When *E. coli* cells in which the *recA* promoter is fused to a *lac* structural gene are irradiated with selected monochromatic wavelengths (254 nm, 313 nm, 334 nm, 365 nm), only 254 nm induces high yields of beta-galactosidase, but there is no induction by any of the NUV wavelengths (Turner and Eisenstark, 1984). Also, lambda prophage induction and Weigle reactivation are observed with FUV but not NUV. Furthermore, prior exposure of the cells either to the selected monochromatic NUV wavelengths *inhibit* the induction of beta-galactosidase by subsequent 254-nm radiation, subsequent 254-nm induction of lambda prophage, Weigle reactivation, and the increase in mutation frequency. These observations are consistent with the hypothesis that NUV blocks subsequent RecA protease action, though other possibilities are not yet ruled out.

Lack of effect of NUV on RecA recombinase

To test whether the recombination function is influenced by NUV, host C600 cells were irradiated with a selected wavelength or a combination of wavelengths (254 nm, 365 nm, or 365 nm then 254 nm) and then infected with lambda *bio11*, a phage mutant that requires a functional RecA-protein recombinase activity to form plaques. Plaque formation is observed after each treatment. Thus, RecA recombinase activity is not blocked by FUV or NUV (indeed, recombination is stimulated).

As an additional test of RecA activity, DNA degradation was measured after irradiation with the same wavelengths, in both a wildtype and a *recA* strain. Since one function of RecA protein is to modulate the RecBC nuclease, a blocking of RecA recombinase activity should result in increased DNA degradation. In the *recA* strain, degradation is observed after 254 nm radiation, but not after 365 nm radiation (Ferron et al., 1972; Turner, 1983), further evidence of a difference between FUV and NUV lesions. DNA degradation is not observed in the wildtype strain following any of the irradiations; thus, the binding activity of RecA protein and its ability to control RecBC nuclease activity is not inactivated by NUV or FUV irradiations. Furthermore, pretreatment with NUV does not interfere with the capacity of RecA to block the action of the RecBC nuclease.

Certain mutants (e.g., *recA*) are more sensitive than wildtype to NUV (Eisenstark, 1970). Since inducible SOS does not account for NUV repair in wildtype, its resistance to NUV is assume to be due to ca. 1000 molecules of constitutive RecA recombinase protein present in each cell (Walker, 1984). Also, since RecA cells are highly sensitive to NUV when in log phase, *but not in stationary phase* (Tuveson and Jonas, 1979), it is assumed that RecA protease is necessary for repair of log-phase cells, but RecA protein is not needed for repair cells in stationary phase.

The photoreceptor for NUV damage

Our studies with phage T7, containing only DNA and protein, emphasize a paradox; i.e., peak absorption is 254 nm for DNA, and 280 nm for protein, but an action spectrum for killing of phage T7 (Ananthaswamy et al., 1979), and *S. typhimurium* (MacKay et al., 1976) show distinct shoulders at 340 nm. The action spectrum for inactivating T7 phage (or bacteria) fits neither the absorption spectrum of DNA nor of protein. Furthermore, if phage particles are irradiated in a nonlethal concentration of H_2O_2, the NUV effect is greatly amplified, with a peak of 340 nm for this synergistic effect (Ananthaswamy et al., 1979). A possible explanation might involve more than one photochemical reaction, the first photoreaction leading to a better NUV chromophore. In a search for a NUV chromophore, we observed that the sulfhydryl in peptide-bound cysteine, in the presence of oxygen (or H_2O_2) is photochemically altered (McCormick et al, 1982). Treatment of 2.5 mM glutathione (a tripeptide containing cysteine) with 2.5 mM H_2O_2 results in the formation of a near-UV chromophore having maximal absorption 25 nm above the absorption of the initial glutathione. From examination of related compounds, it is apparent that the N-acylcysteinamide of the peptide residue is the key element required for generation of the new chromophore. Although we have not yet determined the structure of the new chromophore, we do know that it is not simply the oxidized cysteine. This tripeptide is a useful model molecule for showing that a polypeptide containing cysteine can be altered to generate a new chromophore, one that can absorb NUV and possibly cause biological damage.

Repair of damage to bacteria may involve an additional photoreceptor. The thiolated tRNA molecules in *E. coli* may act as photoreceptors for alarmone synthesis, one of which (pppGppp) may have already been observed (Jagger, 1983).

DNA damage caused by NUV Irradiation

Differences between the biological effects of NUV and FUV (Table 1) indicate that the biochemical lesions are different, and suggest also the involvement of distinctly different alarmones; (Lee et al., 1983). Although a DNA-protein crosslink may be an important NUV lesion, this may be only a preliminary step toward producing a DNA lesion that can be repaired by ExoIII (Sammartano and Tuveson, 1983; Eisenstark and Turner, 1984). Crosslinks are found in phage T7 (Hartman and Eisenstark, 1982) and in mammalian cells (Eisenstark et al., 1982), two locations where the DNA and protein are packaged in close quarters. Our evidence with bacteria is far less striking, but only a small fraction of DNA and protein is tightly bound in these cells. DNA strand breaks (Turner, 1983; Hartman et al., 1979; Hartman and Eisenstark, 1980; Eisenstark et al., 1982; Peak and Peak, 1982; Tuveson et al., 1983) may be secondary consequences of NUV photochemical damage, since the number of DNA breaks do not correlate with the number of lethal events. Further disenchantment with single-strand breaks (SSB) as lethal events comes from knowledge that *xthA* mutants are very sensitive to NUV, but yield few SSB (Turner, 1983). By comparison, the number of DNA-protein crosslinks in phage correlate very well with the number of phage that are inactivated (Hartman et al., 1979). Single-strand breakage by NUV is particularly striking in *Micrococcus radiodurans* (Caimi, 1983), an organism that is very sensitive to NUV, but notoriously resistant to FUV. In evaluating the role of SSB, lack of SSB repair in *polA* mutants following H_2O_2 treatment should be noted (Anathaswamy and Eisenstark, 1977; Yoakum, et al., 1975).

An anomaly remains. Although some experiments show that NUV and H_2O_2 effects may be interchangeable, we have yet to note DNA-protein crosslinks via H_2O_2 alone. Thus, we recognize that NUV effects may not simply be the generation of H_2O_2, for H_2O_2 does lead to numerous SSB (Ananthaswamy and Eisenstark, 1977).

Recovery from NUV damage

A clearer concept of recovery and repair of NUV damage has recently emerged, based on the following observations:

1. By showing that NUV lesions do not lead to SOS induction, and with the knowledge that some genes defective in DNA metabolism lead to NUV sensitivity, we are now aware that *constitutive* levels of certain proteins, particularly RecA recombinase, may still function to repair NUV damage (Turner, 1983). Constitutively produced PolI is clearly necessary for recovery from NUV and H_2O_2, since *polA* mutants are particularly sensitive to these agents.

2. *XthA* mutants are more sensitive to NUV, but no more sensitive to FUV than wildtype, and there are far fewer SSB in the *xthA* mutant (Turner, 1983; Eisenstark and Turner, 1984). This observation supports the view that a major role of ExoIII (the *xthA*

product) is to recognize apurinic or apyrimidinic sites and to nick the DNA at the 3' end of these sites (Weiss, 1981). Although a logical conclusion might be that NUV and H_2O_2 produce apurinic or apyrimidinic sites, there are other possibilities, such as lesions repaired by other enzymatic activities of ExoIII. Also, there could be two separate actions in the case of a DNA–protein crosslink: (i) proteinase action on the polypeptide, leaving an altered base, followed by (ii) action by ExoIII.

3. Throughout our studies, we have considered catalase as a logical "NUV-recovery enzyme," since NUV yields H_2O_2 (McCormick et al, 1976). We now know that a small dose of H_2O_2 can induce catalase and that this extra catalase can protect the cell against highly toxic doses of either H_2O_2 or NUV. As expected, *kat* mutants are sensitive to NUV but not FUV (unpublished). One gene, *katE*, is near *xthA* (both at 38 min). However, the two genes are not likely to be in the same operon, since strains with the deletion *pcn-xthA* are catalase positive; also, *katE* but not *xthA* is inducible (Eisenstark and Turner, 1984). Although neither NUV nor H_2O_2 mimic *all* of the biological effects of the other agent, certain relationships between the two are clear: (i) NUV of cells yields H_2O_2, (ii) either H_2O_2 or NUV will induce catalase synthesis (unpublished); and (iii) induction by one agent will give protection against exposure to the other (unpublished). There are, however, a number of cases in which the effect of H_2O_2 and NUV are not interchangeable. For example, although neither NUV nor H_2O_2 induce the SOS response, only NUV damage blocks SOS (unpublished).

4. We have found that the presence of a plasmid carrying an inducible *sod* gene (Touati, 1983) endows the cell with resistance to NUV, but not to FUV, implicating O_2^- as a lethal radical. This plasmid is a low-copy-number type in a cell that already has a *sod* gene. Although *sod* mutants have been isolated, they are extremely unstable.

5. Constitutive levels of RecA protein and of PolI are needed for repair of NUV lesions, but whether new proteins synthesized following NUV (Peters and Jagger, 1981; Turner 1983; unpublished) are directly involved in repair of DNA is unknown. Our knowledge of recovery from NUV damage may be summarized as follows: The detoxifying enzymes catalase and Mn superoxide dismutase are induced by H_2O_2 and O_2. Constitutive enzymes (e.g., *xthA, recA,* and *polA*) repair DNA damage. The protective role of the *nur* gene product is unknown, as are the roles of the new proteins synthesized following NUV.

The effect of heat and anaerobic metabolism

From our first experiments a dozen years ago, we have been aware of the role of O_2 in NUV damage, and many of our experiments have involved controls of cells grown under N_2 (anaerobically). Among a number of observations, we found that genetic recombination (transduction, RecA-mediated phage recombination) is greatly reduced by oxygen deprivation. This indicates that RecA action needs O_2 particularly as a recombinase (Droffner and Yamamoto, 1983). The involvement of O_2 in PolA activity is of particular interest, especially since *polA* mutants are so highly sensitive to NUV and H_2O_2 (Ananthaswamy and Eisenstark, 1977; Morimyo, 1982).

We have also been aware of the role of heat or heat shock in some of the same metabolic steps that involve NUV damage and recovery (Tyrrell, 1978, 1979; Williams-Hill and Grecz, 1983). Indeed, heat shock in bacteria may shunt cells into utilization of anaerobic pathways (Neidhardt et al., 1983). We have also noted some similarities (as well as differences) in cells that are stressed by NUV, H_2O_2 anaerobiosis, or heat (Lee et al., 1983). We are in constant surveillance as to how this might relate to broader metabolic problems in health and disease, including tumor promotion (Ames, 1983; Levin et al., 1982; Kinsella et al., 1983).

UNKNOWNS

Obviously, stress by NUV or H_2O_2 triggers a number of biochemical activities, but we are left with a number of knowledge gaps, including: (1) determination of the importance of each of the several NUV actions on cell damage and mutation; (2) determination as to whether oxygen radicals, alterations in thiolated tRNAs, or DNA photoproducts start a cascade that leads to recovery; and (3) determination of any common pathways of responses following NUV, FUV, heat shock, or other cellular stresses. Some of these questions should soon be answered as we compare (i) the mechanisms of recovery from NUV and/or H_2O_2 damage, including the roles of the *nur, xthA, polA, recA, sod,* and *kat* gene products, (ii) the new proteins synthesized, (iii) and the new alarmones evoked following each of the stresses. Knowledge of the nature of the DNA lesions would help to understand the biological consequence of such lesions (e.g., blockage of RecA protease without blockage of recombinase function).

ACKNOWLEDGEMENTS

Recent research of the author was supported by Public Health Service Grant No. FD00658 and UMC Research Council No. Bio Med 2652 from N.I.H. I am grateful to Ruth Dalke for handling numerous revisions of this manuscript.

REFERENCES

Ahmad, S. I. (1981). *Photobiochem. Photophys.,* 2, 173.

Ames, B. N. (1983). *Science,* 221, 1256.

Ananthaswamy, H. N., Hartman, P. S., and Eisenstark, A. (1979). *Photochem. Photobiol.,* 29, 53.

Ananthaswamy, H. N. and Eisenstark, A. (1977). *J. Bact.,* 130, 187.

Buzard, R. L. (1980). M.A. thesis. University of Missouri.

Caimi, P. G. (1983). M.A. Thesis, University of Missouri.

Demple, B. and Halbrook, J. (1983). *Nature, 304,* 466.

Demple, B., Halbrook, J., and S. Linn. (1983). *J. Bact.,* 153, 1079.

Droffner, M. L. and Yamamoto, N. (1983). *J. Bact.,* 156, 962.

Eisenstark, A. (1970). *Mutat. Res.,* 10, 1.

Eisenstark, A. (1971). *Adv. in Genetics,* 16, 167.

Eisenstark, A. (1983). *Mutat. Res.,* 122, 267.

Eisenstark, A., Kovacs, S., and Terry, J. (1982). *Jour. Natl. Cancer Inst.,* 69, 177.

Eisenstark, A. and Turner, M. A. (1984). In preparation.

Ferron, W. L., Eisenstark, A., and Mackay, D. (1972). *Biochim. Biophys. Acta,* 277, 651.

Finn, G. J. and Condon, S. (1975). *J. Bact.,* 123, 570.

Hartman, P. S., Eisenstark, A., and Pauw, P. G. (1979). *Proc. Natl. Acad. Sci.,* 76, 3228.

Hartman, P. S. (1981). *Photochem. Photobiol.,* 34, 39.

Hartman, P. S. and Eisenstark, A. (1978). *J. Bact.,* 131, 769.

Hartman, P. S. and Eisenstark, A. (1980). *Mutat. Res.,* 72, 31.

Hartman, P. S. and Eisenstark, A. (1982). *J. Virol.,* 43, 529.

Hassan, H. M. and Fridovich, J. (1978). *J. Biol. Chem.,* 253, 6445.

Jagger, J. (1983). In *Photochemical and Photobiological Reviews.* K. C. Smith (ed.) Plenum Press.

Kelland, L. R., Moss, S. H., and Davies, D. J. G. (1983). *Photochem. Photobiol.,* 37, 617.

Kinsella, A. R., Gainer, H. St. C., and Butler, J. (1983). *Carcinogenesis* 4, 717.

Lee, P. C., Bochner, B. R., and Ames, B. N. (1983). *Proc. Natl. Acad. Sci. USA,* 80, 7496.

Levin, D. E., Hollstein, M., Christman, M. F., Schwiers, E. A., and Ames, B. N. (1982). *Proc. Natl. Acad. Sci. USA* 79, 7445.

Loewen, P. E. (1984). *J. Bacteriol.,* 157, 622.

Mackay, D., Eisenstark, A., Webb, R. B. and Brown, M. S. (1976). *Photochem. Photobiol.,* 24, 337.

McCormick, J. P., Fisher, J. R., Pachlatko, J. P., and Eisenstark, A. (1976). *Science,* 191, 468.

McCormick, J. P., Klita, S., Terry, J., Schrodt, M., and Eisenstark,. A. (1982). *Photochem. Photobiol.,* 36, 367.

Moody, C. S. and Hassan, H. M. (1982). *Proc. Natl. Acad. Sci. USA* 79, 2855.

Morimyo, M. (1981). *J. Bact.,* 152, 208.

Moss, S. H. and Smith, K. C. (1981). *Photochem. Photobiol.,* 33, 203.

Niedhardt, F. C., VanBogelen, R. A., and Lau, E. T. (1983). *J. Bact.,* 153, 597.

Peak, J. G., Peak, M. J., and Foote, C. S. (1982). *Photochem. Photobiol.,* 36, 413.

Peak, M. J. and Peak, J. G. (1982). *Photochem. Photobiol.,* 35, 675.

Peters, M. J. and Jagger, J. (1981). *Nature,* 289, 194.

Sammartano, L. J. and Tuveson, R. W. (1983). *J. Bact.,* 156, 904.

Thomas, G. and Favre, A. (1980). *Eur. J. Biochem.,* 113, 67.

Touati, D. (1983). *J. Bact.,* 155, 1089.

Turner, Mary Ann (1983). Ph.D. Dissertation, University of Missouri.

Turner, M. A. and Eisenstark, A. (1984). *Mol. Gen. Genet.,* 193, 33.

Turner, M. A. and Webb, R. B. (1981). *J. Bact.,* 147, 410.

Tuveson, R. W. and Jonas, R. B. (1979). *Photochem. Photobiol.,* 35, 667.

Tuveson, R. W., Peak, J. G., and Peak, M. J. (1983). *Photochem. Photobiol.,* 37, 109.

Tuveson, R. W. and Violante, E. V. (1982). *Photochem. Photobiol.,* 35, 845.

Tyrrell, R. M. (1978). *Mutat. Res.,* 52, 25.

Tyrrell, R. M. (1979). *Acta Biol. Med. Germ.,* 38, 1259.

Wagner, S. and Snipes, W. (1982). *Photochem. Photobiol.,* 38, 57.

Walker, G. C. (1984). *Microbiological Reviews* 48, 60.

Webb, R. B., Brown, M. S. and Tyrrell, R. M. (1978). *Radiation Research* 74, 298.

Weiss, B. (1981). In *The Enzymes*, Vol. 14, P. B. Boyer, (ed.), Academic Press. p. 203.

Williams. M. and Grecz, N. (1983). *Mutat. Res.,* 107, 13.

Yoakum, G., Eisenstark, A., and Webb, R. B. (1975). In *Molecular Mechanisms for Repair of DNA*, Part B. P. C. Hanawalt and R. B. Setlow, (eds.) Plenum. p. 453.

PART THREE

THE GENOME AND THE REGULATION OF DNA SYNTHESIS

INTRODUCTION

The contributions of the Copenhagen school to the synthesis and function of the bacterial chromosome have their wellspring in physiological experiments. Early experiments with synchronized cells and autoradiography pointed the way for definitive work on the temporal relation between DNA synthesis and the cell cycle. By using age selection with the "baby machine" Helmstetter and Cooper showed that the time needed for the replication apparatus to traverse the chromosome of growing *Escherichia coli* is constant over a wide range of growth rates. Thus, as in the case of components of the Protein-Synthesizing System (PSS), synthesis of DNA is regulated by the frequency of initiation. Work from other laboratories also focused on the importance of the initiation process by finding that certain temperature-sensitive mutants can continue ongoing rounds of replication but cannot initiate new ones under nonpermissive conditions. These studies on regulation have blended well with what has been learned about the structure and organization of the bacterial chromosome, especially in *E. coli*. The result is a reasonably coherent picture of the general features of chromosome structure, biosynthesis, and function. Much remains to be learned since little detail is known. The contributions of this section illustrate the state of our knowledge.

Despite its circularity, the bacterial chromosome has a beginning and an end, at least with respect to its synthesis. This section contains current work on the replication origin, *oriC* in *E. coli,* its adjacent regions, and its terminus. The gene product most studied with regard to initiation, that of the *dnaA* gene, is also treated in some detail.

It has been known in general terms that a mechanistic relationship exists between DNA metabolism and cell division. At long last, it has begin to yield to genetic analysis, as demonstrated in some of the papers that follow. Both physiological and genetic approaches are presented here. The bacterial genome appears to be differentiated with respect to its availability to the transcription apparatus, a topic also covered in this section. Still another contribution presents novel aspects of genetic exchange.

Moselio   Schaechter

Facets of the Chromosomal Origin of Replication *oriC* of *E. coli*

Kaspar von Meyenburg, Flemming G. Hansen, Tove Atlung*, Lars Boe, Ib Groth Clausen, Bo van Deurs[†], Egon Bech Hansen, Birgitte B. Jørgensen, Flemming Jørgensen*, Luud Koppes**, Ole Michelsen, Jørgen Nielsen, Poul Erik Pedersen, Knud V. Rasmussen*, Erik Riise, and Ole Skovgaard

The Technical University of Denmark, Lyngby-Copenhagen
* The University Institute of Microbiology, Copenhagen
† The Panum Institute at The University of Copenhagen
** The University of Uppsala, Sweden

Six years ago the precise localization and structure of the origin of replication of the *E. coli* chromosome was presented at the Cold Spring Harbor meeting by the joint Berlin-Copenhagen group and by Hirota and coworkers (von Meyenburg et al., 1979; Messer et al., 1979; Hirota et al., 1979). It seems timely at this occasion to sum up the continuation of the work on bacterial replication as it developed at the Department of Microbiology of the Technical University of Denmark.

Six years ago many expected—and we were hardly an exception-that one would rather swiftly get an understanding of the mechanism and control of initiation of replication from *oriC*, once having it cloned. However, looking back we must conclude that it has been a fairly slow process, though indeed many new details have become apparent. Indications for this slow progress of further work were contained in the physiological data which we had gathered at that time (von Meyenburg et al., 1979): The presence of additional copies of *oriC* on either minichromosomes or integrated in the chromosome affected growth of the cells only slightly. The effects of minichromosomes with ATP synthase genes (*atp/unc*) on cell growth and division are, as will be decribed below, artifacts due to the high copy number rather than indications of copy control. Furthermore, although the copy number of minichromosomes was much higher than expected, minichromosomes were rather unstable.

Subsequently, the construction of strains with pBR322-oriC chimeras resulted in cells with a total of maybe 50 active oriC sites on the plasmid per site on the chromosome (Hansen et al., 1981a). A direct-control model for E. coli chromosome replication that assumed that the concentration of oriC was felt by the control mechanism (Pritchard et al., 1969; Pritchard, 1978) was made obsolete by this observation, though similar models have been shown to apply to the control of plasmid replication (Nordstrøm et al., 1984). How then do the chromosome and the minichromosomes behave?

GENETIC AND PHYSICAL MAPPING OF THE ORIGIN REGION

The precise mapping of the origin region of the E. coli K-12 chromosome was largely based on the isolation of F' plasmids (Hiraga, 1976; von Meyenburg, et al., 1977) and specialized transducing phages like lambda asn (von Meyenburg et al., 1978, 1979; Miki et al., 1978) and lambda tna (Hansen and von Meyenburg, 1979; Miki et al., 1979). By subcloning and sequencing of DNA segments the arrangement of genes neighboring oriC (Figure 1a,b) and dnaA (Figure 2) was defined (Yasuda and Hirota, 1977; Messer et al., 1978; von Meyenburg and Hansen, 1980; Ream et al., 1980; Oka et al., 1980; Deeley and Yanofsky, 1981; Gay and Walker, 1981a,b; Hansen et al., 1981a,b; Nielsen et al., 1981; Saraste et al., 1981; von Meyenburg et al., 1982; Hansen et al., 1982b; Hansen et al., 1982a; Gellert et al., 1983; Bukh and Messer, 1983; Nielsen et al., 1984).

The singularity of oriC was demonstrated by construction of oriC deletions on a specialized transducing phage lambda asn, which only could be recombined into the chromosome if it contained another replicon giving "integrative suppression" (von Meyenburg and Hansen, 1980). Since then it has been found that the chromosome itself has the potential to replicate independently of oriC if it carries an sdrA mutation (Kogoma and von Meyenburg, 1983) that eliminates RNase H activity (Ogawa et al., 1984) and leads to initiations at other sites on the chromosome (de Massy et al., 1984; Kogoma et al., this volume).

MINICHROMOSOMES AND THE ATP SYNTHASE OPERON

Growth inhibition caused by atp minichromosomes

Certain specialized transducing phages lambda asn were found to establish themselves as minichromosomes when transduced into an asnA asnB recA mutant of E. coli lysogenic for lambda (CM988), allowing the recipient strain to grow in the absence of asparagine (von Meyenburg et al., 1978; 1979). It was this property of the transducing phages (e.g., lambda asn20 and lambda asn105 (Figure 1c)) that we used to define oriC as a site which allowed autonomous replication of the lambda asn genomes in the presence of the lambda cI repressor. Phages not showing this property, for example, lambda asn92, were said to lack oriC. The growth pattern of minichromosome-bearing strains depended on the extent of chromosomal DNA carried by the minichromosome to the left of oriC (von Meyenburg et al., 1979; Hansen et al., 1981b). Minichromosomes lambda asn212 and lambda asn20 allowed good growth (Table 1); the 15-20% reduction in growth

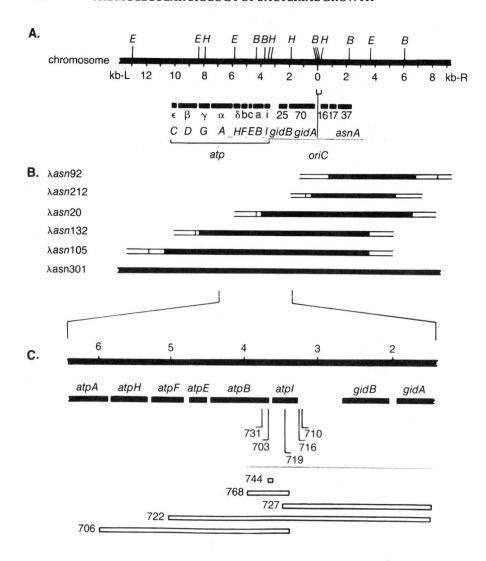

FIGURE 1 (a) Restriction site map and genetic map of the <u>oriC</u> region of the <u>E. coli</u> K12 chromosome and (b) structure of specialized transducing phages λasn (von Meyenburg et al., 1978; Hansen et al., 1981). The zero point of the kb scale is located at <u>ori</u>. Bars indicate position and extent of genes together with the genetic symbols and the polypeptide assignments. Black = chromosomal DNA, white = lambda DNA and unassigned DNA. E: EcoRI, H: Hind III, B: BamHI. (c) <u>atp</u> mutations, deletions (open bars); insertions and point mutations, isolated by selection for fast-growing derivatives of pλasn3 or pλasn301 minichromosome-containing strains. The positions of the mutations and extent of deletions are indicated together with the allele numbers (the mutations <u>atp716</u> and <u>atp719</u> were isolated as Tn10 insertions in λasn20; von Meyenburg et al., 1982a).

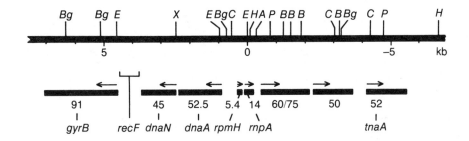

FIGURE 2 Genetic and physical map of the dnaA region. The zero point on the kb scale is located approximately 42 kb from oriC. The position and extent of genes is indicated by bars, genetic symbols, and/or the molecular weight of the polypeptide. The arrows indicate direction of transcription. A: Aval, B: BamHI, Bg: BglII, C: ClaI, H: HindIII, P: PstI, X: XhoI.

rate, as compared to a wildtype control strain, merely reflects the production of minichromosome-less Asn⁻ cells at cell division. Phages carrying additional DNA with part or all of the *atp* operon, for example, lambda *asn3* and lambda *asn105* (Figure 1c), gave rather poor growth (Table 1). This result indicated that the additional chromosomal DNA on these phages adversely affected growth of the host strain or replication of the minichromosomes or both (von Meyenburg et al., 1978). Maintenance of these minichromosomes appeared to be rather poor (von Meyenburg et al., 1979).

The slow growth rate of lambda *asn*-carrying Asn⁺ transductants of strain CM988 provided a genetic approach for the definition of genes or sites responsible for this phenotype, namely the selection of fast-growing mutants. Mutations in lambda *asn3*, *asn301* and lambda *asn105* were found that included insertions, small and large deletions, and some point mutations. Physical and genetic mapping revealed that all were located in or just in front of the *atp* operon (Figure 1c). When transferred into the chromosome, many rendered the cells unable to grow on succinate. In those cases the synthesis of one or several of the subunits of the ATP synthesis was found to be eliminated totally (Hansen et al., 1981b). Few mutations led to only a modest effect on growth on succinate (e.g. the mutations of lambda *asn105-015* and lambda *asn3-10*). In those cases expression of the ATP synthase operon was decreased by 60-80% as measured for the expression of the *atpE* gene, which codes for the smallest of the eight subunits (Figure 3). These findings suggested that the expression of the *atp* genes on the lambda *asn* minichromosomes was responsible for the inhibition of growth and cell division (Hansen et al., 1981b).

Ogura et al (1980) reported the mapping of a copy-control gene, *cop*, in a region that coincided with the beginning of the *atp* operon. Mapping of *cop* was based on the observation that mutations very similar to the ones isolated by us resulted in increased copy number of minichromosomes (in *cis*). However, transfer of our mutations (*atp-706, atp-710, atp-74,* Figure 1c) into the chromosome did not appear to result in any effects on growth beyond what is characteristic of *atp/unc* mutants, i.e., mutants defective in oxidative phosphorylation.

TABLE 1   Growth of *asnA asnB recA1* strains harboring various wildtype and mutant λ*asn* phages integrated or as pλ*asn* plasmids (minichromosomes)*

Phage or plasmid	Size, kb	Chromosomal DNA, kb	Growth rate (doublings/hr)	
			Integrated λ*asn*■	Autonomous pλ*asn*▼
λ*asn92*	40.5	6.0	1.25	—
λ*asn212*	41.1	5.6	1.15	1.0–1.1
λ*asn20*	38.4	9.7	1.15	1.0–1.1
λ*asn3*	48.4	11.3	1.05	0.30–0.33
λ*asn105*	54.0	14.5	1.17	0.63–0.66
λ*asn301*	46.0	29.0	1.22	0.60
λ*asn3-6*● (*atp706*)			1.15	0.9–1.0
λ*asn301-44* (*atp744*)			1.15	0.9

*Growth conditions were selective for Asn⁺ cells (minimal medium AB with 0.2% glucose, 1% casamino acids, and 2 µg/ml thiamine).

■Integrated into *att*λ of strain CM987 together with λ*cI857 S7*.

▼λ*asn* carrying *oriC* established as pλ*asn* minichromosomes after infection of CM988.

●Other mutant phages with mutations (allele number in parentheses) with the same phenotype were λ*asn3-3(atp703)*, λ*asn3-10(atp710)*, λ*asn3-31 (atp731)*, λ*asn301-68(atp768)*, λ*asn105-22(atp722)*, and λ*asn105-27(atp727)*(see Figure 1c).

Therefore, we suggest that the copy–number effects reported by Ogura et al., (1980) were due to relief of inhibition of growth and division by eliminating or decreasing expression of the *atp* operon from the minichromosomes.

Copy numbers of minichromosomes

We had estimated the copy number of lambda *asn* minichromosomes to be 10 (7–13) per genome, independent of the presence or absence of the *atp* operon (von Meyenburg et al., 1979), which is equivalent to about 5 copies per *oriC* in the chromosome. An independent estimate of the copy number of the lambda *asn105* minichromosome that carries the entire *atp* operon was obtained by measuring expression of the *atp* operon. In a CM988/lambda–*asn105* culture synthesis of the c subunit of ATP synthase (the *atpE* gene product) was 4—5 fold higher than in a control strain without minichromosome (Figure 3, lane A, C). Since only the 35% of the cells harboring the lambda *asn105* minichromosome would contribute to the excess synthesis we calculated the level of c–subunit synthesis in the cells carrying a minichromosome to be 8–12 times higher than in the wildtype cells.

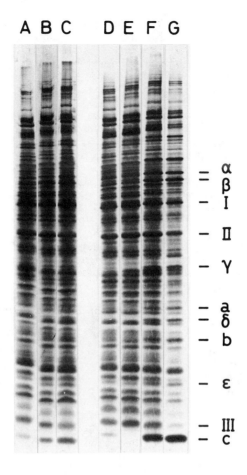

FIGURE 3 Autoradiograms of [^{35}S]methionine-labeled total cell proteins separated by SDS-polyacrylamide gel electrophoresis of: (A) CM987 (wildtype) with λasn92; (B) CM987 with λasn105 integrated in attλ; (C) CM987 with pλasn105 minichromosomes; (D) CM2443 (wildtype); (E) CM1470 (atp706 deletion strain), (F) CM1470 with plasmid pBJC706 resulting in 5-fold overproduction of ATP synthase; and (G) with plasmid pBJC707 giving a 12-fold overproduction of ATP synthase (von Meyenburg et al., 1984). Positions of relevant protein bands are indicated: (a, b, c, α, β, γ, δ, and ε) subunits of the ATP synthase; (I) translation elongation factor EF-Tu; (II) major outer-membrane protein OmpA; (III) lipoprotein.

From the following experiments we conclude that expression of the atp operon is constitutive. At different growth rates the c-subunit was found to represent 0.23-0.30% of total protein and the ratio c-subunit/lipoprotein was rather constant. Furthermore, c-subunit synthesis was increased twofold in a strain diploid for the atp operon (strain CM987 with lambda *asn105* integrated in the lambda attachment site, Figure 3, lane B). Also the determination of beta-galactosidase levels expressed from an *atpB-lacZ* fusion indicated constitutivity (Jorgen Nielsen, unpublished). Therefore, we take the 8-12-fold increase of c-subunit synthesis in the lambda *asn105*-carrying cells to indicate the minichromosome copy number. This value is in agreement with the "old" estimates of copy number (see above). Therefore, the slow growth of the lambda *asn105*-carrying strain could hardly have been due to a low copy number of minichromosomes, as we had assumed initially (von Meyenburg et al., 1978). Together with the analysis of the mutations relieving slow growth, this observation suggested instead that the 8-12 fold increase in ATP synthase level was responsible for growth inhibition. By constructing stable plasmid pBR322 derivatives carrying the eight genes specifying the subunits of ATP synthase (Figure 1), we could demonstrate physiological and morphological effects of the overproduction of ATP synthase (compare lanes D to G in Figure 3; von Meyenburg et al., 1984) that were much the same as those invoked by the lambda *asn105* minichromosome and by an F' carrying *oriC* and *unc/atp* (Masters, 1975; von Meyenburg et al., 1977). These effects are decreased growth rate and inhibition of cell division.

Cytological effects of overproduction of membrane-bound ATP synthase

Electron microscopy of thin sections of cells with a 10-12-fold increased level of ATP synthase (Figure 3, lane G) revealed that unusual vesicles and cisterns were formed in these cells (Figure 4). The vesicles as well as the inner side of the cytoplasmic membrane were covered by a layer of particles 100-120 angstroms thick, which we interpret to represent the F_1 sector of ATP synthase (von Meyenburg et al., 1982b, 1984). The vesicles and cisterns were presumably formed as a result of invagination of the cytoplasmic membrane by insertion of the excess ATP synthase. In wildtype *E. coli* with about 3000 ATP synthase complexes per glucose-grown cell this enzyme covers about 15-20% of the total area of the cytoplasmic membrane. Obviously, there is not sufficient space for all of the excess enzyme if its synthesis is increased more than fivefold. These findings suggest that the Het phenotype (heterogeneity of cell size; von Meyenburg et al., 1977) of strains carrying minichromosomes was the result of the hindrance of cell division by the cytological changes induced by the overproduction of ATP synthase; the 50-65% decrease in growth rate resulting from a 10-12-fold overproduction from a stable pBR322 *atp* plasmid fits quite well with the decrease in growth rate caused by the presence of the lambda *asn105* minichromosomes (Table 1). It is in part a result of the decreased prevalence of the essential protein-synthesizing system (Maaløe, 1979) but may also result from disturbance of membrane functions. Thus, the morphological and physiological effects seen in the minichromosome-containing cells were not an indication of a decreased initiation frequency of chromosome replication, as had been suggested (Masters, 1975; von Meyenburg et al., 1977).

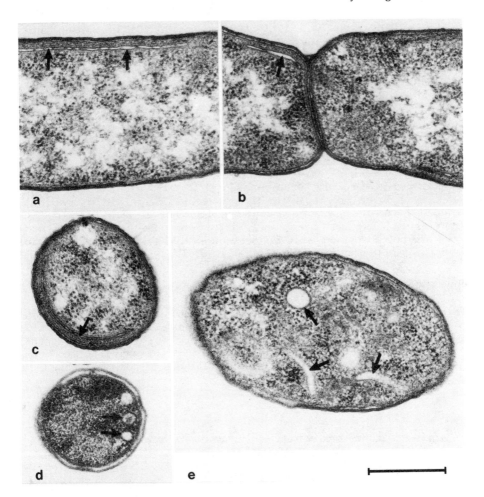

FIGURE 4 Thin-section electron micrographs of cells of strain CM2788 (= CM1470 (pBJC707)) overproducing ATP synthase 12-fold (see Figure 3, lane G). Arrows indicate membrane cisterns and vesicles within the cytoplasm. Typically, the cisterns are found close to the inner (cytoplasmic) membrane of the cells (a-c). Light unstained areas in 5a-c represent sections through the nucleoid (nuclear zone). Bar, 0.5 µm, x 66,000. (From von Meyenburg et al., 1984., reproduced with permission from EMBO Journal.)

THE *dnaA* GENE AND ITS PRODUCT

Map location of the *dnaA* gene

The *dnaA* gene is the only well-characterized gene, whose product is specifically involved in initiation of replication from *oriC* (Hirota et al., 1968; Hirota et al., 1970; Messer et al., 1975; Hansen and Rasmussen, 1977; Marsh and Worcel, 1977; von Meyenburg et al., 1979; Lycett et al., 1980). The DnaA protein appears to act early in the initiation process (Zyskind et al., 1977).

The *dnaA* gene was identified by isolating mutants that were temperature sensitive for initiation of chromosome replication (Hirota et al., 1970; Carl, 1970; Abe and Tomizawa, 1971; Wechsler and Gross, 1971; Beyersmann et al., 1974; Sevastopoulos, 1977). It was subsequently mapped precisely by cloning on the specialized transducing phage lambda *tna* (Hansen and von Meyenburg, 1979), and it is located between the *dnaN* and the *rpmH* genes at 83 min. (Figure 2; Hansen and von Meyenburg, 1979; Miki et al., 1979; Ream et al., 1980; Sakakibara and Mizukami, 1980; Hansen et al., 1982b). The *dnaA* gene codes for a basic polypeptide with a molecular weight of 52.5 kD, as deduced from the nucleotide sequence (Hansen et al., 1982a). The *dnaN* gene (Sako and Sakakibara, 1980) and possibly also *recF* and *gyrB* are cotranscribed with *dnaA*, though the *gyrB* gene appears to have its own promoter (Gellert et al., 1983).

Fine structure genetic mapping of 13 different dnaA(Ts)-mutations

We have used deletion mapping, taking advantage of M13 and lambda *tna* phages carrying segments of the *dnaA* gene, to localize 13 *dnaA*(Ts) mutations. Also lambda *tna* phages carrying different *dnaA*(Ts) mutations were constructed and used to arrive at the fine-structure genetic map of the 13 *dnaA* mutations (Figure 5; Hansen et al., 1984).

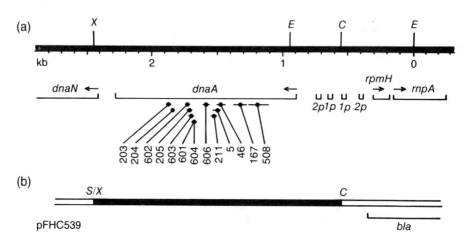

FIGURE 5 (a) Detailed genetic and physical map of the dnaA region. The position and extent of genes and their promoters are indicated by brackets. rpmH and rnpA are cotranscribed, as are dnaA and dnaN. The arrows indicate direction of transcription. The map position of 13 dnaA(Ts) mutations are marked by the respective allele number. (b) The structure of plasmid pFHC539; it consists of the ClaI-XhoI fragment, encompassing the dnaA+ gene and its promoters, inserted in the ClaI and SalI sites of pBR322.

Dominance analysis of dnaA(Ts) mutations

The lambda *tna* phages carrying the *dnaA+* and *dnaA*(Ts) alleles allowed us to analyze the disputed dominance relationship of *dnaA*(Ts) and *dnaA+* alleles (Gotfried and Wechsler, 1977; Zahn et al., 1977; Zahn and Messer, 1979; Murakami et al., 1980; Wechsler, 1980). We tested the 13 different *dnaA*(Ts) alleles (Figure 5) and found all to be recessive to the wildtype *dnaA+* allele (Hansen et al., 1984).

The dominance of *dnaA+* is rather complete. In heterozygous diploid strains *dnaA*(Ts)/*dnaA+* neither the DNA content nor the cell size is in general different from these parameters in wildtype *dnaA+* haploid or diploid cells (Table 2) at 36 C and 41 C; at 36 C the DNA content is decreased and the cell size is increased in the *dnaA*(Ts) strains, while at 41 C replication ceases after termination of rounds of replication in progress. Interestingly, we observed that certain *dnaA*(Ts)/*dnaA+* strains (notably with *dnaA5* and *dnaA46*) were cold

TABLE 2 Growth rate and DNA content of merodiploid *dnaA*(Ts)/ *dnaA+* strains and their haploid counterparts at different temperatures

Strain*	dnaA allele	28°C		36°C		41°C	
		μ▼	DNA▲	μ	DNA	μ	DNA
CM2539	A+	0.85	3.6	1.38	3.4	1.46	3.3
CM2563	A+/ A+	0.86	3.9	1.33	3.8	1.50	3.7
CM2571	A+/ A850	0.82	3.9	1.36	3.7	1.50	3.5
CM2540	A5	0.75	3.1	1.30	1.7	→0	△
CM2575	A5/A+	0.32	5.2	0.96	3.9	1.20	3.4
CM2543	A46	0.86	3.1	1.38	1.9	→0	△
CM2581	A46/A+	0.36	4.8	0.72	4.1	1.18	4.1
CM2559	A167	0.73	3.2	1.30	2.3	→0	△
CM2569	A167/A+	0.62	2.7	1.07	3.9	1.25	3.8

*All strains are *recA1* (Hansen et al., 1984) grown in AB minimal medium (Clark and Maaløe, 1967) with 0.2% glucose, thiamine (5 μg/ml), 1% casamino acids, and tryptophan (20 μg/ml); cultures of haploid and diploid strains were pre-grown overnight at 30°C and 37°C, respectively, and diluted into fresh medium, at the temperatures indicated, to an A_{450} = 0.03.

▼Growth rate (doublings/hr)

▲DNA content (μg DNA per 1 ml of culture at A_{450} = 1.000) determined on duplicate samples harvested at A_{450} = 0.400-0.550 (Brunschede et al., 1977).

△At 41°C the growth rate of the haploid *dnaA*(Ts) strains was constantly decreasing (indicated by → 0), though the precise kinetics differed from strain to strain.

sensitive, exhibiting decreased growth rate at 30 C with an increase of both DNA content and cell size. This phenotype probably belongs to a cold-sensitivity syndrome associated with *dnaA*(Ts) mutations (Wechsler and Zdzienicka, 1975; Kellenberger-Gujer et al., 1978; Atlung, 1981; Hansen et al., 1984).

Effect of high gene dosage of dnaA⁺ and dnaA(Ts) alleles

Earlier experiments had indicated that partial inactivation of the DnaA protein would limit the rate of initiation of DNA replication (Hansen and Rasmussen, 1977). Therefore, by using a high-copy-number plasmid carrying the *dnaA*⁺ gene (pFHC539, Figure 5) we determined whether increased amounts of DnaA protein stimulated DNA synthesis. Plasmid pFHC539 suppressed the temperature sensitivity of the *dnaA*(Ts) mutants, and the DNA content and cell size were restored to values not significantly different from a *dnaA*⁺ wildtype. The presence of pFHC539 in *dnaA*⁺ cells increased the DNA content only marginally although the DnaA protein content was increased about fivefold as estimated by radioactive labeling and 2D-gel electrophoresis (Ole Skovgaard, unpublished). This finding would indicate that the DnaA protein as such is not the rate-determining factor for initiation of chromosome replication. This conclusion was independ-

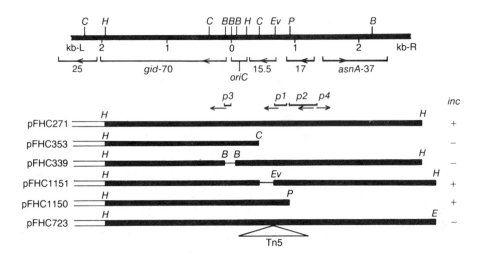

FIGURE 6 Structure of plasmids used to study incompatibility exerted towards the chromosome in dnaA46 strains. The plasmids are aligned with a detailed genetic and physical map of the oriC region. Genes are indicated with their genetic symbol and/or molecular weight. Arrows indicate the direction of transcription from the promoters in the oriC region (p1, p2, p3, p4) as well as in the flanking genes. Restriction enzymes used in the construction of the plasmids were: B: BamHI, C: ClaI, Ev: EcoRV, H: HindIII, P: PvuI. Incompatibility properties are indicated to the right (Inc). + denotes that the plasmid affected growth and chromosome replication in a dnaA46 strain. - denotes that the plasmid had no effect.

ently reached by Churchward et al. (1983). However, we observed a small increase in DNA concentration that was paralleled by a decrease in cell size, indicating a minor stimulation of DNA synthesis by increased level of DnaA protein. Results presented by Atlung et al. (this volume) indicate that an increased concentration of DnaA protein actually stimulated initiation of replication, but without concomitant increase of total DNA synthesis or capacity to synthesize DNA in absence of protein synthesis.

The amount of DnaA protein produced in the strains carrying pFHC539 was considerably less than what would be expected from the 20-fold-increased gene dosage of the *dnaA* gene. This result is an indication of autorepression of *dnaA* gene expression (Atlung et al., this volume). The autorepression appears to be exerted at a 11-bp sequence between the two *dnaA* promoters (Figure 5; Hansen et al., 1982b) that also was found to be present four times in *oriC* and once in the promoter region for the gene encoding the 16-kD protein (Figure 6). This sequence is highly conserved in the *oriC* loci present in different Enterobacteriaceae (Zyskind et al., 1981). The autorepression of *dnaA* gene expression is expected also to affect the expression of the other DNA metabolism gene(s) in the *dnaA* operon (Figure 2).

Plasmids of the same structure as pFHC539 (see Figure 5) carrying any one of the *dnaA*(Ts) mutations *dnaA5, dnaA46, dnaA167, dnaA203, dnaA204,* or *dnaA205* were found to complement the respective *dnaA*(Ts) mutations as well as other alleles at 42 C. This suggested that increased amounts of mutated DnaA(Ts) protein exerted sufficient activity even at 42 C to allow initiation of replication to occur. Some of the *dnaA*(Ts) plasmids (e.g., with *dnaA5* or *dnaA46*) rendered the strains extremely cold sensitive.

THE EFFECTS OF 16-kD GENE TRANSCRIPTION ON *oriC* FUNCTION

Transcription in the oriC region

We used pBR322 as a promoter-cloning vehicle to localize promoters in the *oriC* region (Hansen et al., 1981a). By this *in vivo* approach two promoters to the right (*p1, p2*) and one to the left of *oriC* (*p3*), were defined (Figure 6). The promoters within *oriC,* identified by *in vitro* experiments (Lother et al., 1981) were not revealed in our *in vivo* experiments in which we tested expression of the *tet* gene by measuring tetracycline resistance.

The promoter for the 16-kD-protein gene *p1* gave transcription that enters *oriC* of which—contrary to an earlier interpretation (Hansen et al., 1981a) — only about half is terminated within *oriC.* This transcription into *oriC* caught our attention, because transcription was a well-documented prerequisite for initiation at *oriC* (Lark, 1972, Messer, 1972, von Meyenburg et al., 1979).

16-kD gene transcription and the copy number of oriC plasmids

The 16-kD-gene transcription from *p1* affected the copy number of pBR322-*oriC* chimeras (Hansen et al., 1981a). The copy number of pFHC271 (see Figure 6), which had the intact promoter *p1,* appeared to be four times higher than the one of pBR322. pFHC353, which had the 16-kD-gene promoter *p1* deleted, has only twice the pBR322 copy number. Stuitje and Meijer (1983) observed the same effect of the 16-

kD transcription when monitoring the copy number of minichromosomes. The presence of *oriC* on pBR322 significantly increased the copy number, which we take to indicate that *oriC* was functioning in these plasmids.

Incompatibility between pBR322-oriC chimeras and the chromosome in dnaA(Ts) mutants

Despite the presence of perhaps more than 100 copies of *oriC*, the growth of *dnaA*[+] cells carrying pFHC271 was only slightly affected: a 10% reduction in growth rate, a slightly increased average cell size, and a somewhat heterogeneous distribution in cell size was found. In contrast, the plasmid pFHC271 had deleterious effects in *dnaA*(Ts) strains: growth rate was reduced considerably, the cells were elongated, and, in the absence of selection, plasmidless cells occurred with a high frequency, at 30 C. Plasmid pFHC353 did affect growth of neither *dnaA*[+] nor *dnaA*(Ts) cells with the exception of those with the *dnaA203* or *dnaA204* allele.

The deleterious effects of pFHC271 in *dnaA*(Ts) strains could be suppressed by introducing a *dnaA*[+] plasmid. The effect could also be suppressed by integration of a mini R1 plasmid into the chromosome. These observations demonstrate that the pFHC271 chimeric plasmid compete with the chromosome for the DnaA(Ts) protein. If the chromosome was allowed to initiate by another mechanism or if DnaA[+] protein was provided, the cells were able to grow normally. Also, an excess amount of DnaA46 protein was found to relieve the deleterious effects caused by pFHC271. Also, the adverse effects of pFHC353 in *dnaA203* and *dnaA204* cells were suppressed by integration of an R1 replicon or by supplying DnaA[+] protein.

We localized the site on pFHC271 responsible for those effects in *dnaA*(Ts) strains by constructing different plasmid pFHC271 derivatives. The results are summarized in Figure 6. We interpret the data as follows: Both an intact *oriC* and an intact 16-kD-gene promoter *p1* are required to yield the effects, which were termed incompatibility (Inc), since they appear to be due to a site on the plasmid competing for DnaA(Ts) protein with *oriC* on the chromosome (Inc[+]: pFHC271 and pFHC1150). The 16-kD protein is clearly not needed, since a deletion removing most of its gene (pFHC1151) did not suppress the Inc[+] phenotype. However, Tn5 insertions, which interrupted the transcription from *p1* into *oriC* (e.g., pFHC723), alleviated the incompatibility. Therefore, the site on pFHC271 responsible for the Inc[+] effect in *dnaA*(Ts) mutants could be localized to the region between 0.5 and 0.95 kb-R (Figure 6), i.e., the same region which has formerly been designated *incB* (Yamaguchi et al., 1982; Stuitje and Meijer, 1983).

We replaced the *p1* promoter with several other promoters. The *dnaA* promoters including the 11-bp DnaA protein binding site gave essentially the same response as *p1*. Surprisingly, replacement by the promoters of the *lac* or *trp* operon did not result in the incompatibility, though the levels of transcription are similar to or even greater than that from *p1*.

These observations could be interpreted as follows. Transcription entering *oriC* from the 16-kD promoter leads to an increased "consumption" of DnaA protein. In case of the mutated, only partially active DnaA(Ts) protein (e.g., *dnaA46*) at 30 C this leads to a level of DnaA activity insufficient to initiate at *oriC* at a normal frequency. The observation that incompatibility was provoked only by transcription into *oriC* from *p1* or the *dnaA* promoters, but not from the *lac* or *trp* promoters, suggests that the 11-bp DnaA binding site present in the two former promoters somehow destines the fate of the transcript(ion).

Effect of the 16-kD-gene transcription on chromosome replication

In an approach to study the significance of the 16-kD promoter for initiation of replication at *oriC* in the chromosome a strain was constructed with Tn5 inserted in the 16 kD gene at about 0.45 kb-R (Figure 6). We have estimated the DNA content of this strain (16-kD::Tn5-39) at different growth rates (Figure 7). The insertion led to an increase in DNA content at low growth rates as compared to the wildtype. Introducing plasmid pFHC339, which carries the intact 16-kD gene (Figure 6) into the 16-kD::Tn5-39 strain reduced the high DNA content to a level which was lower than that in the wildtype strain (Figure 7). Also, plasmids pFHC247, carrying only the 16-kD, 17-kD and *asnA* genes, and pLBC45, in which part of the 16-kD gene was deleted, caused a similar decrease of DNA content.

These results suggested that the Tn5 insertion in the 16-kD

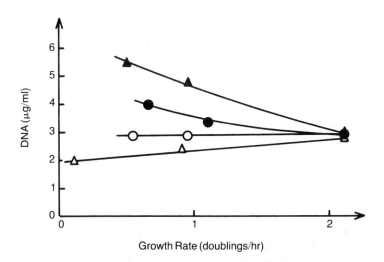

FIGURE 7 Effect of the 16-kD::Tn5-39 mutation on DNA content as a function of growth rate. Closed circles: 'wildtype' (CM3159). Closed triangles: strain CM3164 with insertion Tn5-39 in the 16-kD gene. Open circles: CM3159/pFHC339. Open triangles: CM3164/pFHC339. Strains were grown at 37°C in the following media: AB glucose + casamino acids + uracil; AB + glucose + requirements; AB + succinate + requirements. DNA content is given as µg/ml of a culture at A450 = 1.000.

gene prevented the formation of a repressorlike element; the lack of it would lead to an increased initiation frequency. It would be expressed and supplied in *trans* from a plasmid carrying a DNA-segment to the right of *oriC* including part of the 16-kD gene, probably including promoter *p1*.

Alternatively, the transcription starting from the 16-kD promoter may create a site which could lead to "repression" by immobilizing an initiation factor; if transcription was blocked, this site would not be activated, leading to an increased initiation frequency; if that site is supplied in *trans*, it would, if activated by transcription, again immobilize that initiation factor and thereby lead to a reduced initiation frequency. In conjunction with the results presented above on the incompatibility of pBR322-*oriC* chimeras in *dnaA*(Ts) cells it is tempting to suggest that the DnaA protein is the "initiation factor" (function) being immobilized.

TIMING OF CHROMOSOME AND MINICHROMOSOME REPLICATION

As discussed above, the effects exerted by the region containing the 16-kD and 17-kD genes are fairly subtle: the genes had been found to be dispensable for growth and DNA replication from *oriC* (von Meyenburg and Hansen, 1980), and together with *oriC*, they exhibit an effect that manifested itself as incompatability, as evidenced by increased loss rates of minichromosomes (Stuitje and Meijer, 1983). Therefore, it was interesting to see whether replication of the chromosome was affected by the presence of minichromosomes.

The time interval between consecutive replications of a given DNA segment, the *interreplication time*, was determined for both the minichromosome and chromosomal genes by density-shift experiments, essentially according to Eberle (1968; Luud Koppes, unpublished). The transfer of heavy DNA, pulse-labeled by [^{3}H]thymidine, to the intermediate density upon replication was occurring quite precisely at one doubling time after the density shift of a strain without a minichromosome (Figure 8, strain CM796); the coefficient of variation (CV%) being 12-15% in fair agreement with published data (Eberle, 1968; Newman and Kubitschek, 1978). The presence of minichromosome pλ*asn212* led to a less sharp appearance of chromosomal [^{3}H]DNA of intermediate density (Figure 8), the CV% being 25-28%; the minichromosome itself was found to be re-replicated with kinetics similar to the one of the host chromosome (CV% = 28-33%) (Figure 8). Thus, the presence of the minichromosome increased the spread of the interreplication times of the chromosome, representing a less precise timing of initiation; this is taken to reflect incompatibility between the minichromosome and the chromosome. The spread of the interreplication time of the minichromosome (which is actually the same as its interinitiation time) was very similar to the one of the host chromosome and clearly "nonrandom". Therefore, we can conclude that minichromosomes obey essentially the same timing control as the chromosome. This conclusion appears to be in conflict with results of Leonard et al. (1982); however, in those experiments a minichromosome devoid of the 16- and 17-kD protein genes was analyzed.

CONCLUSIONS

The precise genetic and physical mapping of the *oriC* and *dnaA* region of the *E. coli* K12 chromosome, including the cloning of the *dnaA*

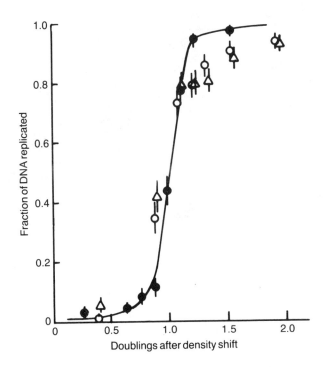

FIGURE 8 Density-shift experiment illustrating sequential replication of chromosomal and minichromosomal DNA. Cells growing exponentially in heavy (^{13}C]glucose, [^{15}N]NH$_4$Cl) were pulse-labeled with [^{3}H]thymidine in heavy (^{13}C]glucose, [^{15}N]NH$_4$Cl) medium and transferred to light medium immediately after the pulse. Relative times of appearance of radioactivity in DNA of hybrid density were determined by CsCl density gradient centrifugation and subsequent assay for specific DNA segments by DNA/DNA hybridization using probes for either the chromosomal atp and rpoBC genes (average/value, open circles) or the λ DNA of the pλasn212 (Figure Ib) minichromosome (triangles) in the asnA⁻asnB strain ER (von Meyenburg et al., 1979). Solid circles represent the atp/rpo data of a control strain without minichromosome (strain ER with λcl857 and the asnB phage λasn53 integrated at attλ = strain CM796).

gene and the construction of a variety of minichromosomes, opened up a diversity of *in vivo* as well as *in vitro* experiments (Messer et al., 1982; Fuller and Kornberg, 1983).

The one of our *in vivo* approaches reported here, that most directly revealed new aspects of the chromosome and minichromosome replication, was the density-shift experiment (previous section). First, this confirmed that initiation of chromosome replication is indeed well timed in *E. coli* (Eberle, 1968; Newman and Kubitschek, 1978), since the spread of interreplication times of specific DNA segments (and inter-initiation times when assaying for the atp segment that neighbors oriC) was small. The presence of a minichromosome diminished this precision of consecutive initiation timing, which in turn appeared to be the same for the minichromosome replication. It indicated (1) that there is interaction between minichromosomes and the chromosome-- probably through competition for events at *oriC* (incompatibility) and (2) that the initiation at *oriC* on the chromosome and on the minichromosomes was subject to the same regulation. It should be noted that these experiments do not directly contribute to the question of timing control in the division cycle.

Let us summarize the elements that affect initiation of replication at *oriC* and that hence may be involved in the interaction between the minichromosomes and the chromosome. These are the following:

1. Transcription from the 16-kD-gene promoter, which may activate a "site" for binding and immobilizing an initiation factor(s). This site appears to be different from *oriC*.

2. The intact and functional *oriC* sequence, which triggers the initiation event and thereby possibly leads to a consumption of DnaA protein.

3. The DnaA protein, which regulates its own synthesis through autorepression, may be the limiting factor for the frequency of initiation both in *dnaA*(Ts) mutants (Hansen and Rasmussen, 1977) and in wildtype cells (Atlung et al., this volume).

The effects of the *atp* operon when carried on minichromosomes on growth and division of *E. coli* must, in the light of the new cytological findings, be interpreted as having resulted from the overproduction of the membrane-bound ATP synthase, which led to disturbance of the cytoplasmic membrane structure (von Meyenburg et al., 1984).

The finding of a high copy number of minichromosomes (von Meyenburg et al., 1979) still remains a mystery. It would be under-standable if a replisome initiating at one minichromosome after finishing a round of replication, which only takes a very short time (von Meyenburg et al., 1979), were able to initiate at another *oriC* site either on the chromosome or on another minichromosome (see also Helmstetter et al., this volume). This idea implies that active replisomes retain the potential for initiation for a while and only lose it when "working" for a long time on the chromosome.

In this context it is worth noticing that the basis for the cessation of initiation of replication at *oriC* after blockage of protein synthesis remains unknown (Maaløe and Hanawalt, 1961). Is it the activity of the DnaA protein that is short-lived, as results of Schaus et al. (1981) might indicate?

We can hardly claim to have arrived at an understanding of the function of 16-kD-gene transcription, DnaA protein and *oriC* and their interaction in the control of initiation of replication at *oriC*. However, we are confident that the findings reviewed here will prove helpful in the development of further *in vivo* and *in vitro* approaches towards this central problem of bacterial physiology.

ACKNOWLEDGEMENTS

We thank Tok Kogoma for constructive criticism and comments on the manuscript. This work was made possible through grants from the Danish Natural Science Research Council, the NOVO Foundation, and the Stiftung Volkswagenwerk.

REFERENCES

Abe, M. and Tomizawa, J-I. (1971). *Genetics*, 69, 1.

Atlung, T. (1981). *ICN-UCLA Symp. Mol. Cell. Biol.*, 21, 297.

Atlung, T., Clausen, E., and Hansen, F. G. (1984). In *Proteins Involved in DNA Replication*, U. Hubscher and S. Spadari (eds.), Plenum, p. 199.

Bachmann, B. J. (1983). *Microbiol. Rev.*, 47, 180.

Beyersmann, D., Messer, W., and Schlicht, M. (1974). *J. Bacteriol.*, 118, 783.

Brunschede, H., Dove, T. L., and Bremer, H. (1977). *J. Bacteriol.*, 129, 1020.

Bukh, H-J. and Messer, W. (1983). *Gene*, 24, 265.

Carl, P. L. (1970). *Mol. Gen. Genet.*, 109, 107.

Churchward, G., Holmans, P., and Bremer, H. (1983). *Mol. Gen. Genet.*, 192, 506.

Clark, D. J. and Maaløe, O. (1967). *J. Mol. Biol.*, 23, 99.

Deeley, M. C. and Yanofsky, C. (1981). *J. Bacteriol.*, 147, 787.

De Massy, B., Fayet, O., and Kogoma, T. (1984). *J. Mol. Biol.*, 178, 277.

Eberle, H. (1968). *J. Mol. Biol.*, 31, 149.

Fuller, R. and Kornberg, A. (1983). *Proc. Natl. Acad. Sci. USA*, 80, 5817.

Gay, N. J. and Walker, J. E. (1981a). *Nucl. Acids Res.*, 9, 2187.

Gay, N. J. and Walker, J. E. (1981b). *Nucl. Acids Res.*, 9, 3919.

Gellert, M., Menzel, R., Mizuuchi, K., O'Dea, M. H., and Friedman, D. I. (1983). *Cold Spring Harbor Symp. Quant. Biol.*, 47, 763.

Gotfried, F. and Wechsler, J. A. (1977). *J. Bacteriol.*, 130, 963.

Hansen, E. B., Hansen, F. G., and von Meyenburg, K. (1982). *Nucl. Acids Res.*, 10, 7373.

Hansen, E. B., Atlung, T., Hansen, F. G., Skovgaard, O., and von Meyenburg, K. (1984). *Mol. Gen. Genet.*, 196, 387.

Hansen, F. G. and Rasmussen, K. V. (1977). *Mol. Gen. Genet.*, 155, 219.

Hansen, F. G. and von Meyenburg, K. (1979). *Mol. Gen. Genet.*, 175, 135.

Hansen, F. G., Koefoed, S., von Meyenburg, K., and Atlung, T. (1981a). *ICN-UCLA Symp. Mol. Cell. Biol.*, 22, 37.

Hansen, F. G., Nielsen, J., Riise, E., and von Meyenburg, K. (1981b). *Mol. Gen. Genet.*, 183, 463.

Hansen, F. G., Hansen, E. B., and Atlung, T. (1982b). *EMBO J.*, 1, 1043.

Hirota, Y., Ryter, A., and Jacob, F. (1968). *Cold Spring Harbor Symp. Quant. Biol.*, 33, 677.

Hirota, Y., Mordoh, J., and Jacob, F. (1970). *J. Mol. Biol.*, 53, 369.

Hirota, Y., Yasuda, S., Yamada, M., Nishimura, A., Sugimoto, K., Sugisaki, H., Oka, A., and Takanami, M. (1979). *Cold Spring Harbor Symp. Quant. Biol.*, 43, 129.

Kellenberger-Gujer, G., Podhajska, A. J., and Caro, L. (1978). *Mol. Gen. Genet.*, 162, 9.

Kogoma, T. and von Meyenburg, K. (1983). *EMBO J.*, 2, 463.

Lark, K. G., Repko, T., and Hoffman, E. J. (1963). *Biochim. Biophys. Acta*, 76, 9.

Lark, K. G. (1972). *J. Mol. Biol.*, 64, 47.

Leonard, A. C., Hucul, J. A., and Helmstetter, C. E. (1982). *J. Bacteriol.*, 149, 499.

Lother, H., Bukh, H-J., Morelli, G., Heimann, B., Chakrabarty, T., and Messer, W. (1981). *ICN-UCLA Symp. Mol. Cell. Biol.*, 22, 57.

Lycett, G. W., Orr, E., and Pritchard, R. H. (1980). *Mol. Gen. Genet.*, 178, 329.

Maaløe, O. and Hanawalt, P. C. (1961). *J. Mol. Biol.*, 3, 144.

Maaløe, O. (1979). In *Biological Regulation and Development*, Vol. I, *Gene Expression*, R. F. Goldberger (ed.), Plenum, p. 487.

Marsh, R. C. and Worcel, A. (1977). *Proc. Natl. Acad. Sci. USA*, 74, 2720.

Messer, W. (1972). *J. Bacteriol.*, 112, 7.

Messer, W., Dankwarth, L., Tippe-Schindler, R., Womack, J. E., and Zahn, G. (1975). *ICN-UCLA Symp. Mol. Cell. Biol.*, 3, 602.

Messer, W., Bergmans, H. E. N., Meijer, M., Womack, J. E., Hansen, F. G., and von Meyenburg, K. (1978). *Mol. Gen. Genet.*, 162, 269.

Messer, W., Meijer, M., Bergmans, H. E. N., Hansen, F. G. von Meyenburg, K., Beck, E., and Schaller, H. (1979). *Cold Spring Harbor Symp. Quant. Biol.*, 43, 139.

Miki, T., Hiraga, S., Nagata, T., and Yura, T. (1978). *Proc. Natl. Acad. Sci. USA*, 75, 5099.

Miki, T., Kimura, M., Hiraga, S., Nagata, T., and Yura, T. (1979). *J. Bacteriol.*, 140, 817.

Newman, C. N. and Kubitcheck, H. E. (1978). *J. Mol. Biol.*, 121, 461.

Nordström, K., Molin, S., and Light, J. (1984). *Plasmid*, 12, 71.

Murakami, A., Inokuchi, H., Hirota, Y., Ozeki, H., and Yamagishi, H. (1980). *Mol. Gen. Genet.*, 180, 235.

Nielsen, J., Hansen, F. G., Hoppe, J., Friedl, P., and von Meyenburg, K. (1981). *Mol. Gen. Genet.*, 184, 33.

Nielsen, J., Jørgensen, B. B., von Meyenburg, K., and Hansen, F. G. (1984). *Mol. Gen. Genet.*, 193, 64.

Nishimura, Y., Caro, L., Berg, C. M., and Hirota, Y. (1971). *J. Mol. Biol.*, 55, 4410.

Ogawa, T., Pickett, G. G., Kogoma, T., and Kornberg, A. (1984). *Proc. Natl. Acad. Sci. USA*, 81, 1040.

Ogura, T., Miki, T., and Hiraga, S. (1980). *Proc. Natl. Acad. Sci. USA*, 77, 3993.

Oka, A., Sugimoto, K., Takanami, M., and Hirota, Y. (1980). *Mol. Gen. Genet.*, 178, 9.

Pritchard, R. H., Barth, P. T., and Collins, J. (1969). *Symp. Soc. Gen. Microbiol.*, 19, 263.

Pritchard, R. H. (1978). In *DNA Synthesis: Present and Future*, I. Molineux and M. Kohiyama (eds.), Plenum, p. 1.

Ream, L. W., Margossian, L., Clark, A. J., Hansen, F. G., and von Meyenburg, K. (1980). *Mol. Gen. Genet.*, 180, 115.

Sakakibara, Y. and Mizukami, I. (1980). *Mol. Gen. Genet.*, 178, 541.

Sakakibara, Y., Tsukano, H., and Sako, T. (1981). *Gene*, 13, 47.

Sako, T. and Sakakibara, Y. (1980). *Mol. Gen. Genet.*, 179, 521.

Saraste, M., Gay, N. J., Eberle, A., Runswick, M. J., and Walker, J. E. (1981). *Nucl. Acids. Res.*, 9, 5287.

Sevastopoulos, C. G., Wehr, C. T., and Glaser, D. A. (1977). *Proc. Natl. Acad. Sci. USA*, 74, 3485.

Sompayrac, L. and Maaløe, O. (1973). *Nature New Biol.*, 241, 133.

Stuitje, A. and Meijer, M. (1983). *Nucl. Acids Res.*, 11, 5775.

von Meyenburg, K., Hansen, F. G., Nielsen, L. D., and Jørgensen, P. (1977). *Mol. Gen. Genet.*, 158, 101.

von Meyenburg, K., Hansen, F. G., Nielsen, L. D., and Riise, E. (1978). *Mol. Gen. Genet.*, 160, 287.

von Meyenburg, K., Hansen, F. G., Riise, E., Bergmans, H. E. N., Meijer, M., and Messer, W. (1979). *Cold Spring Harbor Symp. Quant. Biol.*, 43, 121.

von Meyenburg, K. and Hansen, F. G. (1980). *ICN–UCLA Symp. Mol. Cell. Biol.*, 19, 137.

von Meyenburg, K., Jørgensen, B. B., Nielsen, J., and Hansen, F. G. (1982a). *Mol. Gen. Genet.*, 188, 240.

von Meyenburg, K., Jørgensen, B. B., Nielsen, J., Hansen, F. G., and Michelsen, O. (1982b). *Tokai Exp. Clin. Med.*, 7, 23.

von Meyenburg, K., Jørgensen, B. B., and van Deurs, B. (1984). *EMBO J.*, 3, 1791.

Wechsler, J. A. and Gross, J. D. (1971). *Mol. Gen. Genet.*, 113, 273.

Wechsler, J. A. and Zdziecnika, M. (1974). *ICN–UCLA Symp. Mol. Cell. Biol.*, 3, 624.

Wechsler, J. A. (1980). *J. Bacteriol.*, 144, 856.

Yamaguchi, K., Yamaguchi, M., and Tomizawa, J-I. (1982). *Proc. Natl. Acad. Sci. USA*, 79, 5347.

Yasuda, S. and Hirota, Y. (1977). *Proc. Natl. Acad. Sci. USA*, 74, 5458.

Yuasa, S. and Sakakibara, Y. (1980). *Mol. Gen. Genet.*, 180, 267.

Zahn, G., Tippe-Schindler, R., and Messer, W. (1977). *Mol. Gen. Genet.*, 153, 45.

Zahn, G. and Messer, W. (1979). *Mol. Gen. Genet.*, 168, 197.

Zyskind, J. W., Deen, L. T., and Smith, D. W. (1977). *J. Bacteriol.*, 129, 1466.

Zyskind, J. W., Harding, N. E., Takeda, Y., Cleary, J. M., and Smith, D. W. (1981). *ICN-UCLA Symp. Mol. Cell. Biol.*, 22, 13.

Role of the DnaA Protein in Control of DNA Replication

Tove Atlung, Knud V. Rasmussen, Erik Clausen
The University Institute of Microbiology, Copenhagen

Flemming G. Hansen
The Technical University of Denmark, Lyngby-Copenhagen

INTRODUCTION

Chromosome replication in *E. coli* is regulated primarily at the level of initiation at the origin *oriC*. Initiation takes place, independently of growth rate, at a constant ratio of origin to mass, the initiation mass (Schaechter et al., 1958; Cooper and Helmstetter, 1968; Donachie, 1968; Pritchard et al., 1969), and requires *de novo* protein synthesis (Maaløe and Hanawalt, 1961; Lark et al., 1963). In the replicon model Jacob, Brenner and Cuzin (1963) proposed that a unit capable of autonomous replication carries two specific determinants: a structural gene coding for an initiator protein and a replicator site (or origin) upon which the initiator acts. Later, in an attempt to account for the constant initiation mass Sompayrac and Maaløe (1973) suggested (1) that a fixed amount of initiator must be accumulated for each initiation event to take place and (2) that the gene coding for the initiator is cotranscribed with an autorepressor gene.

Studies of more than 15 conditional lethal *dnaA* mutants [*dnaA*(Ts), see Hansen et al. (1984) for references; *dnaA*(Am), Schaus et al. (1981)] have shown that the DnaA protein is essential for initiation of chromosome replication. Among the known replication factors the DnaA protein is unique, because it is required specifically (Frey et al., 1979) for initiation from *oriC* (Marsh and Worcel, 1977; von Meyenburg et al., 1979; Lycett et al., 1980). Thus, DnaA fits the concept of a specific initiator proposed in the replicon model.

The DnaA protein acts before or during the transcriptional step in the initiation process (Lark, 1972; Zyskind et al., 1977; Fuller and Kornberg, 1983). A number of mutations in *rpoB*, the gene coding for the beta subunit of RNA polymerase, have been shown to suppress *dnaA*(Ts) mutations (Bagdasarian et al., 1977; Atlung, 1981). All *dnaA*(Ts) mutants can be suppressed by *rpoB* mutations, and *rpoB* mutations show allele specificity, i.e., each suppresses some but not all *dnaA*(Ts) alleles (Atlung, 1981; Hansen et al., 1984; T. Atlung, unpublished results), which indicates that the DnaA protein interacts with the RNA polymerase during initiation of replication at *oriC*.

Mutational changes in the DnaA protein and in RNA polymerase can cause overinitiation of DNA replication. In the *dnaA-Cos* mutant, an intragenic pseudorevertant of *dnaA46*, the concentrations of origins and of DNA increase three- to fivefold following a shift to low temperature (Kellenberger-Gujer et al., 1978). Rasmussen et al. (1983) have isolated a mutant (*rpoC907*) that overinitiates two- to threefold when shifted to high temperature. Tanaka et al. (1983) have described several similar mutations in *rpoB* and *rpoC* that cause unconditional overinitiation.

Studies on initiation capacity in the *dnaA46* mutant at intermediate growth temperatures, at which activity of the DnaA protein is limiting for initiation, led Hansen and Rasmussen (1977) to propose that the *dnaA* gene is autoregulated and that when the initiator activity of the DnaA46 protein is heat- inactivated, the protein also loses its activity as an autorepressor.

Thus, physiological studies of the *dnaA* mutants suggested that control of replication might be carried out via an autorepressor-initiator system, as proposed by Sompayrac and Maaløe, with the DnaA protein acting as both autorepressor and initiator.

We now give evidence that the *dnaA* gene is autoregulated, with the DnaA protein acting as an autorepressor at an operator site between the two *dnaA* promoters (Hansen et al., 1982). Expression of a *dnaA-lacZ* fusion was repressed by increased concentrations of DnaA protein and was derepressed by thermoinactivation of the activity of DnaA in Ts mutants. The effect of overproduction of DnaA protein on initiation of chromosome replication was studied using a plasmid in which the *dnaA* gene is under control of the *pL* promoter of phage lambda. We found that induction of synthesis of DnaA protein has no effect on the total DNA concentration but causes amplification of genes close to the chromosomal origin, suggesting that an increased concentration of DnaA protein stimulate initiation at *oriC*.

AUTOREGULATION OF THE *dnaA* GENE

The dnaA gene of E. coli

The *dnaA* gene maps at 83 min on the *E. coli* genetic map and has been cloned and sequenced (Hansen and von Meyenburg, 1979; Hansen et al., 1982). S1 nuclease mapping of *in vivo* transcript starts (Hansen et al., 1982) showed that *dnaA* transcription is initiated from two promoters located 150 and 240 bp upstream from the start of the structural gene (see Figure 1). Between the two promoters, around position -60 of *dnaA2p*, we found an 11-bp sequence that is present four times within *oriC* and once in the -35 region of the 16-kD gene promoter, from which transcription into and through *oriC* originates (von Meyenburg et al., this volume).

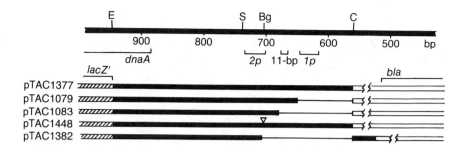

FIGURE 1 Structure of the dnaA"lacZ fusions. Top: Map of the dnaA promoter region showing a few restriction enzyme sites, position of promoters 1p and 2p, the 11-bp oriC homology-DnaA binding site, and the start of the structural dnaA gene. Below is shown the structure of the fusion plasmids used in Table 1. The plasmids are derivatives (kindly supplied by K. Clemmensen and C. Pedersen) of pMLB1034 (Weinstock et al., 1983), a pBR322 replicon carrying the bla gene (resistance to ampicillin), and the lacZ gene with an EcoRI site in the 11th codon allowing in-frame fusion with the dnaA coding sequence. The endpoints of the Bal31 deletions in plasmids PTAC1079 and 1083 have been determined by sequencing. Black bar: DNA from the dnaA promoter region; hatched bar: lacZ DNA; white bar: pBR322 DNA. Thin line: deleted DNA segment. Triangle: 4-bp insertion in the BglII site. pTAC1382 carries a small piece of rnpA DNA. Restriction enzyme sites: Bg: BglII; C: ClaI; E: EcoRI; S: SacII.

Autorepression of the dnaA gene takes place at the 11-bp oriC homology

To facilitate the work on regulation of the dnaA gene, two fusions were constructed. One is an operon fusion between the dnaA promoters and the intact structural tet gene of pBR322 (Atlung et al., 1984), and the other is a gene fusion with the lacZ gene (Figure 1). To test for autoregulation of the dnaA gene by increased concentrations of DnaA protein the tet and lacZ fusion plasmids (both pBR322 replicons) were introduced into strains containing a compatible plasmid (a PACYC184 derivative) carrying either the intact structural dnaA gene and its promoters (pFHC871, which has the same chromosomal DNA as pFHC539, Figure 5, von Meyenburg et al., this volume), or a derivative thereof (pTAC1302) with a 121-bp deletion inside the structural gene (identical to that in pTAC1431, Figure 2). Expression of the fusions was monitored by testing for the level of tetracycline resistance (Tet-r), as described by Hansen et al. (1981), or by measuring beta-galactosidase activity.

In the presence of the dnaA+ plasmid (pFHC871) we observed a twofold reduction both in the level of Tet-r (Atlung et al., 1984) and in the rate of synthesis of the DnaA-beta-galactosidase fusion protein (Table 1) from the plasmid carrying an intact dnaA promoter region. The presence of the control plasmid pTAC1302 had no effect on dnaA gene expression (data not shown), which indicates that the DnaA protein inhibits transcription of its own gene. Tet-r expressed

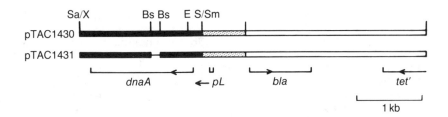

FIGURE 2 Structure of the λpL-dnaA plasmids. The λ pL promoter (from pLc28; Remaut et al., 1981) was inserted in the BglII site in front of the dnaA gene. The remnants of dnaA2p were subsequently deleted to obtain a plasmid with dnaA under exclusive control by λ pL. Finally pTAC1430 was obtained by recloning to pBR327. The analogous plasmid pTAC1431, carrying the 121-bp deletion inside the dnaA gene, was constructed as a control. The position and extent of structural genes and their direction of transcription are given at the top. Black bar: chromosomal DNA; dotted bar: λ DNA; white bar: pBR327 DNA; thin line: deleted DNA segment. Restriction enzyme symbols: Bs: BssHII; E: EcoRI; S: SacII; Sa: SalI; Sm: SmaI; X: XhoI.

from other promoters, e.g., the normal *tet* gene promoter, was unaffected.

Analysis of deletions in the leader region, showed that the regulatory effect was independent of sequences downstream of the transcript start site for *dnaA2p* (Atlung et al., 1984). A 4-bp insertion in the −35 region of *dnaA2p* (pTAC1448; Figure 1) had little effect

TABLE 1   Autoregulation of *dnaA-lacZ* fusion plasmids

Fusion plasmids	Promoter structure	β-gal u/ml x A_{450}	
		pTAC1302 *dnaA121*	pFHC871 *dnaA$^+$*
pTAC1377	Wildtype	700	390
pTAC1079	−1p	570	360
pTAC1083	−1p −11bp	460	480
pTAC1448	4bp	600	600
pTAC1382	−1p −2p	12	ND

Derivatives of strain MC1000 recA1 containing the indicated plasmids were grown exponentially in NY medium (von Meyenburg et al., 1982) containing 200 µg/ml of ampicillin and 20 µg/ml of chloramphenicol. The differential rate of synthesis of β-galactosidase was determined from 6-8 samples according to Miller (1972) in sonicated cells extracts.

ND = not determined.

on transcription in the control strain, but relieved the repression (Table 1). By deleting increasing amounts of DNA upstream from *dnaA2p* using exonuclease Bal31 we found that deletions removing only *dnaA1p* (pTAC1079, Figure 1) had little effect on *dnaA-lacZ* gene expression (Table 1), indicating that this is not the major promoter. Transcription from *dnaA2p* in pTAC1079 was still repressed by surplus DnaA protein. Deletions extending into or beyond the 11-bp *oriC* homology, but leaving the −35 region intact (e.g., pTAC1083), decreased transcription from *dnaA2p* by 20% in the control strain and eliminated the repression by the *dnaA*+ plasmid pFHC871

Physiological studies on autoregulation of the dnaA gene

The differential rate of synthesis of DnaA-beta-galactosidase (henceforth call the fusion protein) per gene copy from the plasmid pTAC1377 was nearly the same at high and low growth rates, suggesting that the DnaA protein is synthesized as a fairly constant fraction of total protein and independently of growth rate.

A higher degree of repression of the fusion protein by the presence of pFHC871 was observed at the low growth rate (Table 2). This is probably caused by a higher concentration of DnaA protein resulting from the increased copy number of pFHC871 with decreasing growth rate. From the plasmid copy numbers and the rate of synthesis of the fusion protein it can be estimated that the presence of pFHC871 leads to a two- to threefold increase in the DnaA protein concentration at the high growth rate and to a six- to eightfold increase at low growth rate. Under these conditions *dnaA* gene expression was 65% and 35%, respectively, of the level found in the haploid strain. Transcription of the *dnaA* gene can only be reduced to approximately 25% of the normal level by excess DnaA protein, as shown by an experiment in which synthesis of the fusion protein from pTAC1377 was followed after induction of a plasmid carrying the *dnaA* gene under control of *pL* of phage lambda (data not shown). We found that the rate of synthesis of the fusion protein was reduced fourfold after the induction of synthesis of DnaA protein and that this rate was then maintained for more than two hours.

Derepression of dnaA gene expression in dnaA(Ts) mutants

The plasmids pTAC1377 and pTAC1448 were introduced into the isogenic *dnaA*+ and *dnaA46* strains CM732 and CM734 (Hansen and von Meyenburg, 1979). Upon a temperature shift from 30 C to 42 C we found that the differential rate of synthesis of the fusion protein from pTAC1377 increased threefold in the mutant strain, while it was unaffected in the *dnaA*+ strain (data not shown). In both strains synthesis of the fusion protein from pTAC1448 was unaffected by the temperature shift. Results similar to those with the *dnaA46* strain were found with strains carrying other *dnaA*(Ts) alleles, including 167, 204, and 508. This showed that at high temperature the autorepressor activity of the in the *dnaA*(Ts) mutants was inactivated, as proposed by Hansen and Rasmussen (1977).

The derepression of *dnaA* gene expression in the *dnaA*(Ts) strains suggests that the level of DnaA protein normally present in

TABLE 2   Effect of growth rate on expression of the *dnaA-lacZ* fusion

Plasmid or genes	Units β-gal x A_{450} at growth rate (per hr)	
	2.3	0.46
pTAC1377 + pTAC1302(*dnaA121*)	700	2650
pTAC1377 + pFHC871(*dnaA*⁺)	460	900
	Copy number/genome equivalent at growth rate listed above	
pTAC1377	10	45
pFHC871 or pTAC1302	7	26
Chromosomal *dnaA* genes	1.6	1.2

Strain MC1000 containing the indicated plasmids were grown in rich medium, NY, or in AB minimal medium (Clark and Maaløe, 1967) supplemented with glycerol. The differential rate of synthesis of β-galactosidase was determined (see Table 1) together with the plasmid DNA content according to Stüber and Bujard (1982). The DNA concentration in MC1000 has been determined in lab-course experiments. The number of chromosomal *dnaA* genes was calculated according to Pritchard and Zaritsky (1970), assuming *C* times of 40 and 60 min for the two growth rates. 1 genome equivalent = 4500 kb. Plasmid pTAC1377 had the same copy number, independently of the *dnaA* allele carried on the pACYC184 plasmid.

a wildtype cell is exerting a fairly high degree of repression on *dnaA* gene expression. The exact level of repression still has to be determined, for example, by using a strain in which the DnaA protein is completely lacking, owing to a Tn10 insertion.

INDUCTION OF *dnaA* PROTEIN CAUSES ABORTIVE OVERINITIATION

Plasmids carrying the *dnaA*⁺ and *dnaAΔ121* genes under control of the inducible promoter *pL* from phage lambda (pTAC1430 and pTAC1431, Figure 2) were constructed to study the effect of overproduction of DnaA on initiation of chromosome replication. The *dnaA* gene expression was found to be under absolute control of *pL*. The two plasmids were transformed into the maxicell strain CSR603 containing the compatible plasmid pNF2690. Figure 3a shows plasmid-encoded proteins synthesized in these maxicells. At 30 C no DnaA protein (52 kD) was made from pTAC1430, while large amounts were synthesized at 42 C. From plasmid pTAC1431 only a very labile truncated DnaA protein (22 kD) was synthesized at 42 C (not visible in this figure).

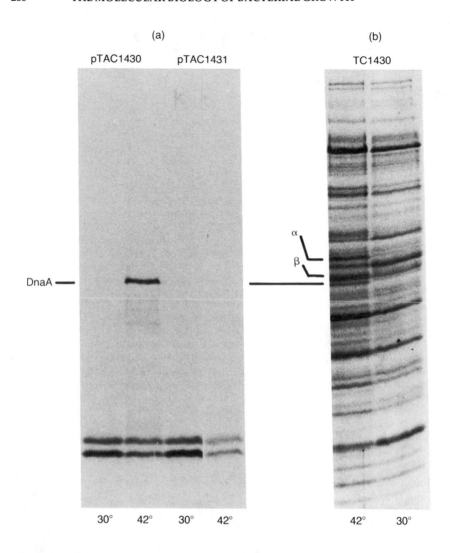

FIGURE 3 Induction of DnaA protein synthesis. (a) Autoradiogram of an SDS polyacrylamide gel (Laemmli, 1970) showing proteins synthesized in maxicells upon induction of λ pL. The maxicell strain CSR603 (Sancar et al., 1979) containing pNF2690 and either pTAC1430 or pTAC1431 was grown at 30°C, UV-irradiated, and incubated overnight with D-cycloserine. Each culture was split in two and labeled with [35S]methionine at 30°C and at 42°C respectively. (b) SDS acrylamide gel, stained with Comassie blue G, showing proteins in total cell extracts of strain TC1430 before and two mass doublings after the heat induction of DnaA protein synthesis. The band designated <u>dnaA</u> was identified through co-migration with [35S]methionine-labeled DnaA protein from the maxicells. Growth conditions were as described in the legend to Figure 4.

Induction of DnaA protein stimulates initiation of replication

For analysis of the effects of induction of DnaA protein synthesis the two *pL*-carrying plasmids were introduced into strain CR34 carrying pNF2690 to supply thermosensitive λ repressor. The resulting strains, TC1430 and TC1431, were grown exponentially at 30 C and then shifted to 42 C for induction of synthesis of DnaA protein. DNA synthesis was monitored by [^{14}C]thymine incorporation. The induction of DnaA protein synthesis in TC1430 had no effect on the DNA/mass ratio (Figure 4). Both at 30 C and at 42 C strain TC1430 had the same DNA concentration as the control strain TC1431 (data not shown).

   Proteins in samples harvested before and following two mass doublings after the shift were separated by SDS gel electrophoresis and visualized by staining (Figure 3b). Surprisingly, a number of proteins, in addition to DnaA and those due to the heat shock, were overproduced upon induction of synthesis of DnaA protein. The two bands just above the DnaA position were identified as the alpha and beta subunits of the ATP synthase by comigration on 2-D O'Farrel gels with radioactively labelled ATP synthase subunits synthesized

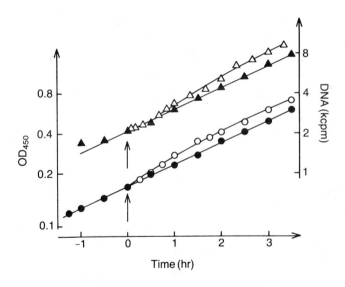

Time (hr)

FIGURE 4   Effect of induction of DnaA protein on growth and DNA synthesis. Strain TC1430 (thyA deo thr leu thi/pNF2690 + pTAC1340) was grown exponentially at 30°C in AB minimal medium supplemented with 0.2% glycerol, 100 µg/ml of leucine and threonine, 1 µg/ml of thiamine, 200 µg/ml of ampicillin, 20 µg/ml of kanamycin, and 10 µg/ml of thymine, 0.05 µCi [^{14}C]thymine/ml. Growth was followed by measuring OD$_{450}$, and DNA synthesis was assayed as accumulation of [^{14}C]thymine in TCA-precipitable material. Part of the culture was shifted to 42°C at the time indicated by the arrow. Closed symbols: 30°C; open symbols: 42°C; circles: OD$_{450}$; triangles: DNA.

upon infection of UV-killed cells by lambda *asn105* (von Meyenburg et al., this volume). A more detailed analysis (Figure 5) showed that the induction of synthesis of DnaA protein had little effect on the rate of synthesis of EF-Tu and of the beta subunit of RNA polymerase (nor on growth, see Figure 4) within the first hour. The rate of synthesis of the ATP synthase alpha and beta subunits and of the c subunit (data not shown) first decreased and then rapidly increased. The induction of DnaA protein resulted in altered rates of synthesis of only a small number of other proteins; some were stimulated, like the ATP synthase subunits, whereas a few others were inhibited.

   The ATP synthase genes constitute one operon, which is transcribed leftwards from a promoter located 3.5 kb to the left of *oriC* (von Meyenburg et al., 1982), and expression of the operon is proportional to gene dosage (von Meyenburg et al., this volume). Therefore, the coordinate overproduction of ATP synthase subunits suggested that the concentration of genes close to *oriC* was increased severalfold upon overproduction of DnaA protein. This hypothesis was tested by direct measurements of gene concentrations by DNA-DNA hybridization. The ratio of *atp* to *lac* genes in the total DNA increased about threefold within 60 minutes after induction of synthesis of the DnaA protein (Figure 6). Thus, the relative gene dosage of the *atp* operon followed kinetics similar to that found for the rate of synthesis of the gene products. Preliminary experiments with other hybridization probes showed that a DNA segment very close to *oriC* was amplified. Furthermore, it was found that the copy number of a minichromosome carrying only 1.5 kb of chromosomal DNA on either side of *oriC* was increased threefold within 45 minutes after overproduction of DnaA protein (Løbner-Olsen and Atlung, unpublished results).

   These results strongly suggest that overproduction of DnaA protein results in additional initiation of replication at *oriC*, though the total DNA concentration of the cells does not change.

Overinitiation following induction of DnaA protein is abortive

Upon inhibition of protein synthesis, replication forks, once initiated, are expected to run to completion, giving rise to an increase in the amount of DNA proportional to the number of origins per chromosome (Maaløe and Hanawalt, 1961; Lark et al., 1963; Pritchard and Zaritsky, 1970). When chloramphenicol (CAM) was added 1 hr after a shift to 42 C, accumulation of DNA ceased within two hours in strains TC1430 and TC1431 (Figure 7). In both strains the DNA content increased 55% relative to that in the exponentially growing culture at the time of CAM addition. No more than 60% residual DNA synthesis was observed, independent of the time of addition of CAM between 5 min and 4 hrs after induction of synthesis of DnaA protein.

   The same results with respect to the DNA/mass ratio, residual DNA synthesis in CAM and increased expression of the *atp* operon, were found on induction of synthesis of DnaA protein from pTAC1430 in another genetic background (TC92) in which overinitiation owing to the *rpoC907* mutation leads to a two- to threefold increase in the DNA/mass ratio (Rasmussen et al., 1983). Thus, overproduction of DnaA protein leads to overinitiation and amplification of DNA segments close to the origin. Unexpectedly, the extra replication forks produced by the overinitiation appear to be abortive or aborted: the overall DNA concentration was unaffected, and only an amount of DNA

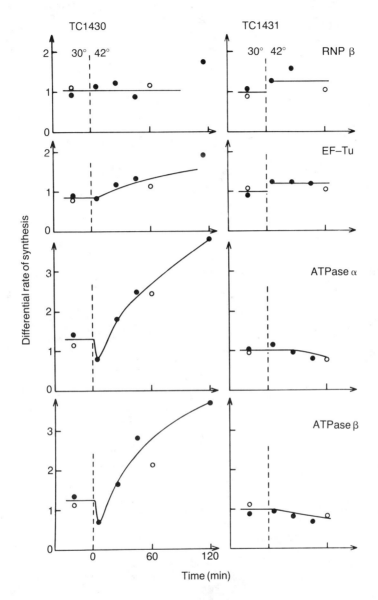

FIGURE 5 Effect of DnaA-protein induction on the differential rate of synthesis of ATP synthase. Strains TC1430 and TC1431 were grown as described in legend to Figures 4. Samples were pulse-labeled with [35S]methionine, mixed with [3H]leucine-labeled reference cells (prepared from TC1430 at 42°C) and the proteins were separated by 2-D O'Farrel gel electrophoresis (O'Farrel, 1975). The ratio of 35S to 3H counts in the spots was normalized to the ratio in the applied sample. These values were, after setting that obtained in the control strain TC1431 at 30°C to 1, plotted against the time of sampling. Values from two independent experiments (solid and open circles) have been combined.

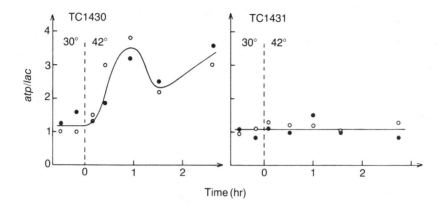

FIGURE 6 Determination of relative gene dosage of <u>atp</u> and <u>lac</u> genes after <u>dnaA</u> induction. Strains TC1430 and TC1431 were grown as in Figure 4. Samples for hybridization were processed according to Kellenberger-Gujer et al. (1978) and loaded onto 0.45-micron BA85 nitrocellulose filters. Probes were prepared from single-stranded M13 recombinant DNA by labeling with [³H]dTTP <u>in vitro</u>. Hybridizations were carried out for 18 hr at 68°C using the conditions for Southern hybridization (Maniatis et al., 1982). The ratio of hybridization by the <u>atp</u> probe relative to the <u>lacZ</u> probe was calculated after subtraction of background on filters loaded with lambda DNA, and plotted against the time of sampling. Data from two independent hybridizations of the same samples (solid and open circles) have been combined.

corresponding to completion of the normal number of replication forks was found after inhibition of protein synthesis. We do not yet know how far from *oriC* the gene amplification extends and whether the amplified segments are stable.

CONCLUSIONS

The dnaA gene is autoregulated

Expression of the *dnaA* gene was shown to be affected by the level of DnaA protein activity. High concentrations of wildtype DnaA protein caused up to a fourfold repression of the synthesis of a *dnaA-lacZ* fusion protein located after the intact *dnaA* promoter region. Synthesis of the same fusion polypeptide was derepressed threefold in a strain carrying a *dnaA*(Ts) mutation (e.g., *dnaA46*). The same pattern of regulation was also found when the *tet* gene of pBR322 was expressed from the *dnaA* promoters. These observations strongly suggest that the DnaA protein regulates its own synthesis and that the autoregulation is at the transcriptional level. The promoter *dnaA2p*, the stronger of the two *dnaA* promoters, is a target for this autoregulation. Transcription from *dnaA1p* is probably not regulated by DnaA protein (Atlung et al., 1984).

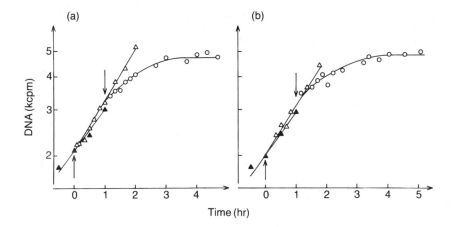

FIGURE 7 Run-out of DNA synthesis after inhibition of protein synthesis. Strains TC1430 and TC1431 were grown as in Figure 4. At time 0 (first arrow) part of the culture was shifted to 42°C. One hr later (second arrow) CAM was added to 200 µg/ml to a part of the 42°C culture. DNA synthesis was followed as incorporation of [^{14}C]thymine. Closed triangles: 30°C; open triangles: 42°C; circles: 42°C + CAM. Panel (a): strain TC1430; Panel (b): strain TC1431.

Two sites in the *dnaA* promoter region appear to be essential for the autorepression. One is located in the –35 region of the *dna2p* promoter. The other coincides with a 9–11 bp sequence that has four homologous counterparts in *oriC* (Hansen et al, 1982). The DnaA protein binds specifically to three 9–11 bp sequences in *oriC* (Fuller and Kornberg, 1983) as well as to that in the *dnaA* promoter region (R. Fuller and A. Wright, personal commun.). Autorepression is independent of sequences downstream from *dnaA2p*, excluding that the possibility that autoregulation is carried out through an attenuator, which seemed likely in view of the effect of *dnaA*(Ts) mutations at the *trp* operon attenuation (Atlung and Hansen, 1983).

Induction of synthesis of DnaA protein leads to initiation of DNA replication

The DnaA protein activity can be limiting for initiation of DNA replication, i.e., when *dnaA*(Ts) mutants are grown at sub-restrictive temperatures (Hansen and Rasmussen, 1977; Tippe-Schidler et al., 1979; Lycett et al., 1980). There are indications that *de novo* synthesis of DnaA protein is required, since initiation stops immediately upon a temperature shift in *dnaA*(Am) mutants containing a temperature-sensitive amber suppressor (Schaus et al., 1981). If DnaA protein activity is a limiting activity for initiation of DNA replication, one would expect that increased concentrations of DnaA protein should result in increased amounts of DNA. However, normal levels of DNA are found in cells containing five times more

DnaA protein than haploid cells (von Meyenburg et al., this volume) and in cells in which synthesis of DnaA protein is induced from the *lac* promoter (Churchward et al., 1983) or from the lambda *pL* promoter. Although the overproduction of DnaA protein from the lambda *pL* plasmid did not lead to a higher content of total DNA, it clearly results in a higher concentration of *oriC* (*atp* operon) DNA, strongly indicating that the newly synthesized DnaA protein stimulated initiation of DNA replication from *oriC*. However, the resulting replication forks must stop close to *oriC*. We do not understand why at least some of the new forks do not proceed further, especially since different mutations can lead to as much as a threefold increase in the normal DNA content (*dnaACos:* Kellenberger–Gujer et al., *1978; rpoB* and *rpoC* mutants: Rasmussen et al., 1983; Tanaka et al., 1983).

Two mass doublings after induction from *pL* the amount of DnaA protein had accumulated to at least 50 times the normal amount, while *oriC* DNA concentration had only increased fourfold. This result probably reflects other limitations imposed on the initiation of DNA replication. The DNA polymerase III holoenzyme might, for example, become a limiting factor under these conditions, since only very few molecules are present in each cell (Kornberg, 1980).

A model for regulation of DNA replication

Our data can be summarized as follows. The DnaA protein is an important factor in initiation of replication. Low activity of DnaA protein causes a low frequency of initiation and thereby a low concentration of the origin and of DNA. High concentration of DnaA protein causes a high frequency of initiation, a high concentration of origin, *but not* a high concentration of DNA, because replication forks stall shortly after initiation. On the basis of these data we propose the following model.

We suggest that regulation of DNA replication is controlled at two levels. At the first level a positively acting initiator protein (DnaA, which is autoregulated in a negative mode) accumulates to a threshold level at which initiation of DNA replication occurs at the chromosomal origin, *oriC*. At the second level, control is exerted *after* initiation of DNA replication. DNA replication forks are kept pausing until a certain mass per initiation complex is reached and then allowed to proceed. During normal bacterial growth these two controls may coincide with time, and the interplay between them results in the precise timing of initiation of DNA replication in the cell cycle. The first level of control in this model bears strong resemblance to the autorepressor initiator model proposed by Sompayrac and Maaløe (1973). The positively acting initiator protein (DnaA) is autoregulated; the only difference from this model is that the DnaA protein itself acts as autorepressor, as suggested by Hansen and Rasmussen (1977). The second level of control in this model might be exerted by an inhibitor with features like those suggested by Pritchard et al., (1969). We suggest that the DnaA protein may have a third function as an inhibitor of fork progression and hence is also involved in the second level of control.

REFERENCES

Atlung, T. (1981). *ICN-UCLA Symp. Mol. Cell Biol.*, 22, 297.

Atlung, T. and Hansen, F. G. (1983). *J. Bacteriol.*, 156, 985.

Atlung, T., Clausen, E., and Hansen, F. G. (1983). In *Proteins Involved in DNA Replication,* U. Hubscher (ed.), Plenum.

Bagdasarian, M. W., Izakowska, M., and Bagdasarian, M. (1977). *J. Bacteriol.,* 130, 577.

Churchward, G., Holmans, P., and Bremer, H. (1983). *Mol. Gen. Genet.,* 192, 506.

Clark, D. J. and Maaløe, O. (1967). *J. Mol. Biol.,* 23, 99.

Cooper, S. and Helmstetter, C. E. (1968). *J. Mol. Biol.,* 31, 519.

Donachie, W. D. (1968). *Nature,* 219, 1077.

Frey, J., Chandler, M., and Caro, L. (1979). *Mol. Gen. Genet.,* 174, 117.

Fuller, R. and Kornberg, A. (1983). *Proc. Natl. Acad. Sci. USA,* 80, 5817.

Hansen, F. G. and Rasmussen, K. V. (1977). *Mol. Gen. Genet.,* 155, 219.

Hansen, F. G. and von Meyenburg, K. (1979). *Mol. Gen. Genet.,* 175, 135.

Hansen, F. G., Koefoed, S., von Meyenburg, K., and Atlung, T. (1981). *ICN-UCLA Symp. Mol. Cell Biol.,* 22, 37.

Hansen, E. B., Hansen, F. G., and von Meyenburg, K. (1982). *Nucleic Acids Res.,* 10, 7373.

Hansen, F. G., Hansen, E. B., and Atlung, T. (1982). *EMBO J.,* 1, 1043.

Hansen, E. B., Atlung, T., Hansen, F. G., Skovgaard, O., and von Meyenburg, K. (1984). *Mol. Gen. Genet.,* 196, 387.

Jacob, F., Brenner, S., and Cuzin, F. (1963). *Cold Spring Harbor Symp. Quant. Biol.,* 28, 329.

Kellenberger-Gujer, G., Podhajska, A. J., and Caro, L. (1978). *Mol. Gen. Genet.,* 162, 9.

Kornberg, A. (1980). *DNA Replication.*

Laemmli, U. K. (1970). *Nature,* 227, 680.

Lark, K. G. (1972). *J. Mol. Biol.,* 64, 47.

Lark, K. G., Repko, T., and Hoffman, E. J. (1963). *Biochim. Biophys. Acta,* 76, 9.

Lycett, G. W., Orr, E., and Pritchard, R. H. (1980). *Mol. Gen. Genet.,* 178, 329.

Maaløe, O. and Hanawalt, P. C. (1961). *J. Mol. Biol.,* 3, 144.

Maniatis, T., Fritsch, E. F., and Sambrook, J. (1982). *Molecular Cloning,* Cold Spring Harbor Laboratories.

Marsh, R. C. and Worcel, A. (1977). *Proc. Natl. Acad. Sci. USA,* 74, 2720.

Miller, J. H. (1972). *Experiments in Molecular Genetics,* Cold Spring Harbor Laboratories.

O'Farrel, P. H. (1975). *J. Biol. Chem.,* 250, 4007.

Pritchard, R. H., Barth, P. T., and Collins, J. (1969). *Symp. Soc. Gen. Microbiol.,* 19, 263.

Pritchard, R. H. and Zaritsky, A. (1970). *Nature,* 226, 126.

Rasmussen, K. V., Atlung, T., Kerszman, G., Hansen, G. E., and Hansen, F. G. (1983). *J. Bacteriol.,* 154, 443.

Remaut, E., Stanssens, P., and Fiers, W. (1981). *Gene,* 15, 81.

Sakakibara, Y. and Yuasa, S. (1982). *Mol. Gen. Genet.,* 186, 87.

Sancar, A., Hack, A. M., and Rupp, W. D. (1979). *J. Bacteriol.,* 137, 692.

Schaechter, M., Maaløe, O., and Kjeldgaard, N. O. (1958). *J. Gen. Microbiol.,* 19, 592.

Schaus, N., O'Day, K., Peters, W., and Wright, A. (1981). *J. Bacteriol.,* 145, 904.

Sompayrac, L. and Maaløe, O. (1973). *Nature New Biol.,* 241, 133.

Stüber, D. and Bujard, H. (1982). *EMBO J.,* 1, 1399.

Tanaka, M., Ohmori, H., and Hiraga, S. (1983). *Mol. Gen. Genet.,* 192, 51.

Tippe-Schindler, R., Zahn, R. G., and Messer, W. (1979). *Mol. Gen. Genet.,* 168, 185.

von Meyenburg, K., Hansen, F. G., Riise, E., Bergmans, H. E. N., Meijer, M., and Messer, W. (1979). *Cold Spring Harbor Symp. Quant. Biol.,* 43, 121.

von Meyenburg, K., Jorgensen, B. B., Nielsen, J., and Hansen, F. G. (1982). *Mol. Gen. Genet.,* 188, 240.

von Meyenburg, K., Hansen, F. G., Atlung, T., Boe, L., Groth Clausen, I., Hansen, E. B., Jorgensen, B. B., Jorgensen, F., Koppes, L., Michelsen, O., Nielsen, J., Pedersen, P. E., Rasmussen, K. V., Riise, E., and Skovggard, O. (1984) This volume.

Weinstock, G. M., ApRhys, C., Berman, M. L., Hampar, B., Jackson, D., Silhavy, T. J., Weiseman, J., and Zweig, M. (1983). *Proc. Natl. Acad. Sci. USA,* 80, 4432.

Zyskind, J. W., Deen, L. T., and Smith, D. W. (1977). *J. Bacteriol.,* 129, 1466.

Initiation of Chromosome and Minichromosome Replication in *E. coli* When the Activity of the *dnaA* Gene Product is Rate-Limiting

Charles E. Helmstetter and Martin Weinberger
Roswell Park Memorial Institute; Buffalo, New York

INTRODUCTION

Initiation of replication at the chromosomal origin in *Escherichia coli*, *oriC*, appears to be directed by a multicomponent complex (Lark and Lark, 1978; Helmstetter et al., 1979; Kornberg, 1980; Blinkowa et al., 1983; Blinkowa and Walker, 1983). Although a number of the components of the complex have been identified and characterized, the determinant that specifies the timing of utilization of the complex, i.e., the initiation frequency regulator, has not been identified.

The purpose of this investigation was to utilize temperature-sensitive mutants of *E. coli* with a defective *dnaA* gene, whose product appears to be a multicopy member of the complex (Blinkowa et al., 1983; Blinkowa and Walker, 1983; Fuller and Kornberg, 1983), in an attempt to answer some questions about the relationship between the function of the initiation complex and the control of initiation frequency. In spite of the fact the DnaA protein may not be the rate-limiting determinant of chromosome duplication (Lycett et al., 1980; Churchward et al., 1983; LaDuca and Helmstetter, 1983; Atlung et al., this volume; von Meyenburg et al., this volume), these mutants can be useful in answering the following fundamental questions. First, can an initiation complex be used more than once, and are the rules governing such reutilization related to the rules governing initiation control? Second, when the number of *oriC* regions in a cell, either on chromosomes or minichromosomes, exceeds the number of available active initiation complexes, is there competition among the origin regions for the complexes, and if so, what determines the choice of origins for initiation?

Our experimental approach to develop answers to these questions stems directly from a fundamental concept developed in Ole Maaløe's laboratory; namely, that information on cellular control processes can be deduced from the manner that cells respond to shifts in their growth states (Maaløe and Kjeldgaard, 1966). Previously, these analyses of transitions generally involved shifts between different nutritional states. However, in our current studies we have analyzed the response of *dnaA* mutants growing at permissive temperature to shifts to higher, less permissive temperatures. We have used comparative analyses of DNA replication in these mutants, with and without resident minichromosomes, to gain some insight into the answers to these questions, and consequently, into the fundamentals of initiation frequency regulation *in vivo*.

MATERIALS AND METHODS

Bacteria and growth conditions. The organisms used were *E. coli* B/r F62 (*dnaA5 thyA his*) (Helmstetter and Krajewski, 1982), *E. coli* B/r F621 (*dnaA5 his recA*) (Leonard et al., 1982), and *E. coli* K12 MM294 (*end thi hsdR supE*), obtained from the Coli Genetic Stock Center. Strain MM294 was made *dnaA* through P1 transduction from B/r F62 and *recA* through mating with K12 KL1699 *recA*, obtained from the Coli Genetic Stock Center. The minichromosome used was pAL2, which consists of a 1.3-megadalton DNA fragment containing *oriC* and a 5.7-megadalton fragment containing a determinant of kanamycin resistance. The construction and properties of the minichromosome have been described in detail previously (Leonard et al., 1982). The cultures were grown in a minimal salts medium (Helmstetter and Krajewski, 1982) containing glucose (0.1%), Casamino acids (0.2, Difco Laboratories, reagent grade), thymine (10 μg/ml), thiamine (1 μg/ml), and histidine (50 μg/ml) when required.
Isotopes. [^{3}H]thymidine was obtained from the New England Nuclear Corp. at a specific activity of 60 to 80 Ci/mmol and used either at this specific activity or diluted with unlabeled thymidine to a specific activity of approximately 16 Ci/mmol. [^{14}C]thymidine and [^{14}C]thymine were obtained from ICN at a specific activity of 56 mCi/mmol. [^{14}C]D-glucose (78.1 atom ^{13}C and NH$_4$Cl (95 atom ^{15}N) were obtained from Merck.
Radioactive labeling. The rates of DNA synthesis in temperature-shifted cultures were measured by pulse-labeling portions of the cultures with [^{3}H]thymidine at frequent intervals. Measurement of incorporation into total cellular DNA was performed as described previously (Helmstetter and Krajewski, 1982). The amount of radioactivity incorporated into pAL2 DNA was determined by pulse-labeling 10-ml portions of the cultures with 10 microCi of [^{3}H]thymidine per ml. After brief exposure to the label, unlabeled thymidine at a final concentration of 100 micrograms per ml was added as a chase for 30 min, and then sodium azide was added at final concentration of 0.1 M in an ice bath. After addition of 1 ml of a stock culture of the same cells containing ^{3}H-labeled pBR322 plasmid DNA, the samples were lysed by the boiling procedure of Holmes and Quigley (1981). The resultant samples were subjected to electrophoresis in 0.7 agarose at 40 to 60 V for 18 to 22 hr in Tris-phosphate buffer (89 mM Tris, 23 mM H$_3$PO$_4$, 2.5 mM Na$_4$EDTA, pH 8.3). After electrophoresis,

fluorography was used to visualize labeled plasmid bands in the gels. The radioactivities in the bands corresponding to supercoiled closed circular (CC) minichromosome DNA and pBR322 DNA were determined by cutting out the bands from the gels and counting in Spectrafluor. The ratio of radioactivities in CC minichromosome to pBR322 DNA was determined in each sample to correct for any variability in the isolation procedure for different samples.

Density shifts. Cells labeled with heavy isotopes in density shift experiments were grown in minimal salts medium containing 0.8 mg/ml [^{15}N]NH$_4$Cl and 0.25 mg/ml [^{13}C]glucose as the sole sources of nitrogen and carbon. Total lysates for CsCl density gradient centrifugation were prepared by resuspending labeled cell pellets in 1.0 ml of 25% sucrose in 50 mM Tris (pH 8.0). Addition of 15 microliters of lysozyme (10 mg/ml) in water was followed by gentle mixing. The cells were placed at 42 C for 45 s and held in an ice bath for 5 min; at that time 322 microliters of 0.25 M EDTA (pH 8.0) and 200 microliters of lysing solution (10% Triton X-100 in 50 mM Tris, 62.5 mM EDTA, pH 8) was added. After incubation for 30 min, 50 microgram of RNase (boiled for 5 min) was added per ml, and the lysates were placed at 37 C for 30 min. Incubation at 37 C continued for another 30 min after the addition of 1 mg Proteinase K per ml. The clear, somewhat viscous lysates were then transferred to an ice bath and sheared with the use of syringes equipped with #23 needles. Solid CsCl (1.27 g/ml) was added to yield a solution with a refractive index of 1.400. Light [^{12}C]-DNA was added as marker to the gradients, which were centrifuged in a Sorval TV865A vertical rotor for 40 to 48 h at 40,000 rpm.

Extracts for density-transfer studies of minichromosome DNA were prepared by the boiling procedure and centrifuged to equilibrium in CsCl-ethidium bromide. Gradients were fractionated and the position of radiolabeled covalently closed pAL2 DNA determined by precipitating 10 microliters per fraction with 5 TCA as described above. Minichromosome-containing fractions were pooled, extracted with TES-CsCl saturated isopropanol to remove ethidium bromide, diluted with a CsCl solution, adjusted to a refractive index of 1.400 and centrifuged as described.

RESULTS

Chromosome replication in dnaA5 mutants of E. coli upon shift from 25 C to higher, less permissive temperatures

When temperature-sensitive, thermoreversible *dnaA* mutants of *E. coli* growing at a permissive temperature are shifted to 41 C, generally identified as a nonpermissive temperature, initiation of chromosome replication ceases, and ongoing rounds of replication continue to completion (reviewed in Helmstetter, et al., 1979; Kornberg, 1980). If the high temperature were totally nonpermissive, such that all the DnaA protein formed in the cell were completely inactive, then new rounds of replication could not be initiated subsequent to the shift. On the other hand, if the fraction of active DnaA protein, from among the few hundred molecules in the cell (Sakakibara and Yuasa, 1982; Atlung et al., this volume), were simply an inverse function of temperature, which is more likely, then a proportion of molecules might be active at the higher temperature. If this were the case, and the *dnaA* protein as well as other components needed for initiation continued to be synthesized at the high temperature, then initiation of new rounds would be anticipated at a later time. This possibility was examined by shifting cultures of *E. coli dnaA5* mutants growing at 25 C to various higher temperatures and pulse-labeling the cultures with [^{3}H]thymidine to determine the rate of DNA replication. Figure 1 shows the results of a few of these experiments with different post-shift temperatures, different genetic backgrounds, and with *recA* mutants containing a minichromosome. Upon shift to the higher temperature the rate of [^{3}H]thymidine incorporation decreased rapidly, consistent with termination of ongoing rounds of replication in the absence of initiation of new rounds. After an interval of continued growth, there was a burst of [^{3}H]thymidine incorporation (Figure 1a,b,e). This pattern is consistent with the predicted initiation of chromosome replication after a delay, during which the cells accumulated sufficient DnaA-dependent initiation complex activity to initiate synthesis. In some cases there was evidence for at least two bursts of incorporation (Figure 1b).

The effect of addition of chloramphenicol on the rate of [^{3}H]thymidine incorporation after shift from 25 to 39 C is also shown (Figure 1c-e). During the first 40 min after the temperature shift addition of chloramphenicol did not significantly alter the initial decrease in [^{3}H]thymidine incorporation, but it prevented subsequent bursts of thymidine incorporation. When chloramphenicol was added at 60 min postshift, a burst of incorporation was induced (Figure 1e). This is similar to previous findings that chloramphenicol induces initiation of chromosome replication in *dnaA* mutants grown at intermediate temperatures (Messer, et al., 1975; Tippe-Schindler et al., 1979; Lycett et al., 1980; LaDuca and Helmstetter, 1983). Thus, approximately 60 min at 39 C was required for sufficient DnaA protein to be present such that complexes were subject to activation by chloramphenicol, presumably due to the enhanced availability of RNA polymerase molecules (Lycett et al., 1980). In the absence of this activation, an additional 30 min was required for formation of functional initiation capacity (Figure 1e).

Density transfer experiments were used to examine DNA replicated during the burst of [^{3}H]thymidine uptake after shift to high temperature. These experiments were performed as shown in Figure 1, except that the culture of B/r F62 was grown in heavy [^{13}C, ^{15}N]-containing medium with [^{14}C]thymine at 25 C, and then shifted to 39 C with addition of 20 microgram of unlabeled thymine per ml. After 95 min, the heavy isotopes were diluted 10-fold by addition of light [^{12}C]glucose and NH$_4$Cl, and the culture was pulse-labeled at intervals with [^{3}H]thymidine. DNA was prepared and subjected to CsCl density gradient centrifugation (Figure 2). The data show that ^{13}C-prelabeled heavy DNA (Figure 2a) was gradually converted to hybrid density. [^{3}H]thymidine was incorporated into DNA of hybrid density throughout the experiment, and at 250 to 280 min (Figure 2f) tritium also appeared in light DNA. The results indicate that a round of replication initiated and completed on the majority of chromosomes during the burst in rate of [^{3}H]thymidine uptake (shown inset in Figure 2d).

The behavior of the cultures in the experiments described in the preceding sections is particularly interesting with regard to the control of chromosome replication. It is well known that incubation of

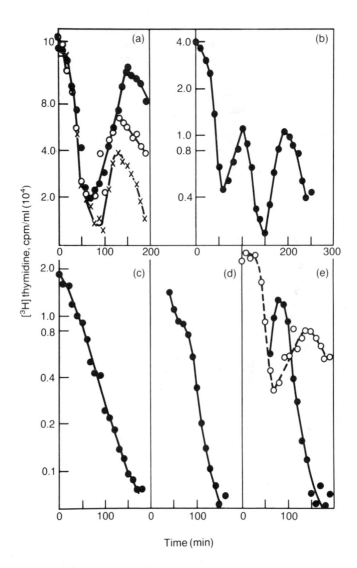

Time (min)

FIGURE 1 Rate of [³H]thymidine incorporation into E. coli dnaA5 mutants after shift from 25°C to higher temperatures. Cultures grown in glucose-casamino acids medium were shifted up in temperature at time zero and 1 ml portions were withdrawn at intervals and pulse-labeled with 1.0 microcuries of [³H] thymidine for 10 min. (a) B/r F62 shifted to 41°C (x), 39°C (open circles) and 37°C (solid circles). (b) K12 MM294 dnaA5 shifted to 41°C. (c-e) A culture of K12 MM294 dnaA5 recA (pAL2) was divided into four parts and shifted to 39°C with addition of 200 μg/ml chloramphenicol (solid circles) at: 0 min (c), 40 min (d), and 60 min (e). The fourth portion, in (e), did not receive chloramphenicol (open circles).

dnaA mutants at intermediate temperatures results in a culture in which the frequency of initiation of chromosome replication is limited by the activity of the DnaA protein (Messer et al., 1975; Tippe‑

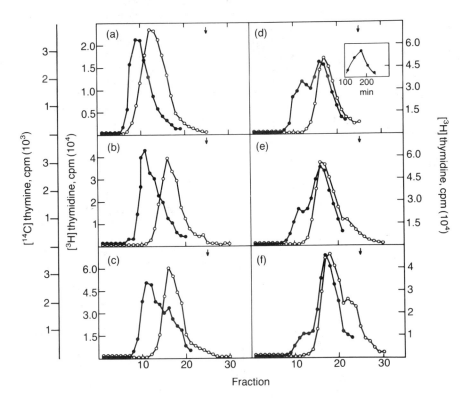

FIGURE 2 Density shift of B/r F62 grown in heavy medium at 25°C and shifted to 39°C. A culture of B/r F62 grown in glucose minimal heavy medium containing 0.2 microcuries/ml [^{14}C]thymine was shifted to 39°C at time 0 (A_{450} = 0.087) concomitant with the addition of 20 μg/ml unlabeled thymine to dilute unincorporated label and effectively stop [$_{14}$C] incorporation. After 95 min at 39°C, heavy label was diluted 10‑fold with the addition of solid [^{12}C]glucose and ^{14}NH$_4$Cl. Starting at 100 min, samples were labeled for 30 min periods with 10 microcuries of [^{3}H]thymidine per ml, lysates were prepared and subjected to CsCl density gradient centrifugation. Prior to centrifugation 2000 cpm of [^{14}C]‑labeled light DNA was added to each gradient. The position of this marker DNA in each gradient is designated with an arrow. Labeling periods were (a) 100‑130 min, (b) 130‑160 min, (c) 160‑190 min, (d) 190‑220 min, (e) 220‑250 min, and (f) 250‑280 min. Solid circles, ^{14}C; open circles, ^{3}H. The inset in (d) shows the rate of [^{3}H]thymidine incorporation in a density shift experiment performed as described in Figure 1.

Schindler et al., 1979; Lycett et al., 1980; Sakakibara and Yuasa, 1982; LaDuca and Helmstetter, 1983). This is also true in the experiments we have described, with one important addition: after shift from permissive temperature to the higher temperatures, not only is the DnaA protein rate-limiting for initiation of replication, but the initiations take place in one or more synchronous waves. At temperatures at which continued growth might be expected, the initiations must gradually become asynchronous since *dnaA* cultures grown for long periods of time at intermediate temperatures do not show synchronous initiation, except possibly as they approach the stationary phase (LaDuca and Helmstetter, 1983). Since the activity of the DnaA protein is the determinant of initiation frequency in the cultures, and this product is part of the initiation complex, we can conclude that activity of the initiation complex is limiting in the cells and determines the timing of initiation of chromosome replication. Given this concept, plus the fact that initiation takes place in synchronous waves, we can investigate whether resident minichromosomes compete for the limiting active replication complexes, and possibly whether a complex can be used more than once.

Effect of minichromosome maintenance on DNA replication in B/r F621 (pAL2) after a temperature shift

In order to examine the competition between regions containing *oriC* for the limiting active initiation complexes at high temperatures, the temperature shifts were performed in parallel on *E. coli* B/r F621 cultures that either contained or lacked pAL2 minichromosomes. In these experiments, [^{3}H]thymidine incorporation, shown in Figure 3 for three different postshift temperatures, monitors the rate of chromosome replication, because minichromosome DNA corresponds to less than 1% of the total DNA. There was no dramatic alteration in the incorporation pattern due to minichromosome maintenance, except for the possibility of a slight, 10-min delay in the midpoint of the increase in incorporation of ^{3}H in the minichromosome-containing cultures.

The next question in this study concerned the timing of replication of the minichromosomes after shift to high temperature. These experiments were performed in the same manner except that [^{3}H]thymidine incorporation in minichromosome DNA was identified by agarose gel electrophoresis. Upon shift to high temperature, uptake of radioactivity into minichromosome DNA ceased (Figure 4b), as we and others have shown previously (von Meyenburg et al., 1978; Lother et al., 1981; Leonard et al., 1982). However, there was a subsequent burst of synthesis of pAL2 DNA at about the time initiation of chromosome replication began (Figure 4a). The period of incorporation into minichromosome DNA was short compared to the duration of incorporation of label into total DNA. In a density-transfer experiment performed as described in Figure 2, the [^{3}H]thymidine incorporated during the burst of pAL2 replication appeared in minichromosomes of hybrid density. This result indicates that most minichromosome molecules replicated once during this period. The significance of these findings with respect to competitive interactions and replication complex reutilization will be discussed in detail in the Discussion.

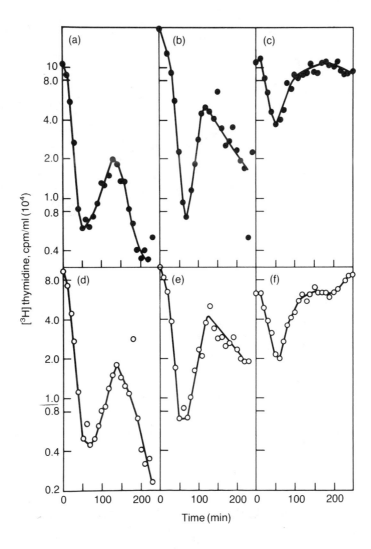

FIGURE 3   Rate of [³H]thymidine incorporation into <u>E.coli</u> B/r F621 and F621 (pAL2) after shift from 25°C to higher temperatures. (a,b,c) B/r F621 shifted to 41°C, 37°C, and 33°C, respectively. (d,e,f) B/r F621 (pAL2) shifted to 41°C, 37°C, and 33°C. Samples (1 ml) were withdrawn at intervals and exposed to 1.0 microcuries of [³H]thymidine for 6 min. Experiments at the same temperature were performed simultaneously with two separate cultures in glucose-casamino acids minimal medium.

DISCUSSION AND CONCLUSIONS

The goal of this work was to answer certain fundamental questions concerning the behavior of the complex that is involved in initiation

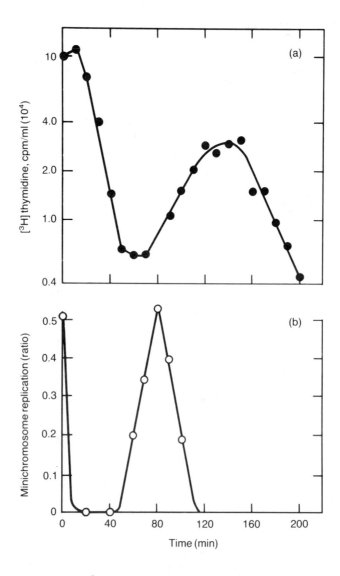

FIGURE 4   Rate of [³H]thymidine incorporation into total DNA and minichromosome DNA in B/r F621 (pAL2) after shift from 25°C to 41°C. (a) A culture of F621 (pAL2) in glucose-casamino acids medium was shifted to 41°C at time zero and samples (1 ml) were removed at intervals and exposed to 1.0 microcuries of [³H]thymidine for 6 min. Total acid precipitable incorporation is shown. (b) A separate culture was shifted to 41°C and 10 ml samples were exposed to 10 microcuries/ml of [³H]thymidine for 6 min. Unlabeled thymidine was added, at 100 µg/ml final concentration, as a chase. 1 ml of a culture containing ³H-labeled pBR322 plasmid DNA was added to each sample and the ratio of radioactivities in CC pAL2 DNA to CC pBR322 DNA, determined by agarose gel electrophoresis, is shown.

of *E. coli* chromosome replication *in vivo*. Can an initiation complex be reutilized? Do *oriC*-containing replicons compete for complexes when they are in limiting supply? The experimental situation we have analyzed, namely, DNA replication in thermoreversible *dnaA* mutants shifted to high temperatures, with and without resident minichromosomes, is uniquely suited to such studies. This is because the activity of DnaA protein, a component of the initiation complex, is rate-limiting for initiation, and most importantly, chromosomes and minichromosomes initiate replication in synchronous waves under these conditions.

Before discussing the findings in relation to the questions posed, it is necessary to define the concept of initiation complex reutilization, and to consider how competition for complexes would manifest itself in our experiments. A replication complex will be considered to be reutilized when it (or its components in a reorganized form) is capable of initiating a second round of replication at *oriC* without additional macromolecular synthesis to form new components of the complex, to form a second complex, or to release an inhibitor of the complex. The question of reuse of initiation complexes *in vivo* has been discussed many times with regard to the control mechanism for the frequency of initiation of chromosome replication (e.g., Lark and Lark, 1978; Helmstetter et al., 1979; Kornberg, 1980; Atlung et al., this volume). Since the interval between successive initiation events requires protein synthesis, it has often been suggested that once a complex is used to initiate a round of replication, a new round cannot begin until a new active complex has been formed. However, this is an open question for at least two reasons. First, reutilization of the complex for initiation of chromosome replication may, for initiation to take place, require an interaction between the complex, *oriC,* and an envelope site in the cell (Craine and Rupert, 1978; Nicolaidis and Holland, 1978; Wolf-Watz and Masters, 1979; Nagai et al., 1980; Hendrickson et al., 1982; Jacq et al., 1983). It is not unreasonable to suggest that this interaction with the envelope limits expression of initiation complex activity at chromosomal origins, whereas the complexes themselves can be reutilized. Second, control of extrachromosomal *oriC* function in minichromosomes is not regulated as stringently as chromosomal *oriC* function, since minichromosomes replicate in the absence of protein synthesis, at a gradually decreasing rate, for about one generation after initiation of chromosome replication has ceased (von Meyenburg et al., 1978; Leonard et al., 1982). This suggests that initiation complexes are in excess in the cell and can be utilized to initiate minichromosome replication, or that complexes can be reutilized for replication of these molecules.

Next we will consider how competition for initiation complexes might be detected. If there were competition in our experiments in that the complex activity was rate-limiting for initiation, then it would most likely be manifested in two ways. If the *oriC* on chromosomes and minichromosomes were equally competitive and complexes could not be used more than once, then the magnitude of the burst of [^{3}H]thymidine incorporation at high temperature, which corresponds essentially to initiation and elongation of chromosomes, would be significantly reduced in the cells containing minichromosomes. Alternatively, if the same situation obtained, but the complexes used for initiation and replication of minichromosomes could be reutilized, then there would be a delay in the increase in [^{3}H]thymidine incorporation into chromosomal DNA. The length of the delay would depend upon two things: (1) the time a complex was occupied in replicating a minichromosome, which could vary from a few seconds, if it left the molecule upon completion of the initial polymerization event, to a few minutes if it were not released until completion of supercoiled monomer formation (Leonard et al., 1982); and (2) the ratio of minichromosomes to chromosomes in individual cells, which averages about 5 for pAL2. Since the time occupied by a complex in replicating minichromosomes is unknown, it is impossible at this point to calculate the extent of the possible delay, but eventually all complexes would be occupied in chromosome replication if the minichromosome resident time were short compared to the time for a round of chromosome replication. Thus, the magnitude of the burst in [^{3}H]thymidine incorporation would probably be unaffected by the presence of minichromosomes in this case.

Now we will consider our findings with regard to the questions of reutilization and competition. When the *dnaA* mutants were shifted from permissive temperature to temperatures previously called nonpermissive, initiation of chromosome and minichromosome replication was prevented until an interval of time had passed, which required protein synthesis; then, there was a burst of minichromosome replication coincident with initiation of chromosome replication. At 39 C to 41 C, this round of chromosome replication was completed, and additional protein synthesis was required for the next synchronous initiation. This is an experimental situation in which the activity of DnaA protein in initiation complexes is rate-limiting for initiation of both chromosome and minichromosome replication, and therefore, the concentration of active complexes determines the timing of initiation. All other requirements for initiation, including any envelope sites, are in excess, because multiple initiations took place upon shift to permissive temperatures in the absence of protein synthesis (Evans and Eberle, 1975; Eberle and Forrest, 1982; Helmstetter et al., 1984). These results are most consistent with the conclusion that once a replication complex has initiated a round of chromosome replication, it cannot be reutilized until further protein synthesis has taken place. The interval between successive rounds of replication at high temperature must have involved resynthesis (or reassembly) of labile components, or synthesis of a new complex containing sufficient active DnaA protein.

Reutilization of the initiation complex after minichromosome replication, and the possible competitive interactions between minichromosomes and the chromosome are more difficult to evaluate. The burst of minichromosome replication coincident with initiation of chromosome replication, and the formation of molecules of hybrid density during the burst in a density shift experiment, could be interpreted as consistent with the absence of complex utilization for more than one round of minichromosome replication. However, transfer of a complex from a completed minichromosome to a chromosome could have taken place, thereby removing complexes from availability for continued minichromosome replication. As analyzed above, this situation should produce a delay in initiation of chromosome replication in some cells. There was a slight delay of about 10 min in the rise in [^{3}H]thymidine incorporation in cells containing a minichromosome, but the final peak heights were very similar. Thus, there could be reutilization of complexes that initiate minichromosome replication and competition for their use, if the resident time for a complex on a minichromosome is short and/or the chromosomal *oriC* has a competitive advantage that offsets the higher copy number of minichromosomes.

REFERENCES

Blinkowa, A., Haldenwang, W. G., Ramsey, J. A., Henson, J. M., Mullin, D.A., and Walker, J.R. (1983). *J. Bacteriol.,* 153, 66.

Blinkowa, A. and Walker, J.R. (1983). *J. Bacteriol.,* 153, 535.

Churchward, G., Holmans, P., and Bremer, H. (1983). *Mol. Gen. Genet.,* 192, 506.

Craine, B. and Rupert, C. S. (1978). *J. Bacteriol.,* 134,193.

Eberle, H. and Forrest, N. (1982). *Mol. Gen. Genet.,* 186, 66.

Evans, I. M. and Eberle, H. (1975). *J. Bacteriol.,* 121, 883.

Fuller, R. S. and Kornberg, A. (1983). *Proc. Natl. Acad. Sci. USA,* 80, 5817.

Helmstetter, C. E. and Krajewski, C. A. (1982). *J. Bacteriol.,* 149, 685.

Helmstetter, C. E., Krajewski, C. A., Leonard, A. C. and Weinberger, M. (1985). Submitted to *J. Bacteriol.*

Helmstetter, C. E., Pierucci, O., Weinberger, M., Holmes, M., and Tang, M.-S. (1979). In *The Bacteria,* L. N. Ornston and J. R. Sokatch (eds.) Academic Press, vol. VII, p. 517.

Hendrickson, W. G., Kusano, T., Yamaki, H., Balakrishnan, R., King, M., Murchie, J. and Schaechter, M. (1982). *Cell,* 30, 915.

Holmes, D. S. and Quigley, M. 1981. *Anal. Biochem.,* 114, 193.

Jacq, A., Kohiyama, M., Lother, H. and Messer, W. (1983). *Mol. Gen. Genet.,* 191, 460.

Kornberg, A. (1980). *DNA Replication,* and *1982 Supplement to DNA Replication,* W. H. Freeman.

LaDuca, R. J. and Helmstetter, C. E. (1983). *J. Bacteriol.,* 154, 1371.

Lark, K. G. and C. A. Lark (1978). *Cold Spring Harbor Symp. Quant. Biol.,* 43, 537.

Leonard, A. C., Hucul, J. A., and Helmstetter, C. E. (1982). *J. Bacteriol.,* 149, 499.

Lother, H., Buhk, H-J., Morelli, G., Hermann, B., Chakraborty, T., and Messer, W. (1981). *ICN-UCLA Symp. Molec. Cell. Biol.*, 22, 57.

Lycett, G.W., Orr, E., and Pritchard, R. H. (1980). *Mol. Gen. Genet.*, 178, 329.

Maaløe, O. and Kjeldgaard, N. O. (1966). *Control of Macromolecular Synthesis,* W.A. Benjamin.

Messer, W., Dankwarth, L., Tippe-Schindler, R., Womack, J. R., and Zahn, G., (1975). In *DNA Synthesis and its Regulation.* M. Goulian and P. Hanawalt (eds.) *ICN-UCLA Symp. Mol. Cell. Biol.,* vol. 3. W. A. Benjamin, p. 602.

Nagai, K., Hendrickson, W., Balakrishnan, R., Yamaki, N., Boyd, D., and Schaechter, M. (1980). *Proc. Natl. Acad. Sci. USA,* 77, 262.

Nicolaidis, A. A. and Holland, B. I. (1978). *J. Bacteriol.,* 135, 178.

Sakakibara, Y. and Yuasa, S. (1982). *Mol. Gen. Genet.,* 186, 87.

Tippe-Schindler, R., Zahn, G., and Messer, W. (1979). *Mol. Gen. Genet.,* 168, 185.

von Meyenburg, K., Hansen, F. G., Riise, E., Begmans, H. E. N., Meijer, M., and Messer, W. (1978). *Cold Spring Harbor Symp. Quant. Biol.,* 43, 121.

Wolf-Watz, H. and Masters, M. (1979). *J. Bacteriol.,* 140, 50.

The Role of RNase H in Initiation of DNA Replication in *E. coli*

Tokio Kogoma, Harvey Bialy, Nelda L. Subia, Ted A. Torrey, and Gavin G. Pickett
The University of New Mexico, Albuquerque

Kaspar von Meyenburg
The Technical University of Denmark, Lyngby-Copenhagen

INTRODUCTION

Stable DNA-replication (*sdrA/rnh*) mutants, capable of continued DNA replication in the absence of protein synthesis (SDR), are devoid of ribonuclease H (RNase H) activity (Ogawa et al., 1984). The *rnh-224* mutation (formerly *sdrA224;* Torrey et al., 1984) has been shown to render the cells able to dispense with the *oriC* site and the DnaA[+] protein (Kogoma and von Meyenburg, 1983). In the absence of *oriC+*, *rnh* mutants initiate rounds of DNA replication in at least four discrete regions of the chromosome (de Massy et al., 1984). In *rnh⁻ oriC+ dnaA+* strains, these sites are also used as the origins of replication, though initiation is predominantly in the *oriC* region (de Massy et al., 1984). Thus, it appears the *rnh* mutations, i.e., the absence of RNase H, allows certain chromosomal nucleotide sequences (collectively termed *oriK*), which are normally repressed, to be used as replication origins. It has also been shown that in a partially purified enzyme system, RNase H inhibits *dnaA+*-independent initiation (Ogawa et al., 1984). These observations led to the suggestion that RNase H confers specificity to the *dnaA+*-dependent initiation at *oriC* by preventing use of alternative sequences (*oriK*) as initiation sites for replication (Ogawa et al., 1984). However, it has not been demonstrated nor excluded that RNase H has a more direct role in the *dnaA+*-dependent initiation at *oriC*, for example, in the processing of a primer RNA, as shown in ColE1 plasmid replication *in vitro* (Itoh and Tomizawa, 1980).

In this presentation, we summarize the properties of *rnh* (*sdrA*) mutants and discuss the results of experiments that suggest the absence of a direct involvement of RNase H in DNA initiation at *oriC*. We also show that the level of RNase H does not change appreciably during or after induction of stable DNA replication in *rnh⁺* strains.

STABLE DNA REPLICATION

Based on the kinetics of acquisition and loss of immunity to thymineless death during and after amino acid starvation, respectively, Maaløe and Hanawalt (1961) hypothesized that "protein and/or RNA synthesis is required to initiate but not to sustain DNA replication." The autoradiography experiments that followed provided evidence consistent with several predictions drawn from the hypothesis (Hanawalt et al., 1961). A series of density-shift experiments, which had been initiated earlier with Maaløe's "explosive send-off" (Lark et al., 1963), were designed and performed by Lark and co-workers to show that amino acid starvation results in the alignment of the DNA replication cycles of individual cells (Lark et al., 1963). This was a direct demonstration of the chromosome-synthetic behavior predicted from the Maaløe-Hanawalt hypothesis. Subsequently, this requirement for protein and RNA syntheses was shown to be a unique property of the *dnaA⁺*-dependent initiation at the *oriC* site (von Meyenburg et al., 1979).

Despite the ample evidence demonstrating this requirement, its biochemical nature is not understood; only instability of one or more factors necessary for an initiation event is suggested. In keeping with this expectation, when Kogoma and Lark (1970, 1975) uncovered certain conditions in which this strict requirement for protein synthesis can be circumvented, the term stable DNA replication (SDR) was coined to describe the ability to continue DNA replication in the absence of protein synthesis, e.g., in chloramphenicol (Cam). A genetic study examining the effects of several mutations at the *recA* and *lexA* loci led to the proposal that SDR is an SOS function (Kogoma et al., 1979). Consistent with this proposal, it was demonstrated that most of the SOS-inducing treatments also induce SDR and that iSDR (induced stable DNA replication) is considerably more resistant to irradiation with ultraviolet light than normal replication (Kogoma et al., 1979). iSDR appears to be mutagenic and is not only dependent on RecA⁺ for its induction, like other SOS functions, but also for its maintenance, i.e., loss of RecA⁺ activity results in cessation of iSDR (Lark and Lark, 1979).

STABLE DNA REPLICATION MUTANTS

In an attempt to gain some insights into the genetic basis of this phenomenon, we isolated mutants that express SDR constitutively (cSDR) (Kogoma, 1978). The mutations map at a locus (*sdrA*) located at 4.5 min on the *E. coli* genetic map and are recessive to the wildtype allele (Kogoma et al., 1981; Torrey et al., 1984). Independently, Lark et al. (1981) isolated an Sdr^C mutant and mapped the mutation at another locus (*sdrT;* 99 min). Whereas the *sdrT* mutant depends on *recA⁺* not only for the cSDR but also for reproductive DNA replication (i.e., DNA replication during the cell cycle), *sdrA* mutants exhibit *recA⁺* dependence only for cSDR. Thus, *recA⁻* mutations are not lethal for *sdrA* mutants (Torrey and Kogoma, 1982). *sdrA* mutations were found to render cells sensitive to nutritionally rich media, e.g., L-broth (the Srm⁻ phenotype) (Torrey et al., 1984).

More recently, *sdrA* mutants were found to be devoid of RNase H activity (Ogawa et al., 1984). The demonstration that a cloned small fragment of the *E. coli* genome, which contains only the *rnh⁺* gene, complements all the phenotypes of *sdrA* mutants led to the conclusion that *sdrA* is allelic to *rnh* (Ogawa et al., 1984). Despite the *in vitro* demonstration of the requirement of ColE1-type plasmid replication for RNase H (Itoh and Tomizawa, 1980; Masukata and Tomizawa, 1984), *sdrA* (*rnh*) mutations were found to have no detectable effects on the *in vivo* replication of this type of plasmid (Kogoma and Subia, 1983). Similar observations were also made with independent isolates of *rnh⁻* (Ogawa and Okazaki, 1984; Naito et al., 1984). This apparent paradox can be resolved by the observation that in the absence of RNase H the requirement for DNA polymerase I can be bypassed (Kogoma and Subia, 1983; Kogoma, 1984). This DNA polymerase I-independent plasmid replication (the Pir⁻ phenotype) involves a transcription event in the *ori* region of the plasmid and DNA polymerase III (Kogoma, 1984). Thus, the absence of RNase H activity appears to invoke an alternative mechanism for plasmid replication that does not involve either the normally required DNA polymerase I or RNase H. Plasmid molecules replicating via the alternative mechanism exist as highly concatemeric forms, i.e., multimers of the unit plasmid joined in a head-to-tail fashion (the Pcf⁻ phenotype) (Kogoma and Subia, 1983).

Taken with the additional phenotypes described below, these observations clearly point to the importance of RNase H in several aspects of nucleic acid metabolism in *E. coli*. The diverse set of phenotypes displayed by *rnh* mutants is summarized in Table 1.

INDUCTION OF SDR DOES NOT INVOLVE INACTIVATION OF RNASE H

The induction of SDR in *sdrA⁺* (*rnh⁺*) strains is *recA⁺*-dependent. The identity of *sdrA* with the structural gene for RNase H (*rnh*) prompted us to determine if treatments inducing SDR, e.g., UV irradiation and incubation with nalidixic acid (Kogoma et al., 1979), resulted in a loss of RNase H activity, and, therefore, if iSDR was mediated at the biochemical level by the *rnh* locus. As shown in Table 2, there was no loss of RNase H activity, measured *in vitro*, immediately or after 60 min in the presence of Cam following UV irradiation or nalidixic acid treatment.

We conclude from these results that the induction of iSDR does not proceed via a direct inactivation of RNase H, such as might be envisioned if an activated RecA protease cleaved the enzyme. However, inducing treatments might elicit the production of a soluble inhibitor of RNase H. We have investigated this possibility in two ways: first, by preparing extracts from induced cells by a gentle nondetergent procedure and adding them to partially purified RNase H preparations; second, by examining the SDR inducibility of cells containing an *rnh⁺* plasmid, pSK760, which overproduced the enzyme by 8- to 15-fold (Kanaya and Crouch, 1983; this work, data not shown). We reasoned that if a soluble inhibitor was produced upon induction, then the overproduced RNase H should titrate this inhibitor, thus raising the threshold level of inductive treatments. Neither of these approaches gave any indication of the presence of an RNase H inhibitor following induction. Extracts prepared from induced cells do not inhibit RNase

TABLE 1 Phenotypes of different RNase H-negative(rnh) mutants [c,f,g]

Phenotype	Mutations			
	$sdrA$	$dasF$	rnh	$herA$
Sdr[c] (constitutive SDR)	+[a,b]	+[c,d]	+[c]	ND
Das⁻ ($dnaA$(Ts) suppression)	+[e]	+[c,d]	+[c]	ND
Dos⁻ ($\Delta oriC$ suppression)	+[e]	+[c,d]	+[c]	ND
Srm⁻ (sensitivity to rich media)	+[d]	+[d]	+[c]	ND
Her⁻ (host factor affecting ColE1 replication)	ND	ND	+[g]	+[g]
Pir⁻ ($polA^+$-independent plasmid replication	+[h,i]	ND	ND	ND
Pcf⁻ (plasmid concatemer formation)	+[h,j]	ND	+[j]	ND

+ = phenotype demonstrated; ND = not determined.

a, Kogoma, 1978; b, Torrey and Kogoma, 1982; c, Ogawa et al., 1984; d, Torrey et al., 1984; e, Kogoma and von Meyenburg, 1983; f, Ogawa and Okazaki, 1984; g, Naito et al., 1984; h, Kogoma and Subia, 1983; i, Kogoma, 1984; j, Subia and Kogoma, 1985.

H, and the SDR inducibilities of cells with and without the rnh^+ plasmid are indistinguishable (data not shown). Thus, it appears that the inactivation of RNase H, directly or indirectly, is not the signal that leads to iSDR.

RNASE H CONFERS SPECIFICITY TO dnaA⁺-DEPENDENT INITIATION AT *oriC*

As mentioned above, cSDR is dependent on $recA^+$. However, $sdrA$ mutants can continue DNA replication in the total absence of RecA⁺ function, e.g., in $recA$-deletion strains, unless protein synthesis is inhibited (Torrey and Kogoma, 1982). This observation suggested that the ordinary replication ($recA^+$-independent) is in operation in $sdrA$ mutants and that inhibition of protein synthesis results in both the cessation of the ordinary replication and the activation of the cSDR capability, which can be expressed only in the presence of RecA⁺ protein. This view presented the possibility that cSDR is a new type of replication distinct from that dependent on $dnaA^+$ and $oriC^+$ (Kogoma et al., 1981). Consistent with this idea, we were able to isolate extragenic suppressor mutations (rin) that allow cSDR to occur despite the absence of RecA protein (Torrey and Kogoma, 1982). On the basis of these results we proposed a hypothesis ("Switch Model") that E.

TABLE 2   Specific activity of RNase H before and after induction

Strain		Treatment	RNase H[i] units/mg protein
AQ2	(rnh^+)[a]	None	28
AQ4	($rnh2$)[a]	None	≤0.8
AQ377	($rnh102$)[b]	None	≤1.1
ON152	($rnh91$)[c]	None	≤1.0
AB1157	(rnh^+)[d]	None	46
AB1157		35 J/m^2 UV[e]	54
AB1157		35 J/m^2 UV, 60 min Cam[f]	80
AB1157		No UV, 60 min Cam[f]	75
AB1157		Nalidixic acid[g]	40
AB1157		Nalidixic acid, 60 min Cam[h]	56

[a] Kogoma, 1978;   [b] Kogoma et al., 1981;   [c] Ogawa and Okazaki, 1984; [d] Kogoma et al., 1979.

[e]5 ml of exponentially growing cultures at 4-5 x 10^8/ml were harvested from supplemented M9 media, washed once with M9 buffer, and resuspended in the same buffer. UV irradiation was for 15 sec at 2.3 J/m^2/sec, after which the cells were diluted 1:1 in 2x supplemented M9 medium, and incubated for 40 min at 37° in the dark. Occurrence of iSDR was monitored by the incorporation of [^{3}H]thymidine into acid-insoluble form, as described by Kogoma (1978).

[f]Cells, treated as in e, were incubated for an additional 60 min in the presence of 150 µg/ml chloramphenicol (Cam) before extraction of RNase H.

[g]Exponentially growing cells were incubated with 40 ug/ml of nalidixic acid for 45 min at 37°, washed by filtration with 10 volumes of warm M9 buffer, and incubated for 30 min at 37° before determination of RNase H activity.

[h]Cells, treated as in g, were incubated for an additional 60 min in the presence of Cam before extraction of RNase H.

[i] Cells were collected by centrifugation and RNase H extracted as described by Carl et al., (1980). Enzyme activity was assayed using [^{32}P] øX174 DNA-RNA (a gift of R.J. Crouch) as substrate, as described by Kanaya and Crouch (1983). Units of activity are nmol acid-insoluble nucleotide/15 min at 37°. Protein was determined by the method of Lowry et al., (1951).

coli has the potential to express a normally repressed initiation system if the sdrA gene is mutated (Kogoma et al., 1981; Torrey and Kogoma, 1982). In this hypothesis sdrA was viewed as a gene coding for a repressor that prevents a switch from the dnaA$^+$-dependent initiation mechanism to the alternative initiation pathway.

The hypothesis was directly supported by the subsequent demonstration that *sdrA* mutants could dispense with the DnaA protein and *oriC* site (Kogoma and von Meyenburg, 1983). Under *oriC⁻* conditions initiation can occur in at least four discrete regions of the chromosome of *sdrA* mutants (de Massy et al., 1984). It appeared, therefore, that inactivation of the *sdrA* gene product, in turn, activates certain nucleotide sequences (*oriK*) to be used as replication origins. It was further substantiated by the discovery that *sdrA* is the structural gene (*rnh*) encoding RNase H and that the presence of RNase H suppresses *dnaA⁺*-independent replication in an *in vitro* initiation system (Ogawa et al., 1984). These observations led to the proposal that RNase H functions as a specificity factor by preventing the use of certain nucleotide sequences other than the *oriC* sequence as replication origins (Ogawa et al., 1984).

IS RNASE H DIRECTLY INVOLVED IN *dnaA⁺*-DEPENDENT INITIATION AT *oriC*?

Initiation of a round of chromosome replication during the cell cycle (reproductive replication) involves the *oriC* site and the DnaA protein. The fact that genetic manipulations inactivating the *dnaA* gene or removing the *oriC* site from the chromosome are possible only in integratively suppressed strains (von Meyenburg and Hansen, 1980) and in *rnh⁻* mutants (Kogoma and von Meyenburg, 1983) indicates that the requirement of the ordinary initiation system for these two genes is absolute. We call this initiation system the *dnaA⁺ oriC⁺* initiation pathway. Since *recA* mutations have no effect on the reproductive replication of the wildtype (*rnh⁺ dnaA⁺ oriC⁺*) strains, the *dnaA⁺ oriC⁺* initiation pathway is independent of RecA function. In contrast, cSDR in *rnh-224* mutants (the *sdrA⁻* pathway) can be completely inhibited by certain *recA* mutations; however, reproductive replication in these mutants is unaffected (Torrey and Kogoma, 1982), suggesting the presence of normal replication. Since *rnh-224* mutant cells contain no detectable amounts of RNase H activity (less than 0.05% that of the wildtype; Ogawa et al., 1984), this observation implies that the *dnaA⁺ oriC⁺* pathway can function in the absence of active RNase H. However, such a conclusion is valid only if *E. coli* cells harbor no other replication systems, apart from the *dnaA⁺ oriC⁺* and *sdrA⁻* pathways, that operate under these conditions.

The first indication that this is indeed the case was provided by the observation that *rnh* mutants exhibited cSDR in the absence of the *oriC* site or the DnaA protein (Kogoma and von Meyenburg, 1983). The observation indicated that the *sdrA⁻* pathway is independent of *dnaA⁺* and *oriC⁺*, and suggested that the *sdrA⁻* pathway operates as a reproductive replication system in place of the *dnaA⁺ oriC⁺* pathway under these conditions. This suggestion predicted that the reproductive replication in *rnh⁻ dnaA*::Tn10 and *rnh⁻ oriC*(deletion) strains should become *recA⁺*-dependent. To test this prediction, we constructed *rnh⁻ dnaA*::Tn10 *recA*(Ts) and *rnh⁻ oriC(deletion) recA*(Ts) mutants and examined them for the colony formation and DNA synthesis at 42 C, the restrictive temperature for the RecA mutant protein (Figures 1 and 2). The results show that *rnh⁻ dnaA*::Tn10 *recA*(Ts) cells are unable to form colonies at 42 C, though at 30 C the *recA*(Ts) triple mutant can grow as well as the *recA⁺* counterpart (Figure 1). A very similar result was obtained with *rnh⁻ oriC(deletion) recA*(Ts) triple mutants (not shown). DNA synthesis in *rnh⁻ dnaA*::Tn10 *recA*(Ts) and *rnh⁻ oriC*(deletion) *recA*(Ts) continued in the presence of Cam at 30 C and ceased at 42 C within 2 hours after the addition of Cam (Figure 2(b) and 2(d)). Temperature shift had no effect on DNA synthesis in the *recA⁺*

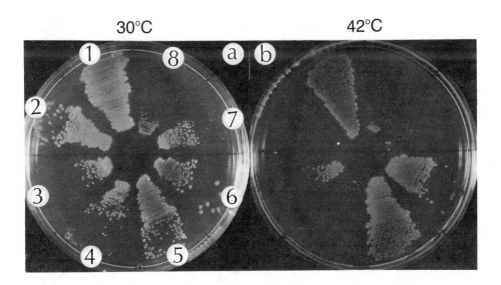

FIGURE 1 Effects of the recA200 mutation and dnaA⁺-containing plasmids on the colony formation of rnh dnaA mutants at 42°C. Eight strains of different combinations of rnh, dnaA, and recA alleles (otherwise isogenic) with or without plasmid (see the table below for the genotypes) were grown on supplemented M9G plates at 30°C overnight. Single colonies were spread with toothpicks on two supplemented M9G plates. One (a) was incubated at 30°C for 56 hours and the other (b) at 42°C for 36 hours.

Sector number	Strain	Genotype (alleles)				Plasmid
		rnh	*dnaA*	*oriC*	*recA*	
1	AQ1993	+	+	+	+	None
2	AQ2016	+	A5	+	+	None
3	AQ2096	224	850::Tn10	+	+	None
4	AQ2118	224	850::Tn10	+	200	None
5	AQ2201	224	850::Tn10	+	200	pFHC506
6	AQ2216	224	850::Tn10	+	200	pFHC539
7	AQ2239	224	850::Tn10	+	200	pDKT143
8	AQ2203	224	850::Tn10	+	200	pBR322

counterparts. Thus, cSDR is temperature sensitive in these *recA*(Ts) strains. Figure 2 also shows that DNA replication is temperature sensitive in these mutants even in the absence of Cam at 42 C, indicating that the reproductive replication becomes dependent on *recA+* when the *dnaA+ oriC+* pathway is blocked by the *dnaA*::Tn10 or *oriC*(deletion) mutation (Figures 2(a) and 2(c)).

These observations indicated that the *sdrA−* pathway can be adopted for reproductive replication if the *dnaA+ oriC+* pathway is not operational and that there is no other replication system available in these mutants other than these two initiation pathways. Thus, we infer that the *dnaA+ oriC+* pathway can function without active RNase H in *rnh− recA− (dnaA+ oriC+)* strains.

This inference predicts that introduction of the DnaA protein should restore temperature-resistant DNA replication to *rnh− dnaA*::Tn10 (*oriC+*) *recA*(Ts) cells, but not to *rnh− oriC*(deletion) *recA*(Ts) cells. Plasmid pFHC539 is a pBR322 derivative, constructed by F.G. Hansen, carrying an intact *dnaA+* gene. Cells harboring this plasmid have been shown to contain at least 5 times more DnaA protein than haploid *dnaA+* cells (von Meyenburg et al., this volume). pFHC506 is a similar construct except for the deletion of part of the regulatory site including a promoter (*1p*) and the DnaA–protein binding sequence (E.B. Hansen et al., 1982; F.G. Hansen, personal comm.). Figure 1 shows that introduction of either plasmid could render *rnh− dnaA*::Tn10 (*oriC+*) *recA*(Ts) mutant cells temperature resistant. pBR322 or pDKT143, a pBR322 construct similar to pFHC506 but lacking the entire *dnaA+* gene, did not suppress the temperature sensitivity, indicating that the introduced *dnaA+* gene was responsible for the rescue effect (Figure 1). In contrast, none of these plasmids restored temperature resistance to the *oriC*(deletion) counterpart (not shown). Thus, the rescue by the cloned *dnaA+* gene occurs only when the *oriC* region is present. It was previously shown that initiation occurs predominantly in this region of the chromosome in *rnh− dnaA+* cells, although *oriK* sites in other regions are also used as replication origins in the absence of RNase H (de Massy et al., 1984).

REPLICATION OF MINICHROMOSOMES IN *rnh* MUTANTS

The above results allow us to conclude that the *dnaA+ oriC+* pathway can function in the absence of active RNase H only if the observed *recA+*-independent, *dnaA+*-dependent initiation occurs specifically at *oriC*. To investigate this further, we examined replication and maintenance of "minichromosomes", i.e., *oriC* plasmid and p*λasn20*, in *rnh* mutants. The p*λasn20* minichromosomes (von Meyenburg et al., 1979) and plasmid pOC24 (Messer et al., 1979) replicate solely from *oriC*. The replication of minichromosomes strictly requires active DnaA protein in *rnh+* strains. The results summarized in Table 3 show that pOC24 transformed *rnh* mutants in the presence of active DnaA activity (CM3440; AQ2020 at 30 C) but not under *dnaA−* conditions (AQ2096; AQ2020 at 42 C). A similar result was obtained with p*λasn20* minichromosome after infection of *rnh* mutants with *λasn20* (data not shown). Plasmid pTKQ27, a derivative of pBR322 (Kogoma, 1984), transformed all strains at all temperatures. Thus, the loss of RNase H activity did not change the *dnaA+* dependence of minichromosome replication, indicating a lack of nucleotide sequences (*oriK*) on the minichromosomes that can be activated, as a result of the loss of

RNase H, for use as initiation sites in a *dnaA*+-independent manner, i.e., via the *sdrA*⁻ pathway.

Both types of minichromosomes are quite unstable; that is, even

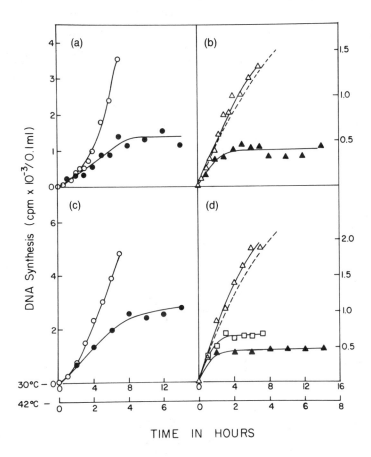

FIGURE 2 DNA synthesis in AQ2118 and AQ2120 at 30°C or 42°C in the presence and absence of chloramphenicol. AQ2118 (rnh224 dnaA850::Tn10 oriC⁺ recA200) (a,b) and AQ2120 (rnh224 dnaA5 oriC del1071 recA200)) (c,d) were grown in M9G medium supplemented with casamino acids (0.2%) at 30°C to titers of 0.83 x 10⁸ and 1.5 x 10⁸ cells/ml, respectively. Portions of the cultures were added to [³H]thymine (10 µCi/8 µg/ml) and incubated in the absence of Cam (a,c) at 30°C (open circles) and 42°C (solid circles) or in the presence of Cam (150 µg/ml) (b;d) at 30°C (open triangles) and 42°C (solid triangles). An additional portion of the AQ2120 culture was incubated in rifampicin (100 ug/ml) at 30°C (squares). 0.1 ml samples were withdrawn, and radioactivity in the acid-insoluble fraction determined as described previously (Kogoma, 1978). Dashed line: DNA synthesis of comparable cultures of the recA⁺ counterparts of AQ2118 and AQ2120 at 42°C in the presence of Cam. Note the two-fold difference in time scale for 30°C and 42°C cultures.

TABLE 3   Transformation of *rnh dnaA* mutants with pOC24 and pTKQ27

Strain	Relevant genotype	Temperature	Transformants per µg DNA of	
			pOC24	pTKQ27
CM3438	*rnh+*	30°	7.28	75.36
		42°	18.98	85.44
CM3440	*rnh224*	30°	4.40	24.00
AQ2096	*rnh224 dnaA850::*Tn10	30°	<0.01	6.16
AQ2020	*rnh224 dnaA5 ∆oriC*	30°	5.12	24.00
		42°	<0.01	18.40

when grown under selective growth conditions, minichromosome-less cells constitute a sizable portion of the populations in single colonies and in liquid cultures (von Meyenburg et al., 1979). The fractions of pλ*asn20* minichromosome-less cells in *rnh+* (CM3438) and *rnh⁻* (CM3440) cell populations were found not to differ significantly (Table 4). The stability of plasmids such as these minichromosomes lacking a partitioning mechanism is largely determined by the copy number (Summers and Sherratt, 1984). The above results suggest that the lack of RNase H activity does not grossly alter the copy number of minichromosomes.

DISCUSSION

That RNase H activity is required for DNA initiation to occur exclusively at *oriC* was demonstrated by both *in vivo* (Kogoma and von Meyenburg, 1983; de Massy et al., 1984) and *in vitro* experiments (Ogawa et al., 1984). However, the results described above do not indicate a direct involvement of RNase H activity in the initiation process; *dnaA+*-dependent initiation at *oriC* appears to occur in the absence of active RNase H. We conclude that RNase H does not directly participate in the biochemical process leading to initiation, though the possibility that another enzyme with a similar function may replace RNase H under certain conditions cannot be ruled out.

The *rnh224* (*sdrA224*) allele used in these experiments above endows the cells with no detectable levels of RNase H activity as measured *in vitro* (less than 0.05% that of *rnh+* cells; Ogawa et al., 1984). In our discussion we have assumed that the residual RNase H activity, if any, has no physiological significance. Our conclusion is of course only as valid as the specificity and sensitivity of the measurements. Recently, Kanaya and Crouch (1984) reported experiments in which genetic manipulations that replaced the chromosomal *rnh+* gene by an *rnh* allele insertionally inactivated by a transposon did not yield viable cells if an active *rnh* allele was not present elsewhere in the cells. They concluded that RNase H was essential for the growth of *E. coli*. However, it is possible that RNase H activity is necessary only under certain growth conditions, e.g.,

TABLE 4   Stability of minichromosomes in rnh^+ and rnh^- strains

Strain	Analysis of single colonies		
	Kan-r	Kan-r + Kan-s	% Kan-r
CM3438 (rnh^+)	162	359	45.1 ± 7.3
CM3440 (rnh^-)	177	374	47.1 ± 10.5

CM3438 and CM3440 were infected with $\lambda asn20$ (zzz::Tn5), plated on M9G-casamino acid plates containing kanamycin (Kan), and incubated overnight at 30°C. Single colonies of the Kan-r transductants were spread on plates not containing Kan and allowed to form colonies at 30°C. These secondary single colonies were analyzed for the Kan-r and Kan-s phenotypes at 30°C.

in nutritionally rich media. We have observed that *rnh* mutations render cells with certain genetic backgrounds extremely sensitive to broth (Torrey et al., 1984; von Meyenburg and Kogoma, unpublished).

Initiation via the *sdrA* pathway is sensitive to rifampicin (Figure 2(d); Kogoma, 1978), suggesting that active transcription is involved. It is possible that in the absence of RNase H activity a DNA-RNA hybrid consisting of a template DNA strand and the RNA transcript in certain regions of the chromosome may be stabilized by RecA (and perhaps also SSB) protein binding to the displaced single-stranded DNA region. This "R-loop" structure, similar to the D loop thought to exist on certain replicating DNA molecules (Kasamatsu et al., 1971; Kornberg, 1980) and in early stages of homologous recombination (Dressler and Potter, 1982), may be used by a replisome to begin a round of DNA replication. In the context of this model RNase H is thought to exert its specificity function by preventing formation of potential initiation structures (R loops), thus ensuring exclusive initiation at *oriC*.

The observed lack of change in the RNase H levels after treatments that result in iSDR (Table 2) suggests that factors other than RNase H are involved, provided that iSDR is initiated by the same mechanism as cSDR. Alternatively, iSDR may involve a separate initiation mechanism. In contrast to the UV-resistant and mutagenic property of iSDR, the cSDR that results from mutational inactivation of RNase H (rnh^-) is no more resistant to UV radiation than normal replication, nor is it significantly error-prone (Kogoma and Rask, unpublished). On the other hand, the cSDR conferred by *sdrT* mutations appears to be mutagenic (Lark et al., 1981); *sdrT* mutant cells contain as much RNase H activity as *sdrT*$^+$ cells (Bialy, Lark and Kogoma, unpublished). Therefore, it may be that mutagenic SDR (i.e., iSDR and *sdrT*-conferred cSDR) results from an alteration in the DNA replication machinery, as suggested earlier (Kogoma and Lark, 1975). Compatible with this possibility, it has been observed that direct thermal activation of a heat-activable RecA (*tif-1*) protease, in the presence of Cam, results in the appearance of SDR capability in a *lexA71*::Tn5 *recA441* (*tif1*) strain (Witkin and Kogoma, 1984).

In closing, it must be pointed out that despite the advances in our understanding of this unique phenomenon made possible by the isolation of *sdrA* (*rnh*) mutants, the precise molecular nature of the requirement for protein synthesis for initiation of DNA replication in *E. coli,* first uncovered by Maaløe and Hanawalt in 1961, remains elusive.

ACKNOWLEDGEMENTS

We would like to thank Flemming Hansen for plasmids, Robert Crouch for his generous gift of RNase H substrate, and David Bear for his assistance in preparing more. Excellent technical assistance by Hans Snyder is greatly appreciated. This work was supported by a NATO Grant for International Collaboration in Research to K. v. M. and T. K.; grants (GM 22092 and Minority Biomedical Science Program Grant PRO 8139) from the National Institutes of Health and a grant-in-aid for research from the University of New Mexico to T. K.; grants from the Danish Natural Science Research Council and the NOVO Foundation to K.v.M. T. K. wishes to dedicate this paper to Prof. T. Yanagita, who introduced him to the Copenhagen view of bacterial growth.

REFERENCES

Carl, P. L., Bloom, L., and Crouch, R. J. (1980). *J. Bacteriol.,* 144, 28.

de Massy, B., Fayet, O., and Kogoma, T. (1984). *J. Mol. Biol.,* 178, 227.

Dressler, D., and Potter, H. (1982). *Ann. Rev. Biochem.,* 51, 727.

Hanawalt, P. C., Maaløe, O., Cummings, D. J., and Schaechter, M. (1961). *J. Mol. Biol.,* 3, 156.

Hansen, E. B., Hansen, F. G., and von Meyenburg, K. (1982). *Nucl. Acids Res.,* 10, 7373.

Itoh, T., and Tomizawa, J. (1980). *Proc. Natl. Acad. Sci. USA,* 77, 2450.

Kanaya, S., and Crouch, R. J. (1983). *J. Biol. Chem.,* 258, 1276.

Kanaya, S., and Crouch, R. J. (1984). *Proc. Natl. Acad. Sci. USA,* 81, 3447.

Kasamatsu, H., Robberson, D. L., and Vinograd, J. (1971). *Proc. Natl. Acad. Sci. USA,* 68, 2252.

Kogoma, T. (1978). *J. Mol. Biol.,* 121, 55.

Kogoma, T. (1984). *Proc. Natl. Acad. Sci. USA,* 81, in press.

Kogoma, T., and Lark, K. G. (1970). *J. Mol. Biol.,* 52, 143.

Kogoma, T., and Lark, K. G. (1975). *J. Mol. Biol.,* 94, 243.

Kogoma, T., and Subia, N. L. (1983). In *Mechanisms of DNA*

Replication and Recombination, N. R. Cozarelli (ed.). Alan Liss, Inc. pp. 337-349.

Kogoma, T., and von Meyenburg, K. (1983). *EMBO J.,* 2, 263.

Kogoma, T., Torrey, T. A., and Connaughton, M. J. (1979). *Mol. Gen. Genet.,* 176, 1.

Kogoma, T., Torrey, T. A., Subia, M. L., and Pickett, G. G. (1981). In *The Initiation of DNA Replication,* D. S. Ray (ed.), Academic Press. pp. 361-374.

Kornberg, A. (1980). *DNA Replication.* W. H. Freeman.

Lark, K. G., and Lark, C. A. (1979). *Cold Spring Harbor Symp. Quant. Biol.,* 43, 537.

Lark, K. G., Repko, T., and Hoffman, E. J. (1963). *Biochim. Biophys. Acta,* 76, 9.

Lark, K. G., Lark, C. A., and Meenen, E. A. (1981). In *The Initiation of DNA Replication,* D. S. Ray (ed.), Academic Press. pp. 337-360.

Lowry, O. H., Rosebrough, N. J., Farr, A. L., and Randall, R. J. (1951). *J. Biol. Chem.,* 193, 265.

Maaløe, O., and Hanawalt, P. C. (1961). *J. Mol. Biol.,* 3, 144.

Masukata, H., and Tomizawa, J. (1984). *Cell,* 36, 513.

Messer, W., Meijer, M., Bergmans, H. E. N., Hansen, F. G. von Meyenburg, K., Beck, E., and Schaller, H. (1979). *Cold Spring Harbor Symp. Quant. Biol.,* 43, 139.

Naito, S., Kitani, T., Ogawa, T., Okazaki, T., and Uchida, H. (1984). *Proc. Natl. Acad. Sci. USA,* 81, 550.

Ogawa, T., and Okazaki, T. (1984). *Mol. Gen. Genet.,* 193, 231.

Ogawa, T., Pickett, G. G., Kogoma, T., and Kornberg, A. (1984). *Proc. Natl. Acad. Sci. USA,* 81, 1040.

Summers, O. K., and Sherratt, D. J. (1984). *Cell,* 36, 1097.

Subia, N. L., and Kogoma, T. (1985). Submitted.

Torrey, T. A., and Kogoma, T. (1982). *Mol. Gen. Genet.,* 187, 225.

Torrey, T. A., Atlung, T., and Kogoma, T. (1984). *Mol. Gen. Genet.,* 196, 350.

von Meyenburg, K., and Hansen, F. G. (1980). *ICN-UCLA Symp. Mol. Cell. Biol.,* 19, 137.

von Meyenburg, K., Hansen, F. G., Riise, E., Bergmans, H. E. N., Meijer, M., and Messer, W. (1979). *Cold Spring Harbor Symp. Quant. Biol.*, 43, 121.

Witkin, E. M. and Kogoma, T. (1984). *Proc. Nat. Acad. Sci. USA*, 81, in press.

The Terminus Region of the Chromosome

Peter Kuempel and Joan Henson
University of Colorado, Boulder

Our laboratory has been interested for a long time in topics related to bacterial growth and the events of the cell cycle of *Escherichia coli*. Like many others who have worked in this field, we were first intrigued with bacterial growth due to the classic studies of Ole Maaløe and the "Copenhagen School." For one of us (P.L.K.), this interest was nurtured considerably by a postdoctoral year spent in Ole's lab (1965). Since that time, a flood of literature concerning bacterial growth has appeared. For a recent synthesis of all these separate details into an integrated picture of growth and division of *E. coli*, it is refreshing to note that one of the best sources available is still a product of the "Copenhagen School" (Ingraham, Maaløe and Neidhardt, 1983).

PREVIOUS STUDIES OF THE TERMINUS

Our current interest is the terminus region of the chromosome. We first became interested in this region as the result of some autoradiographic experiments that demonstrated that *E. coli* chromosome replication was initiated in a bidirectional fashion, and that the two forks traveled at approximately the same rate (Prescott and Kuempel, 1972; Kuempel et al., 1973a). We extended these autoradiographic studies to determine if it was possible to detect replication forks meeting elsewhere on the circular chromosome at the end of the replication cycle. A large number of such patterns was observed, and the frequency of these events demonstrated that the forks were meeting in the region directly opposite the origin (Kuempel et al., 1973b). Upon examining the then-current genetic map of that region, we were fascinated by the presence of a large region (about 6 min) that contained very few genetic loci and across which mapping by means of bacteriophage P1 cotransduction was not possible. Although we have now studied this region for a number of years, our fascination with it still continues.

A fair amount is now known about the terminus region. A cotransduction map exists for this region, as a result of the isolation of various transposon insertions (Bitner and Kuempel, 1981, 1982; Fouts and Barbour, 1982), and other genetic loci are slowly being assigned to the region. Genetically, there is nothing unusual about this region, except for the scarcity of identified loci, and as demonstrated by M. Masters (this volume), a number of proteins are encoded there. A restriction map of the terminus region is also available (Bouché, 1982), and this map has been a tremendous asset for further analyses of the region.

An interesting property of the terminus is that it inhibits replication forks (Kuempel et al., 1977; Kuempel and Duerr, 1979; Louarn et al., 1977, 1979). The recent experiments of Bouché et al. (1982) indicate that forks traveling in a direction clockwise with respect to the genetic map were inhibited in the region between 180 and 240 kb on the Bouché map (1982). Forks traveling counter-clockwise were inhibited in the region between 300 and 240 kb, and the forks usually met at approximately 240 kb. These coordinates are shown on the genetic map in Figure 1.

Circular chromosomes that replicate in a bidirectional fashion do not necessarily have a terminus that inhibits replication forks. Some examples are SV40 (Lai and Nathans, 1975) and the Cairns form of lambda (Valenzuela et al., 1976). This suggests that the inhibitory sites present in the terminus of the *E. coli* chromosome have evolved

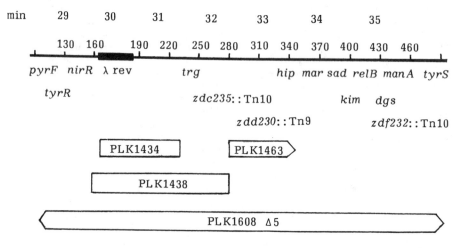

FIGURE 1 Map of the terminus region of the <u>E. coli</u> chromosome. The coordinates in min (minutes) and kb (kilobase pairs of DNA) are from Bachmann (1983) and Bouche (1982). Genetic markers whose location in the Bouche map are known are <u>pyrF</u> (Donovan and Kushner, 1983), <u>nirR, trg, kim,</u> and <u>manA</u> (Bouche et al., 1982), <u>zdc235::Tn10</u> (Henson and Kuempel, 1983) and <u>dgs</u> (unpublished experiments). The <u>hip</u> (Moyed and Bertrand, 1983) and <u>mar</u> (George and Levy, 1983) loci have recently mapped in the terminus region. The deletion strains are described in the text.

to provide some function. An intriguing hypothesis is that the terminus contains features that are important at the end of the replication cycle or for coordinating chromosome replication with cell division. Inhibiting replication in the terminus region would insure that these functions would be used optimally, since the replication cycle would always end in this region.

POTENTIAL FUNCTIONS OF THE TERMINUS

What functions that are important for the cell cycle might be encoded in the terminus region? A number of clues concerning these potential functions come from studies of low copy number plasmids. These plasmids are normally quite stable, even though they might be present at a copy number as low as 1 or 2. However, altered plasmids that are very unstable can be obtained, and analyses of these plasmids have demonstrated that several different functions are important for plasmid stability. It would be expected that the bacterial chromosome would have functions that are at least equally sophisticated.

Partition sites

One function that is presumably encoded in the chromosome of *E. coli* that may be in the terminus region, is a *par* (partitioning) site. These sites act in *cis*, and they specify the equipartition of chromosomes to daughter cells (Nordstrom et al., 1980; Meacock and Cohen, 1980; Miki et al., 1980). The *par* site of pSC101 contains 270 bp (Meacock and Cohen, 1980), and it apparently does not encode a protein (Miller et al., 1983). The *par* region of phage P1 and the F plasmid are larger, and in addition to a site that acts in *cis* they contain genes for proteins that are essential for the function of these sites (Austin and Abeles, 1983a, 1983b; Austin and Wierzbicki, 1983; Ogura and Hiraga, 1983a). These proteins can act in *trans*. It is noteworthy that *par* sites can be inserted into heterologous plasmids and thereby confer stability on previously unstable plasmids (Meacock and Cohen, 1980; Austin and Abeles, 1983a; Ogura and Hiraga, 1983a).

It is usually thought that *par* sites function by means of an attachment to the cell membrane. This is a variation of the original replicon model of Jacob et al. (1963), and the partitioning of daughter chromosomes would be achieved by membrane growth between the *par* sites. It has been demonstrated with R1 and pSC101 that DNA of *par*+ plasmids is associated with the outer membrane fraction, whereas DNA of *par*‾ plasmids is not (Gustafson et al., 1983). The replication origin region of the *E. coli* chromosome is also associated with the outer membrane (Kusano et al., 1984), and this region could contain the *par* site.

Par sites can also function at a variety of different positions in a chromosome (Meacock and Cohen, 1980), and it is possible that a *par* site of the *E. coli* chromosome is in the terminus region. Such a location would considerably simplify the partitioning of chromosomes in cells growing in rich medium (Cooper and Helmstetter, 1968) or in low concentrations of thymine (Pritchard and Zaritsky, 1970). In these conditions, a new cycle of replication is initiated before the preceding cycle has been completed; this can lead to partially replicated chromosomes that contain one copy of the terminus region and more than four copies of the origin. Once the replication cycle is completed, it would be simpler to partition from the newly doubled terminus than from the multiple origins. There is no evidence that the terminus region of the *E. coli* chromosome is associated with the cell envelope, but such evidence has been obtained for *Bacillus subtilis* (Hye et al., 1976; Sargent and Bennet, 1982).

Decatenation of daughter chromosomes

Before daughter chromosomes can be effectively partitioned to daughter cells, it is also necessary that any chromosomes that are physically joined together be separated from each other. One way that daughter chromosomes can be linked together is by means of interlocked (or catenated) circles. For example, Sundin and Varshavsky (1981) have determined that two daughter chromosomes of SV40 can be interlocked with catenation linkage numbers as high as 20. They suggest that unwinding of the last 10 or 20 turns of the parental double helix during replication leaves an equal number of interlocking turns joining the two daughter chromosomes. These interlocking turns could then be removed by a type II topoisomerase, such as the *E. coli* DNA gyrase. It has been demonstrated that this enzyme can decatenate interlocked chromosomes (Mizuuchi et al., 1980; Kreuzer and Cozzarelli, 1980).

In a chromosome as large as that of *E. coli,* the interlocking turns present at the end of the replication cycle would probably be confined to the terminus region. This would occur if the growing daughter chromosomes formed separate domains that were connected by the remaining unreplicated DNA. DNA gyrase shows specificity for particular sites, and the decatenation of daughter chromosomes would be facilitated if sites of this type were enriched in the terminus region (Fisher et al., 1981; Morrison and Cozzarelli, 1981). DNA gyrase does seem to be involved in the decatenation of daughter chromosomes in *E. coli:* when cells containing a temperature-sensitive DNA gyrase were incubated at a nonpermissive temperature, nucleoids isolated from such cells had a doublet shape and increased amount of DNA (Steck and Drlica, 1984). Exposure of the nucleoids to gyrase *in vivo* or *in vitro* caused a reversion to normal singlet morphology.

Site-specific recombination

Another process that can leave daughter chromosomes joined together is recombination, which produces a single circular dimer that cannot be partitioned to daughter cells. Studies with the circular prophage form of P1 have demonstrated that the resolution of such structures is very important for plasmid stability. The plasmid encodes a site-specific recombination enzyme (*cre*) that functions at the *loxP* site, and this rapidly converts circular dimers to monomers (Austin et al., 1981). P1 mutants lacking the *cre* system are very unstable. A similar system is important for the stable maintenance of ColE1 (Summers and Sherratt, 1984). If a similar system is present in the *E. coli* chromosome, it would seem advantageous to have the recombination site located in the terminus. As a result, the site would only function at the end of the replication cycle when two copies of the site would be present. A system of this type could also function to decatenate interlocked chromosomes.

Cell division control

In addition to being potentially involved in chromosome separation and partitioning, the terminus region could also be involved in coordinating cell division with chromosome replication. Cell division usually occurs 20 minutes after the completion of a replication cycle (Cooper and Helmstetter, 1968), and a number of attempts have been made to determine how these processes are coordinated. Many experiments have demonstrated that if DNA synthesis is inhibited, the subsequent cell division is also inhibited. This observation could be taken to indicate that in some way replication of the terminus signals cell division, but it is now understood that these experiments used conditions that induced the SOS response (Little and Mount, 1982). Part of this response is the synthesis of the *sfiA* (or *sulA*) gene product which acts on the product of the *sfiB* (or *sulB* or *ftsZ;* Lutkenhaus, 1983) gene to inhibit cell division.

Although induction of the SOS response leads to a potent inhibition of cell division in stressed cells, this system does not seem to be involved in the normal control of cell division. For example, *sfiA* mutants are not abnormal with respect to cell size or number of anucleate cells, and these mutants also exhibit a normal response to a shift of growth medium (Huisman et al., 1983). A new locus, *sfiC,* has recently been identified, and its expression also inhibits cell division by acting on the *sfiB* protein. The *sfiC* locus does not seem to be involved in the normal control of cell division, either, since *sfiA sfiC* mutants show normal control of cell division (D'Ari and Huisman, 1983).

There is some evidence that cells possess still another system for controlling cell division, and this system could be the one that normally functions at the end of a replication cycle when completed daughter chromosomes are present. The strains that were used in these experiments were either *sfiB* or *sfiA sfiC* mutants, so induction of the SOS response would not inhibit cell division. When DNA synthesis was inhibited, cell division did not stop immediately stop. However, division was inhibited after a period of 20 to 30 minutes (Huisman et al., 1980; Burton and Holland, 1983). It has been proposed that cell division was inhibited once there were no more completed daughter chromosomes to partition to daughter cells, and the cells contained only partially replicated chromosomes.

A control system that coordinates cell division with the presence of plasmid chromosomes that can be partitioned to daughter cells has recently been shown to be specified by genes in the F plasmid (Ogura and Hiraga, 1983b). When plasmid replication was inhibited and the plasmid copy number dropped to one per cell, cell division was inhibited. Two genes for this control system have been identified in the F plasmid, and the system appears to operate independently of *recA* activity. All of the components essential for this system might be encoded in the F plasmid. However, it seems likely that this system also uses components from the host cell. Further analysis of this system could provide important information about the elusive *E. coli* system that coordinates the termination of replication with cell division.

DELETIONS IN THE TERMINUS REGION

Considering the nature of the functions mentioned above, how could one map and study them in a structure as complex as the *E. coli* chromosome? Some of these functions might be necessary for cell survival, but others probably enhance survival without being essential. It should at least be possible to obtain mutations in loci of this latter type. However, there does not seem to be a simple way to select for mutations in most of these loci, and mutant phenotypes might even be hard to detect. Our approach has been to obtain deletions of various parts of the terminus region, the rationale being that if a region can be deleted, the presence of that region is not necessary for cell survival. Strains harboring various deletions can then be examined to determine if any aspect of chromosome separation, partitioning, or cell division is affected.

Obtaining deletions has been simplified by use of various transposon insertions that have been isolated in the terminus region. Tn10 insertions have been particularly useful, since the types of deletions caused by this transposon have been well characterized (Kleckner et al., 1979; Ross et al., 1979). In addition, it is possible to select directly for tetracycline-sensitive (Tet-s) cells (Bochner et al., 1980). These cells often contain deletions that start at the inside end of one of the IS10 elements of the transposon, and extend through the genes for tetracycline resistance and the other IS10 and into the chromosome.

Figure 1 shows some of the deletions that we have obtained, and the relevant features of the terminus region that we have used to characterize these deletions. Prophage lambda *reverse* has been important in the isolation of most of these deletions, and it is shown integrated at its appropriate location. All of these deletions, except for the one isolated in strain PLK1608, have no noticeable effect on either cell morphology, viability, or growth rate. If some aspect of cells harboring these deletions was affected, it was not apparent in these simple tests, and more sophisticated analyses are required.

The deletion in strain PLK1434, and others similar to it, have been recently described (Henson et al., 1984). These deletions cause fusion fragments that contain DNA homologous to both the left end of lambda *reverse* and the right side of the *trg* region. In addition, the deletions caused the loss of chromosomal DNA normally present between lambda *reverse* and *trg-2*:Tn10. We have not yet cloned all the DNA present between the end points of these deletions, so we cannot be certain that all of this DNA is absent. However, based on the data we have obtained, it seems unlikely that any of this DNA is still present.

The deletion in strain PLK1438 has also been well characterized. We have identified the fusion fragment in this strain, and have also demonstrated that various chromosomal fragments normally present between lambda *reverse* and *zdc-235*::Tn10 were absent. This deletion removed 102 kb of chromosomal DNA, which is approximately 2.5 min on the genetic map. Based on the results of Bouché et al. (1982), the deletions in PLK1438 and PLK1434 removed the region in which clockwise-traveling replication forks are inhibited and the region in which replication forks usually meet. We are now testing these strains to determine if replication forks are still inhibited. Our hypothesis is that the region in which inhibition occurs and where the forks meet is shifted toward the right side of the terminus region in these strains.

The deletion in strain PLK1463, and other strains similar to it, extended to the right of *zdc-235*::Tn10 (Henson and Kuempel, 1983).

These deletions extended through *zdd-230*::Tn9, and consequently removed a minimum of 40 kb of chromosomal DNA. We have not yet identified the endpoints of these deletions, but deletions of this interval were readily obtained. Approximately 15% of the Tet-s derivatives that we have tested were also sensitive to chloramphenicol.

The largest deletion that we have obtained is that present in PLK1608. This strain was isolated among temperature-resistant derivatives of a very stable lysogen containing temperature-sensitive lambda *reverse*. We have determined that this strain was missing DNA homologous to cloned fragments from the following regions of the Bouché map: 163-188, 223-230, 243-253, 289-311, 380-400, 417-430, 440-455 kb. We have not yet identified the end points of this deletion, or the fusion fragment. Genetically, the strain was *pyrF+ tyrR nirR trg sad manA tyrS+*. The deletion could be transduced into other strains using *zdg-232*::Tn10 as a selectable marker, and all of the *manA* recipients also became *sad trg tyrR*. This was a cotransduction of loci that are normally separated by 7 min.

The presence of this deletion, which we call deletion 5, had a dramatic effect on the phenotype of the cells. The cells grew at one half the normal rate in several media and only one half of the cells formed colonies. The cell morphology was very heterogeneous, and there were minicells, normal-sized cells, and filaments up to 60 micrometers long. Autoradiographic examination of cells labeled with [^{3}H]thymine demonstrated that 15% of the cells lacked DNA. The DNA distribution in the longer cells was also heterogeneous. Some of these cells contained the DNA distributed along the entire length of the filament, whereas others contained DNA in discrete regions (Figure 2).

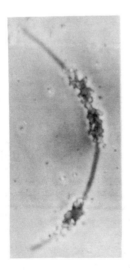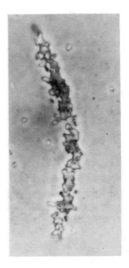

FIGURE 2 Autoradiography of strain PLK1648, which contains deletion 5. Cells were grown for 12 generations in [^{3}H]thymine before preparation. The cells shown here were approximately 30 μm long.

The deletion in strain PLK1608 is one of the few we have obtained that removed DNA from the region between kb 340 and kb 400. We have another deletion that has removed this region, as well as most of the rest of the terminus region. This strain also grew slowly, and the cells had a very heterogeneous morphology. It is possible that deletion of a site between 340 and 400 kb causes this phenotype. However, it is also possible that this phenotype is only caused by very large deletions. We are presently isolating smaller deletions of this interval for further study. Regardless of the outcome of those experiments, it appears that most, if not all, of the terminus region is nonessential for cell growth, though the region contains functions that enhance cell growth.

ACKNOWLEDGEMENTS

This research has been supported by grants from the American Cancer Society (MV-137A), the National Institute of General Medical Sciences (GM32968), and the Biomedical Research Grant Program at the University of Colorado.

REFERENCES

Austin, S. and Abeles, A. (1983a). *J. Mol. Biol.*, 169, 353.

Austin, S. and Abeles, A. (1983b). *J. Mol. Biol.*, 169, 373.

Austin, S. and Wierzbicki, A. (1983). *Plasmid*, 10, 73.

Austin, S., Ziese, M., and Sternberg, N. (1981). *Cell*, 25, 729.

Bachmann, B. J. (1983). *Microbiol. Rev.*, 47, 180.

Bitner, R. and Kuempel, P. (1981). *Mol. Gen. Genet.*, 184, 208.

Bitner, R. and Kuempel, P. (1982). *J. Bacteriol.*, 149, 529.

Bochner, B., Huang, H., Schieven, G., and Ames, B. (1980). *J. Bacteriol.*, 143, 926.

Bouché, J. (1982). *J. Mol. Biol.*, 154, 1.

Bouché, J., Gelugne, J., Louarn, J., Louarn, J. M., and Kaiser, K. (1982). *J. Mol. Biol.*, 154, 21.

Burton, P. and Holland, I. B. (1983). *Mol. Gen. Genet.*, 190, 128.

Cooper, S. and Helmstetter, C. (1968). *J. Mol. Biol.*, 31, 519.

D'Ari, R. and Huisman, O. (1983). *J. Bacteriol.*, 156, 243.

Donovan, W. P. and Kushner, S. R. (1983). *Gene*, 25, 39.

Fisher, L. M., Mizuuchi, K., O'Dea, M. H., Ohmori, H., and Gellert, M. (1981). *Proc. Natl. Acad. Sci. USA*, 78, 4165.

Fouts, K. and Barbour, S. (1982). *J. Bacteriol.*, 149, 106.

George, A. M. and Levy, S. B. (1983). *J. Bacteriol.*, 155, 541.

Gustafsson, P., Wolf-Watz, H., Lind, L, Johansson, K., and Nordstrom, K. (1983). *EMBO J.*, 2, 27.

Henson, J. and Kuempel, P. (1983). *Mol. Gen. Genet.*, 189, 506.

Henson, J., Kopp, B., and Kuempel, P. (1984). *Mol. Gen. Genet.*, 193, 263.

Huisman, O. and D'Ari, R. (1981). *Nature,* 290, 797.

Huisman, O., D'Ari, R., and George, J. (1980). *Mol. Gen. Genet.*, 177, 629.

Huisman, O., Jacques, M., D'Ari, R., and Caro, L. (1983). *J. Bacteriol.*, 153, 1072.

Hye, R., O'Sullivan, A., Howard, K., and Sueoka, N. (1976) In *Microbiology 1976,* Am. Soc. Microbiology, p. 83.

Ingraham, J., Maaløe, O., and Neidhardt, F. (1983). *Growth of the Bacterial Cell,* Sinauer.

Jacob, F., Brenner, S., and Cuzin, F. (1963). *Cold Spring Harbor Symp. Quant. Biol.,* 28, 329.

Kleckner, N., Reichardt, K., and Botstein, D. (1979). *J. Mol. Biol.,* 127, 89.

Kreuzer, K. and Cozzarelli, N. (1980). *Cell,* 20, 245.

Kuempel, P. and Duerr, S. (1979). *Cold Spring Harbor Symp. Quant. Biol.,* 47, 563.

Kuempel, R., Duerr, S., and Seeley, N. (1977). *Proc. Natl. Acad. Sci. USA,* 74, 3927.

Kuempel, P., Maglothin, P., and Prescott, D. (1973a) In *DNA Synthesis In Vitro,* R. Wells and R. Inman (eds.), University Park Press, p. 463.

Kuempel, P., Maglothin, P., and Prescott, D. (1973b). *Mol. Gen. Genet.,* 125, 1.

Kusano, T., Steinmetz, D., Hendrickson, W., Murchie, J., King, M., Benson, A., and Schaechter, M. (1984). *J. Bacteriol.,* 158, 313.

Lai, C. and Nathans, D. (1975). *J. Mol. Biol.,* 97, 113.

Little, J. and Mount, D. (1982). *Cell,* 29, 11.

Louarn, J., Patte, J., and Louarn, J. M. (1977). *J. Mol. Biol.,* 115, 295.

Louarn, J., Patte, J., and Louarn, J. M. (1979). *Mol. Gen. Genet.,* 172, 7.

Lutkenhaus, J. (1983). *J. Bacteriol.,* 154, 1339.

Meacock, P. and Cohen, S. (1980). *Cell,* 20, 529.

Miki, T., Easton, A., and Rownd, R. (1980). *J. Bacteriol.,* 141, 87.

Miller, C., Tucker, W. Meacock, P., Gustafsson, P., and Cohen, S. (1983). *Gene,* 24, 309.

Mizuuchi, K., Fisher, L., O'Dea, M., and Gellert, M. (1980). *Proc. Natl. Acad. Sci. USA,* 77, 1847.

Morrison, A. and Cozzarelli, N. (1981). *Proc. Natl. Acad. Sci. USA,* 78, 1416.

Moyed, H. and Bertrand, K. (1983). *J. Bacteriol.,* 155, 768.

Nordstrøm, K., Molin, S., and Aagaard-Hansen, H. (1980). *Plasmid,* 4, 215, 332.

Ogura, T. and Hiraga, S. (1983a). *Cell,* 32, 351.

Ogura, T. and Hiraga, S. (1983b). *Proc. Natl. Acad. Sci. USA,* 80, 4784.

Prescott, D. and Kuempel, P. (1972). *Proc. Natl. Acad. Sci. USA,* 69, 2842.

Pritchard, R. and Zaritsky, A. (1970). *Nature,* 226, 126.

Ross, D., Swan, J., and Kleckner, N. (1979). *Cell,* 16, 721.

Sargent, M. and Bennett, M. (1982). *J. Bacteriol.,* 150, 623.

Steck, T. and Drlica, K. (1984). *Cell,* 36, 1081.

Summers, D. and Sherratt, D. (1984). *Cell,* 36, 1097.

Sundin, O. and Varshavsky, A. (1981). *Cell,* 25, 659.

Valenzuela, M., Freifelder, D., and Inman, R. (1976) *J. Mol. Biol.,* 102, 569.

Weiss, A., Smith, M., Iismaa, T., and Wake, R. G. (1983). *Gene,* 24, 83.

Is the Chromosome of *E. coli* Differentiated Along its Length With Respect to Gene Density or Accessibility to Transcription?

Millicent Masters*, Peter D. Moir**, Renate Spiegelberg*, J. H. Pringle*, and Carl W. Vermeulen†

* *Edinburgh University, Scotland*
** *The University Medical School, Nottingham*
† *College of William and Mary, Williamsburg, Virginia*

INTRODUCTION

The chromosome of *Escherichia coli* is a single circular molecule of DNA 1.4 mm in length. Although this DNA molecule is well over 1000x longer than the cell in which it resides, it is so thin that it need occupy only 10-20% of the cellular volume. Electron micrographs of cells show the nuclear body as a compact structure of close to this predicted size. Such compactness requires that the DNA be tightly packed and intricately folded, and indeed in the final chapter of "Control of Macromolecular Synthesis" Maaløe and Kjeldgaard point out that there would be no room for ribosomes within the folded chromosome (nor have they been observed there). How then does the DNA manage to direct protein synthesis? If transcription were to occur deep within the nucleoid, translation would have to take place at a site remote from the transcribed gene, but this is not the case. Since in bacteria translation and transcription are closely coupled in both time and space, the two processes must occur at the only place where both can occur: at the interface between nucleoid and cytoplasm.

Realizing that all parts of the chromosome cannot be exposed at the nuclear surface simultaneously and anxious to explain how, despite this, the *lac* gene can be available to transcription/translation at all times, Maaløe and Kjeldgaard suggested that the DNA is in constant motion such that each gene is exposed at the nuclear surface every few seconds. This would certainly be necessary if all genes, like the *lac* gene, were to be available for transcription at all times. An alternative organization of the chromosome might have genes that need to be constantly available for transcription located permanently near the nuclear surface and those that do not would be consigned to the interior. Such a structure could be relatively static.

We were inclined to favor the latter view as a result of several observations, none in themselves compelling, but each suggesting independently that the chromosome might well be differentiated along its length into relatively active and passive regions. The first such observation concerned the distribution of mapped genes in *E. coli*. Edition 5 of the linkage map (Bachmann et al., 1976) showed that the mapped genes were not distributed equally along the length of the chromosome but instead tended to be clustered together into gene-dense regions. This pattern of gene distribution has remained a feature of later editions of the linkage map (Figure 1). Edition 7 (Bachmann, 1983) lists over 1000 genes. Taking the length of the *E. coli* chromosome as 4000 kilobase pairs (Kb) and an average protein as having a molecular weight of 40,000, the genome of *E. coli* can be estimated to have the capacity to code for more than 3000 proteins. Thus, the mapped fraction of the genes is about 30% of the total, and these remain clustered to form gene-dense areas, suggesting that the chromosome possesses at least some degree of differentiation. The second relevant observation was made while we were studying the frequency of recombination after P1 transduction. We noticed that recombination frequency varied in a way which indicated that it occurred more efficiently in gene-dense regions (Masters, 1977; Newman and Masters, 1980). This observation suggested that these regions might be more accessible to the enzymes that promote recombination, either because of their location within the cell (such as at the surface of the nucleoid) or because some feature of their primary or secondary structure renders them better substrates for recombination, possibly a concomitant of more frequent functioning as transcriptional templates. The third observation was that substantial sections (60 Kb, 40 Kb) of at least one gene-sparse region can be deleted without adversely affecting the growth of the cell (Henson et al., 1984; Kuempel and Henson, this volume). Thus, at least one region lacking mapped genes can also now be described as lacking essential genes. Finally, a substantial body of evidence has now accumulated that suggests that the bacterial nuclear body possesses a defined physical structure in the form of the 100-or-so independent domains of supercoiling that have been so elegantly demonstrated (for review see Pettijohn, 1976). These domains could well serve as the means of organizing the nucleoid into defined inner and outer regions.

Taken together, these observations have led us to hypothesize that the bacterial chromosome is differentiated along its length into regions that differ in the frequency with which they interact with the macromolecules that recombine and transcribe DNA. Verifying or falsifying this idea is not a simple task, and one must of necessity start with and test what are certainly oversimplified hypotheses. Our first approach has definitely been in this category. We decided to ask whether a region particularly sparse in mapped genes contains DNA that is capable of directing the synthesis of polypeptides, that is, to ask whether it has any genes at all. The answer obtained is in the affirmative: the region we chose to study is capable of directing the

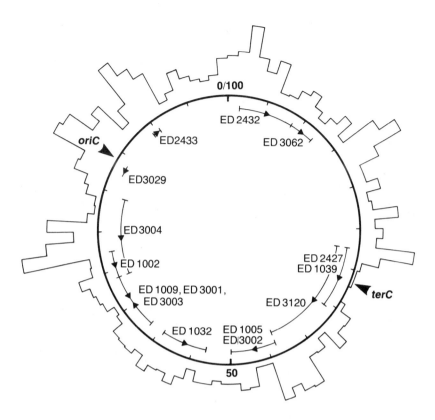

FIGURE 1   Distribution of genes on the <u>E. coli</u> chromosome. This drawing is based on Edition 7 of the linkage map (Bachman, 1983). The loci in each minute interval (with the exception of the bracketed loci of uncertain position) were counted and represented as a bar of proportionate length placed over the appropriate interval. The smallest bar (e.g., that at 31 min) represents a single gene. The regions within which the F' is integrated to form the various Hfr strains used are also shown.

synthesis of polypeptides. Furthermore, as already mentioned, subsequent work by Kuempel et al. (this volume) has shown that the region in question can be deleted without adversely affecting cell growth. Thus, this DNA, sparse in mapped loci, contains genes, but they are not essential for cell growth in laboratory conditions. We have yet to discover whether the products of these genes are made during normal cell growth.

Our second approach was to address the question of expression by asking whether *lac* genes when located at different positions on the chromosome would be equally capable of expression. One might expect that if gene-sparse regions were buried in the interior of the DNA mass, insertions in these regions might be poorly expressed. Our experiments, when considered from this point of view, have considerable limitations. However, the results we have obtained indicate that the *lac* system is expressed equally well at a number of different locations. Thus, neither of these approaches has yielded results that support the idea that there is any easily detectable heterogeneity in the coding abilities or transcriptional competence of different parts of the chromosome.

POLYPEPTIDES ENCODED IN THE TERMINUS REGION OF THE CHROMOSOME

In order to ask whether gene-sparse regions code for polypeptides we decided to study the region in which fewest genes have so far been identified. This is the region, diametrically opposite to the replication origin, in which termination of bidirectional replication occurs. The entire 6-min segment between minutes 29 and 35 on the genetic map contain only 6 mapped loci and, in addition, two defective prophages with homology to phage lambda (*rac, kim*). The region is ca. 250 Kb long and contains sufficient DNA to code for about 200 proteins of average size. Thus, it certainly must be regarded as gene-sparse.

Our approach has been to clone DNA from this region into a vector (pBR325) convenient for the study of polypeptide synthesis directed by the cloned DNA. Our task was made easier by the availability of a plasmid, pTH51, that contained DNA complementing *trg,* one of the few mapped loci in the terminus region. This plasmid had been constructed by subcloning a Pst fragment containing the *trg* gene from a larger plasmid in the Carbon-Clarke collection (Harayama et al., 1982). Bouché's invaluable extended restriction map of the terminus region (Bouché, 1982) provided us with the information that the chromosomal Pst fragment contained in pTH51 would be divided between two HindIII fragments, one of 23 Kb, the other of 11 Kb. Therefore, we constructed a HindIII library of chromosomal DNA and identified plasmids containing fragments homologous to the Pst fragment in pTH51 by colony hybridization. These contained inserts of the predicted size and were called pPM1000 and pPM2000. Following a similar strategy, pPM2000 was used to identify pPM4000 in an EcoRI gene bank. Thus, we isolated a total of 40 Kb (or 1 minute) of DNA from the terminus region to study further. Various subclonings were made eventually yielding the collection of plasmids shown in Figure 2.

Each of the plasmids was transformed into DS410, a minicell-producing strain. Minicells were prepared and plasmid-coded proteins labelled with [^{35}S]methionine, separated on one-dimensional poly-acrylamide gels, and identified by autoradiography. We were, with reasonable certainty, able to assign the coding sequences of 10 polypeptides to particular segments of the DNA we had cloned. Together these account for about 50% of the coding capacity of the 33 Kb of DNA we have studied in detail so far (Figure 2). We consider this estimate to be a minimum one, because poorly expressed peptides, small peptides, or peptides obscured by the abundantly produced *amp* or *cmp* products could easily be missed. Thus, our result is well within the range obtained by Neidhardt et al. (1983), who found that for seventeen plasmids of the Clarke-Carbon library, expression varied between 36% and 123% of coding capacity (with an average of 70%).

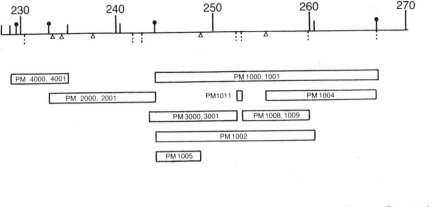

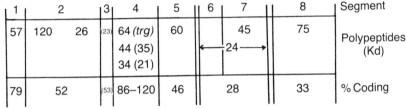

FIGURE 2   Polypeptides encoded by the terminus region of the chromosome. The upper line shows the regions we have cloned. The numbers (in Kb) refer to the Bouché (1982) map. Our restriction mapping confirms that of Bouché except that we find an extra Pst site at Kb253. EcoRI (solid vertical lines), Pst (dashed vertical lines), HindIII (solid dots), Sal (triangles). The chromosomal segments contained in the plasmids we have isolated are shown beneath the restriction map. Segments with two numbers have been cloned in both orientations. Only those located to the right of Kb243 have so far been analyzed in detail. The table showing the polypeptides encoded by segments 1-8 is colinear with the restriction map. The identification of polypeptides whose sizes are shown in parentheses is still uncertain.

BETA-GALACTOSIDASE PRODUCTION IN STRAINS WITH TRANSPOSED *lac* GENES

In order to determine whether *lac* expression would vary with gene position we used a collection of Hfr strains made by Paul Broda (Broda et al., 1972; Masters and Broda, 1971). These were made by transferring an F'ts114 *lac* into a strain with a *lac* deletion and seeking Lac+ derivatives at 42 C. These proved to be a collection of Hfr strains in which the F'*lac* had integrated at a number of different points on the chromosome. The sites of integration were determined using conjugation (see Figures 1 and 3). In most cases this could not be done sufficiently precisely to say with certainty whether the insertion has occurred in a gene-dense or gene-sparse region (Figure 1).

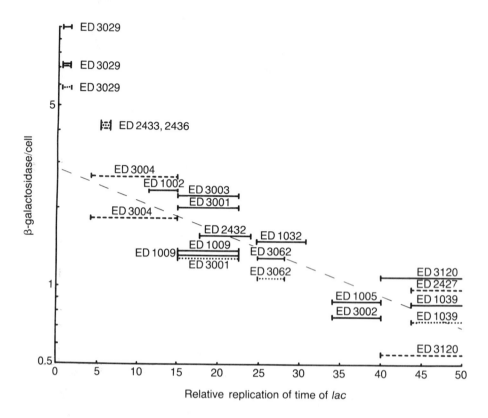

FIGURE 3 β-Galactosidase synthesized by transposition Hfrs. Cultures of each strain were grown in L-broth supplemented with 1 mM IPTG. Overnight cultures were diluted and grown to $OD_{540} < 0.2$ and rediluted to fewer than 3×10^5 cells/ml. When cell counts had risen to 1.5×10^6, samples were taken at 10 minute intervals for counting, sizing, enzyme analysis, and determination of percent Lac+ cells. β-Galactosidase was assayed as described by Miller (1972). Data were only included from those samples in which the cells were > 95% Lac+ and that remained constant in size throughout the experiment. Cell numbers were used to measure growth rate. Enzyme levels were corrected to a notional interdivision time of 20 min by using the formula $fG_x = 2[(\underline{C}(1 - \underline{x}) + \underline{D})/\underline{T}]$, in which $\underline{G}_x$ is the number of genes per cell at position $\underline{x}$ at the interdivision time $\underline{T}$. $\underline{C}$, the time required for chromosome replication, and $\underline{D}$, the interval between termination and division, were taken as 40 min and 20 min, respectively. $\underline{G}_x$ was calculated for the notional and observed growth rates and enzyme/cell multiplied by the ratio of the two. The corrected value is plotted against a horizontal bar that defines the period within which the $\underline{lac}$ gene might be supposed to replicate. The solid, dashed, and dotted bars represent different experiments. The length of the bar indicates how accurately the site of insertion of the F' has been determined. The horizontal axis has been divided into 50 replication units corresponding to the genetic map folded over on a line drawn between origin and terminus.

The amount of beta-galactosidase per cell was measured in broth cultures of these strains grown in the presence of the inducer IPTG. We found that in order to avoid catabolite effects, it was necessary to measure enzyme levels in cultures with very low cell concentrations (about 10^6 cells/ml). At higher cell concentrations the enzyme level per cell increased continuously with cell density, presumably as a result of cAMP accumulation, and could not be used as a measure of gene content.

In Figure 3 we present the results obtained for 17 separately isolated Hfr strains, inserted at a minimum of eight different sites on the chromosome. The length of each bar defines the region within which the F' must be integrated; those that include 50 minutes span the terminus. All strains did not grow at the same rate, so the value of enzyme/cell was corrected for the consequent variation in the value of genes/cell by adjusting all data to a growth rate of three divisions per hour. At this growth rate there should be four origins per terminus, so a position-dependent fourfold variation in enzyme/cell could be anticipated if the fully induced level of expression were proportional to the number of structural genes in the cell. The line drawn through the data in Figure 3 is drawn to have the expected slope of 4. A straight line is expected on this log-linear plot, because the variation in gene frequency from origin to terminus is exponential.

As can be seen, enzyme synthesis in all Hfr strains resulting from insertions between 7.5 and 50 minutes on the replication map is close to the line, the greatest deviations being about 20%. The *lac* genes carried on the F plasmids inserted within the gene-sparse terminus region are not expressed particularly poorly, and ED3003, which produces the most enzyme/gene in this group, is not inserted in a demonstrably gene-dense region. Thus, we can conclude that, in this group of Hfr strains expression under comparable growth conditions depends mainly on the copy number of the gene.

Hfr ED3029 and the methionine-dependent pair ED2433 and ED2436 (these probably result from a single insertion event) behave differently; they express the transposed *lac* gene at levels higher than would be anticipated when compared with the other strains. Both are integrated close to the replication origin in markedly gene-dense regions, and thus it is possible that the increased expression from *lac* could result from such regions being favored for transcription. However, other factors may be important and are perhaps the more probable causes of the higher levels of expression that are observed. ED3029 grows much more slowly than the other strains (generation time = 60-80 min on broth) and, as a result of the insertion, may be deficient in a gene product important to normal growth. This deficiency could result in an over-production of cAMP or CRP protein, even at low cell densities, and a consequent enhanced expression of *lac*. The overproduction of beta-galactosidase by ED2433/36 is more difficult to explain away convincingly, since its growth is not impaired by the insertion into a *met* gene. Since we know that the F' is inserted within an active gene, it is possible that an external promoter is responsible for the extra transcription. As yet, there is no evidence for such an interpretation, and it remains possible that the level of expression we see is a correlate of the high gene density of the region.

Using Hfr strains for this sort of experiment was convenient for several reasons (not the least of which is that they were readily available), but this approach also has severe limitations. First, the inserted DNA is so large that the environment within which the *lac* gene is functioning may be considered to be that of F rather than of the chromosome. Second, insertion of such a large piece of DNA may influence chromosome structure in its vicinity in a way that obscures any effects dependent upon gene environment. A better test of the effect of gene position on expression would involve accurately mapped transpositions of a small DNA fragment insulated by flanking transcriptional terminators from exogenously initiated transcription. Construction of a family of such strains is not a trivial undertaking, and we have not yet proceeded with such an approach.

CONCLUSIONS

The experiments we report here do not support the idea that the chromosome is differentiated into regions varying in their accessibility to transcription or containing genes at different densities. The observations mentioned at the beginning of this article still require explanation, particularly the one indicating that a large section of a gene-sparse region can be deleted with no apparent ill effects on the cell. This suggests the possibility that gene-sparse regions are perhaps repositories for nonessential genes. The terminus region is known to contain two defective prophages with homology to phage lambda, and two additional regions with homology to lambda have been identified. An excisable element termed *e14* is located at 25 min on the map in an otherwise gene-sparse region. It may well be that a significant portion of the unmapped DNA of *E. coli* consists of such nonessential phage or plasmid remnants. Thus, a possible picture of the chromosome is one of regions dense in essential genes interspersed with regions composed of nonessential or infrequently expressed genes.

ACKNOWLEDEGMENTS

We should like to acknowledge the helpful discussion of W. D. Donachie and the technical assistance of C. Henry. We thank Mark Hanks and Ian Oliver for help with figure preparation.

REFERENCES

Bachmann, B. J. (1983). *Microb. Rev.,* 47, 180.

Bachmann, B. J., Low, K. B., and Taylor, A. L. (1976). *Bact. Rev.,* 40, 116.

Broda, P., Meacock, P., and Achtman, M. (1972). *Mol. Gen. Genet.,* 116, 336.

Bouche, J. P. (1982). *J. Mol. Biol.,* 154, 1.

Harayama, S., Engstrom, P., Wolf-Watz, H., Iino, T., and Hazelbauer, G. (1982). *J. Bacteriol.,* 152, 372.

Henson, J. M., Kopp, B., and Kuempel, P. L. (1984). *Mol. Gen. Genet.,* 193, 263.

Masters, M. (1977). *Mol. Gen. Genet.,* 155, 197.

Masters, M. and Broda, P. (1971). *Nature,* 232, 137.

Miller, J. H. (1972). *Experiments in Molecular Genetics,* Cold Spring Harbor Laoratory, p. 352.

Neidhardt, F. C., Vaughn, V., Phillips, T. A., and Bloch, P. L. (1983). *Microbiol. Rev.,* 47, 231.

Newman, B. J. and Masters, M. (1980). *Mol. Gen. Genet.,* 180, 585.

Pettijohn, D. G. (1976). *CRC Crit. Rev. Biochem.,* 4, 175.

Genes Required for Cell Division in *Escherichia coli*

William D. Donachie
Edinburgh University, Scotland

Mistreatment of *E. coli* cells in any of a large number of different ways can lead to inhibition of cell division. Therefore, it is not surprising that mutations in many genes can also lead to complete or partial blocking of division (Slater and Schaechter, 1974). The task of finding which of these genes are involved in cell division in unperturbed cells seemed daunting in 1974. Nevertheless, a few foolhardy people continued to search for such genes and now, ten years later, the prospect seems immensely more encouraging. I want to suggest the possibility that the number of genes that are primarily involved in cell division is in fact quite small and that most (and perhaps all) of these are found within a single gene cluster, the so-called "major morphogene cluster" (Donachie, Begg, and Sullivan, 1984).

This view has arisen as the result of working for a number of years with various mutants showing defects in cell division. It seems plausible that division is blocked or impaired in many of these mutants by the activation of emergency repair systems that also result in inhibition of primary division genes. For example, treatments or mutations that block DNA replication or that damage DNA in certain ways are now known to induce a whole set of genes required for the "SOS" response (see Little and Mount, 1982, for a review). The products of most of these genes appear to be required to repair DNA damage, but the product of the *sfiA* (*sulA*) gene acts as a specific inhibitor of cell division. It is important to note that this gene plays no role whatsoever in the regulation of division in undamaged cells (Huisman, Jacques, D'Ari, and Caro, 1983). The existence of this normally repressed gene as part of the SOS regulon is thought to be responsible for inhibition of cell division in mutants with lesions in genes that are primarily involved in steps in DNA replication or the synthesis of specific DNA precursors. Inhibition of division by SfiA protein continues until this protein is inactivated by the action of the La protease, the product of the *lon* gene (Charette et al., 1981; Chung and Goldberg, 1981; Mitsuzawa and Gottesman, 1983). In consequence, *lon*⁻ mutants are very prone to produce division-inhibited, *long* cells when treated in any of the myriad ways that succeed in evoking a transient SOS response. There is a similar system (the "adaptive response") that is induced by DNA-alkylating agents (see Walker, 1984 for review) and, although I do not know whether this system also blocks division, I would be mildly surprised if it did not. There is a third multigene regulon that is turned on by a variety of stress situations (including temperature change and also DNA damage itself) and that also appears to induce a division block. This is the "heat shock" regulon, which is described by Neidhardt elsewhere in this book. Indeed, it was only when I heard Neidhardt describe the heat shock response of *E. coli* in Tuscaloosa that I became properly aware both of what a very interesting and important phenomenon it is and of its likely involvement in cell division. Among the mutations that are found in screening for cell division mutants are *dnaK* mutations (K.J. Begg, unpublished). These mutants stop dividing soon after a shift to high temperature (42 C in our experiments) but continue to grow exponentially and to replicate and segregate chromosomal DNA normally. The filamentous cells formed by these mutants at high temperature have a quite characteristic and unique appearance, with phase-dark inclusions and irregular protrusions, but they are nevertheless able to resume normal growth and division after a shift back to lower temperatures. As explained by Neidhardt, the *dnaK* gene is part of the HTP regulon, and it appears to be required to turn off the HTP response once more after its transient induction by a temperature shift. We have ourselves noticed that *E. coli* cells exhibit a transient inhibition of division shortly after a shift to 42 C, at a time corresponding to the period of maximum HTP induction, and we therefore assume that this inhibition is a part of the HTP response and that the failure of *dnaK* mutants to switch off this regulon is responsible for their appearance in hunts for division mutants. The role of the *dnaK* gene in the HTP response would therefore be formally equivalent to that of the *lon* gene in the SOS response. (The mechanism of action could of course be totally different.)   The SOS regulon is repressed by the LexA protein, and mutations in the *lexA* gene can therefore give rise to mutants that are temperature-sensitive for division (*tsl* mutations). The HTP regulon is under the control of the *htpR* locus and, though the product of this gene appears to be a positively acting, sigma-like protein required for the transcription of HTP genes (Grossman et al., 1984), one could imagine how mutations at that locus also could give rise to temperature-sensitive "division" mutants. Thus, it is very exciting to find that some of the temperature-sensitive division mutants that we find result from mutations ("*ftsE*" and "*ftsS*") that map extremely close to the *htpR* locus (G.P.C. Salmond, personal communication).

Many of the temperature-sensitive "division" mutants described in the literature also have other defects in growth, and many die at the restrictive temperature (see Donachie et al., 1984). Thus, it seems quite plausible to assume that at least some of these mutant strains are secondarily blocked in division as the result of the induction of the HTP system or the SOS system in the course of their death agonies at 42 C. We have therefore concentrated our attention on those mutants that are completely blocked in cell division at 42 C but that continue to grow normally and replicate and segregate their DNA normally for prolonged periods at the restrictive temperature for division. Furthermore, we require that viability is completely maintained when the cells are shifted back to the permissive temperature. When we map the mutations that give such a phenotype then we find that, apart from *dnaK, ftsE,* and *ftsS* mutations, the vast majority lie within the major morphogene cluster (K.J. Begg, unpublished). The extent of this gene cluster is not known with certainty, but it stretches for about 20 kilobases or about 0.5% of the total genome and contains at least 14 genes. Each of these genes has some role in the growth and function of the cell envelope (see Donachie et al., 1984 for a review). Seven of the genes are known to control individual reactions in the net synthesis of peptidoglycan (forming the shape-maintaining sacculus of the cell envelope), but five (or possibly six) appear to be required exclusively for cell division. These genes are *ftsI (pbpB, sep), ftsQ, ftsA, ftsZ (sfiB, sulB), envA,* (and possibly *azi*).

We do not know much about the biochemical reactions involved in forming the septum and converting it into two new cell poles. However, physiological studies on doubly mutant strains carrying a mutation in one of the above "primary" division genes, together with a mutation in a gene from another smaller cluster of morphogenes that govern the rod-shape of normal *E. coli* cells (Stoker et al., 1983) are informative. These studies indicate that the individual gene products act at successive stages in the initiation, formation, and completion of the septum (K.J. Begg and W.D. Donachie, in preparation). Thus, the FtsZ protein appears to be required for the very earliest step in the initiation of septation, the FtsI and FtsQ proteins for the next stages, FtsA protein for septum completion, and the EnvA product for the final splitting of the septum to form cell poles. The proteins coded for by these genes and the complete DNA sequence for at least three of the genes (*ftsI, ftsQ, ftsA;* Nakamura et al., 1983; Robinson et al., 1984) are known; however, enzymatic activities have been ascribed to only two of the proteins. Ishino et al. (1981) have shown that penicillin-binding protein (PBP) 3, the FtsI protein, has transpeptidase and transglycosylase activities on peptidoglycan substrates *in vitro,* while *envA* mutants are reported to be deficient in another peptidoglycan enzyme, N-acetyl muramyl L-alanine amidase (Wolf-Watz and Normark, 1976). It seems most likely that the products of the remaining genes in this group will also prove to have enzymatic activity on peptidoglycan.

One of the most fascinating features of cell division is that it is periodic. Therefore, we have been interested in regulation of expression of the primary division genes. The transcriptional organization of the genes in the major morphogene cluster is most interesting. For example, despite the continuous run of genes of closely related function, the cluster is not organized as a classical operon. Thus, every gene can be cut out from the cluster and expressed in isolation (i.e., inserted into a cloning vector of some kind) from its own promoter. Nevertheless, the direction of transcription of each cluster gene (known in 9 of the 14 genes) is the same, and no strong transcriptional terminators have been detected between neighboring genes. Indeed, the promoters of the *ftsZ* gene

actually lie within the coding sequence of the neighboring *ftsA* gene
(Sullivan and Donachie, 1984), while the promoter of *ftsA* in turn lies
within the coding sequence of its upstream neighbor *ftsQ* (Robinson
et al., 1984). Hence, at least some of the transcriptional units in the
cluster overlap. We do not yet know what transcripts are produced
from the intact cluster *in vivo,* but we do know that normal cell
division results even when any of the genes is expressed exclusively
from its own contiguous promoter (i.e., from a cloned fragment of
the cluster). If normal cell division requires regulation of expression
of these genes, then such regulation must be able to take place in
the cloned fragments. Accordingly, we have studied the regulation of
transcription from cloned promoters fused to either *galK* or *lacZ* coding
segments (to provide easy assays for transcription). In particular,
we have studied the regulation of transcription from one of the
promoters (lying within *ftsA*) of the *ftsZ* gene. Expression of *ftsZ* is
particularly interesting, because it is the earliest known step in cell
division. It is also exciting to find that the FtsZ protein is in fact
the target for the SfiA product, i.e., for the specific inhibitor of
division that is produced on induction of the SOS response
(Lutkenhaus, 1983).

We have found two ways that the frequency of transcription
from this *ftsZ* promoter may be regulated (Donachie et al., 1983,
1984). Transcription appears to be negatively affected by the presence
of the FtsI, FtsQ, FtsA, and FtsZ gene products. For example,
transcription from this promoter is increased approximately 11-fold in
cells that have an amber mutation in *ftsA*. Inhibition of cell division
itself is not sufficient to affect transcription, because neither mutations
in *ftsE* (mapping near *htpR*) nor blockage of division by induction of
the SOS response (with UV or thymine starvation) has any detectable
effect on transcription from the *ftsZ* promoter. However, the relative
rate of transcription is strongly dependent on growth rate of the
cells. The concentration of protein (beta-galactosidase or galactokinase)
produced under the control of this promoter is about five times higher
in cells growing slowly than in fast-growing cells. (This effect is
independent both of the gene to that the promoter is fused and of
whether the fusion is carried on a multicopy plasmid vector or is
present as a single chromosomal copy in a phage λ vector.) Because
the average size of cells also depends on the growth rate (Schaechter,
Maaloe and Kjeldgaard, 1958) we are able to calculate that the number
of *ftsZ* transcripts per cell remains approximately the same at all
growth rates. I particularly like this conclusion, because it is what
one might expect if each cell has to produce the same number of FtsZ
molecules in each cycle in order to make one septum.

At first sight, the apparent "autoregulatory" control of *ftsZ*
expression (by the Fts proteins of the major cluster) would not be
expected to give rise to a constant number of FtsZ protein molecules
per cell, but rather to a constant concentration. However, we could
assume that the *ftsZ* promoter (and perhaps those of the other division
genes in the cluster) is repressed by low concentrations of the FtsI,
FtsQ, FtsA, and FtsZ proteins present during growth between
divisions and that the periodic use of relatively large quantities of
these proteins during septum formation results in the periodic
derepression of the gene(s). The total range of expression that we
have observed for this *ftsZ* promoter is approximately 50-fold, and
consequently the amount of transcription taking place during septation

could be very much greater than the total amount of transcription of *ftsZ* during the rest of the cell cycle. If the amount of FtsZ protein required for septation is approximately the same in cells of different sizes (i.e., with different growth rates), then the average number of *ftsZ* transcripts produced per cell would be approximately independent of cell size, as observed. This model assumes that *ftsZ* proteins are sequestered or inactivated during septum formation and although we do not know whether this is true for FtsZ protein itself, it does seem that this is the case for the FtsA protein (Donachie et al., 1979; Tormo and Vicente, 1984).

Models like this provide a way in which cell division genes may be turned on and off periodically as the cell initiates and completes cell division. What they do not tell us is the event that actually switches cell envelope growth from elongation to localized septation. In our model, the genes are passively derepressed by a change in the cellular environment and then automatically repressed once more when this change is complete. The periodic change in cell state that initiates cell division remains as elusive as ever.

ACKNOWLEDGEMENTS

This essay is mostly about genetics but little of this work would have been interpretable without the knowledge of the physiology of cell growth and division that has stemmed in such large part from the work of Ole Maaløe and his colleagues and collaborators. I am grateful to Ole for inviting me to visit and work at the Microbiology Institute in Copenhagen in past years. I know that it did me a great deal of good, and I can only hope that that in itself is worthwhile.

I would also like to take this opportunity to thank those colleagues and friends who have worked with me here in Edinburgh and who are together largely responsible for the output of this lab. My special thanks therefore go to Ken Begg, Graham Hatfull, Varda Kagan-Zur, Patrick Kelly, Dan Kenan, Joe Lutkenhaus, the late Nozomu Otsuji, Arthur Robinson, George Salmond, Renate Spiegelberg, Neil Sullivan, and Miguel Vicente.

REFERENCES

Charette, M., Henderson, G. and Markowitz, A. (1981). *Proc. Natl. Acad. Sci. USA, 78,* 4728.

Chung, C. H. and Goldberg, A. L. (1981). *Proc. Natl. Acad. Sci. USA, 78,* 4931.

Donachie, W. D., Begg, K. J., and Sullivan, N. F. (1984). In *Microbial Development,* R. Losick and L. Shapiro (eds.). Cold Spring Harbor Publications.

Donachie, W. D., Sullivan, N. F., Kenan, D. J., Derbyshire, S. V., Begg, K. J., and Kagan-Zur, V. (1983). In *Progress in Cell Cycle Controls,* J. Chaloupka, A. Kotyk and E. Streiblova (eds.). Czechoslovak Academy of Sciences, Prague, pp. 28-33.

Donachie, W. D., Begg, K. J., Lutkenhaus, J. F., Salmond, G. P.

C., Martinez-Salas, E., and Vicente, M. (1979). *J. Bacteriol.,* 140, 388.

Grossman, A. D., Zhou, Y-N., Gross, C., Hellig, J., Christie, G. E., and Calendar, R. (1984). *J. Bacteriol.,* in press.

Huisman, O., Jacques, M., D'Ari, R., and Caro, L. (1983). *J. Bacteriol.,* 153, 1072.

Ishino, F. and Matsuhashi, M. (1981). *Bioch. Bioph. Res. Comm.,* 101, 905.

Little, J. W. and Mount, D. W. (1982). *Cell,* 29, 11.

Lutkenhaus, J. F. (1983). *J. Bacteriol.,* 154, 1339.

Nakamura, M., Maruyama, I. N., Soma, M., Kato, J., Suzuki, H., and Hirota, Y. (1983). *Mol. Gen. Genet.,* 191, 1.

Mitsuzawa, S. and Gottesman, S. (1983). *Proc. Natl. Acad. Sci. USA,* 80, 358.

Robinson, A. C., Kenan, D. J., Hatfull, G. F., Sullivan, N. F., Spiegelberg, R., and Donachie, W. D. (1984). *J. Bacteriol.,* 160, 546.

Schaechter, M., Maaløe, O., and Kjeldgaard, N. O. (1958). *J. Gen. Microbiol.,* 19, 592.

Slater, M. and Schaechter, M. (1974). *Bacteriol. Rev.,* 38, 199.

Stoker, N. G., Broome-Smith, J. K., Edelman, A., and Spratt, B. (1983). *J. Bacteriol.,* 155, 847.

Sullivan, N. F. and Donachie, W. D. (1984). *J. Bacteriol.,* 158, 1198.

Tormo, A. and Vicente, M. (1984). *J. Bacteriol.,* 157, 779.

Walker, G. C. (1984). *Microbiol. Rev.,* 48, 60.

Wolf-Watz, H. and Normark, S. (1976). *J. Bacteriol.,* 128, 580.

The Effect of DNA Chain Elongation Rate on Bacterial Cells

Robert H. Pritchard
University of Leicester, England

INTRODUCTION

Among the well-recognized contributions of the Copenhagen school to microbial physiology was their early appreciation that valuable insights could be obtained into the control of cell growth and the coordination of the synthesis of macromolecules by studying the way cultures cope with transitions from one steady-state rate of growth to another (Maaløe and Kjeldgaard, 1966). An analogous transition, between conditions allowing one rate of chain elongation and another in DNA synthesis, can be equally revealing but is less widely known. This type of transition was called a "step-up" or "step-down" (Pritchard and Zaritsky, 1970) by analogy with the term "shift-up" used to describe transitions between different growth rates.

To illustrate how step transitions have been exploited in studies on cell growth I will refer to another theme that recurs among the papers of the Copenhagen school. This is the idea that the rate-determining step in macromolecular synthesis is the initiation step. In the case of chromosome replication in bacteria this led to the argument that it should be seen as having two components--an initiation step that was rate determining, and a chain elongation phase that was not (Maaløe, 1961).

A difficulty that prevented immediate acceptance of this idea was that dichotomous replication had not been recognized, or even conceived of. It was therefore not obvious how a cell growing in a medium permitting, say, one doubling per hour and apparently needing the whole of this time to replicate its DNA (all cells incorporating radioactive thymine) could nevertheless increase its rate of DNA synthesis to accommodate a faster growth rate, unless it increased the rate of chain elongation.

The discovery of dichotomous replication (Oishi et al., 1964; Pritchard and Lark, 1964) resolved this difficulty, but this implication of multifork replication was not widely appreciated for some time, because thinking about replication in bacteria had, in the meantime, been colored by the commitment of the scientific community to the replicon hypothesis (Jacob et al., 1963).

THYMINE LIMITATION AND CHAIN-ELONGATION RATE

If the rate of DNA synthesis is determined by the frequency of initiation of rounds of chromosome replication, it follows logically that the rate of chain elongation should not be rate-determining. This proposition is testable by varying the rate of chain elongation and asking whether this affects the rate of DNA synthesis.

Obviously the rate of chain elongation may be important for other reasons, and alterations from the natural rate may not be benign. Zaritsky and I therefore set out in 1968 to answer the following questions. Can we reduce the rate of chain elongation in growing cultures of *E. coli*? If we can, does it affect the overall rate of DNA synthesis? And, does it affect the growth rate or any other easily measured cell parameters?

The method we chose to try to influence the rate of chain elongation was thymine limitation of *thy* strains of *E. coli*. This immediately posed an interesting question. If reducing the rate of chain elongation does not reduce the growth rate, how can we know if a particular thymine concentration is rate-limiting or not? Indeed how do we know that the standard concentration of thymine used to cultivate particular *thy⁻* strains is not already rate-limiting for chain elongation?

When we looked through the published literature on *thy⁻* strains, we came across an interesting paper by Friesen and Maaløe (1965). They reported that when several *thy⁻* strains were pulse-labeled with [³H]thymine, the amount of label incorporated into DNA increased with increasing concentrations of thymine. There was no increase in the growth rate, and they showed that the mutants were not leaky. They were unable to offer an explanation for this paradox but it seemed to us to indicate that all concentrations of thymine being used were rate-limiting for chain elongation but not rate-limiting for growth. The increased incorporation with increased concentration of thymine in the pulse we surmised to be a transient effect, caused by an acceleration in the rate of chain elongation. Therefore, we set out to investigate this phenomenon in more detail. Figure 1 illustrates and summarizes what we found. On the left is a steady-state exponential culture of *E. coli* 15T⁻ growing in one thymine concentration. On the right is the same culture after addition of more thymine to give ten times the previous concentration but the same specific activity. Clearly the first concentration was not rate-limiting for mass increase, but it was for DNA synthesis, which accelerated within seconds of the addition of extra thymine. However, within about 60 minutes the rate of DNA synthesis had fallen back to match the growth rate.

We described this experiment as a step-up experiment, because we deduced that the addition of extra thymine reduced the step time for addition of deoxynucleotides at replication forks. We also deduced that the fall back to the pre-step-up rate of incorporation reflected the fact that the steady-state rate was determined solely by the

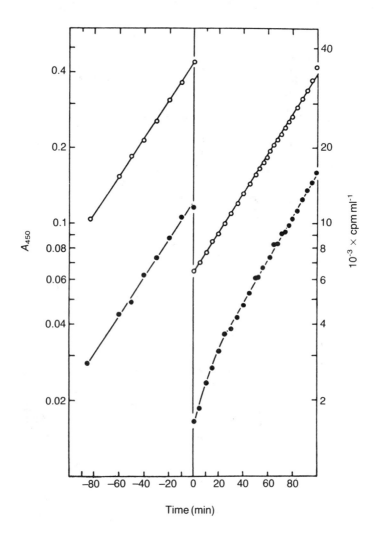

FIGURE 1 Increase of absorbance (open circles) and of DNA (solid circles) in a steady-state culture of E. coli 15T⁻ after a step-up. Minimal medium was used and the concentrations of thymine were 0.4 μg/ml before time 0, and 5.0 μg/ml (same specific activity) after dilution into fresh prewarmed medium at time 0. (From Zaritsky and Pritchard, 1970.)

frequency of initiation of rounds of replication, which was itself determined by the growth rate, which did not change. In other words, the overall rate of DNA synthesis is determined by the average number of replication forks and the rate of nucleotide addition at each fork. If we increase the rate of nucleotide addition, there is an immediate increase in the rate of DNA synthesis, but there is also a progressive reduction in the number of forks per chromosome until at equilibrium there is no net change in the rate of DNA synthesis. Friesen and Maaloe in their pulse-labeling experiments saw the first effect but not the second.

Although this and other experiments showed that changing the step time by about a factor of two was benign, there were significant affects on the composition of cells. In high thymine the steady-state concentration of DNA (DNA per unit of absorbance) was higher than it is in low thymine (Figure 1). We deduced that this was because initiation occurs at a constant average mass per chromosome origin, which is not affected by thymine concentration, but the amount of DNA accumulated in a given period by progression of a replication fork from the origin of replication will be greater in high thymine than in low thymine.

The relationship between DNA and mass was expressed quantitatively as follows (Pritchard and Zaritsky, 1970):

$$\overline{G}/\overline{M} = \frac{\tau}{kC \ln 2} (1 - 2^{C/\tau}) \tag{1}$$

in which $\overline{G}$ is the average number of genome equivalents per cell, $\overline{M}$ is the average cell mass, τ is the doubling time, C is the transit time of a replication fork from origin to terminus, and k is a constant.

It can be seen that the change in DNA concentration resulting from a change in the step time or from a change in growth rate are essentially equivalent. A shift-up changes the C/τ ratio, because C is relatively unaffected by changes in growth rate (Cooper and Helmstetter, 1968) A step-down changes the ratio in the same direction, because decreasing the rate of chain elongation does not affect τ. A step-down transition, unlike a shift-down, can be made as readily as a step up (Figure 2).

Although a step-down or a step-up changes the concentration of DNA, the effect is not expected to be uniform for different genes. If thymine concentration does not affect the initiation concentration (the number of origins in unit cell mass at the time of initiation), the average concentration of a gene near the origin will be unaffected. For all other locations there will be a change in gene concentration, the magnitude increasing with distance from the origin. The relationship can be expressed quantitatively as follows (Chandler and Pritchard, 1975):

$$\overline{F}_x/\overline{M} = \frac{2^{-Cx/\tau}}{k} \tag{2}$$

in which $\overline{F}_x$ is the average number of copies of a gene located x per cent of the distance between origin and terminus.

The expected pattern of change is shown in Figure 3(a), and a direct experimental verification of this pattern has been made using not thymine limitation, but a comparison of rep^+ and rep^- strains of E. coli that have different DNA rates of chain elongation (Lane and Denhardt, 1974). In our original work we used a more indirect method to show that there was a gradient of change in gene concentration, namely, to compare the differential rate of synthesis of enzymes specified by genes located at different positions on the E. coli map. The kind of result we obtained is given in Figure 4, which seems to

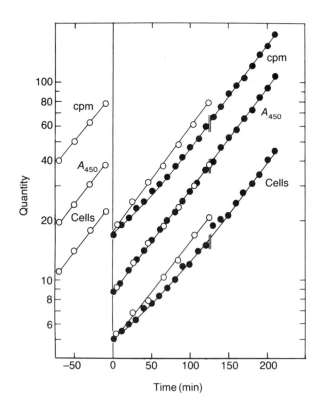

FIGURE 2 Increase in absorbance, DNA, and cell number in a steady-state exponential culture of E. coli K12 thy⁻ after a step-down. Minimal medium was used and the concentrations of thymine were 2.0 µg/ml before time 0, and 0.4 µg/ml after dilution into fresh prewarmed medium containing no thymine (solid circles). A dilution was also made into medium containing the same concentration of thymine as a control (open circles). Breaks in curves indicate time of a twofold dilution to maintain absorbance below 0.4. (From Zaritsky and Pritchard, 1973.)

show that the rate of synthesis of beta-galactosidase is modulated by the concentration of thymine in the growth medium. However, calculation shows that the rate is constant if expressed as enzyme activity per gene copy. In an extensive analysis of this kind we showed that the magnitude of the effect of thymine on enzyme output was proportional to the distance of the gene from the chromosome origin, with zero change in the case of tryptophanase encoded by a gene located close to the origin (Chandler and Pritchard, 1975).

DNA CONCENTRATION AND GROWTH RATE

Two obvious conclusions can be drawn from these observations. One is that DNA concentration is not rate limiting for growth--in our

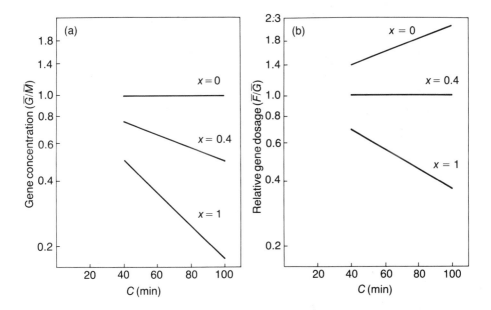

FIGURE 3 Changes in gene concentration and relative gene dose expected from changes in C. (a) Gene concentration. The relative number of copies of a gene in unit mass is shown for a gene at the origin ($\underline{x}$ = 0), at the terminus ($\underline{x}$ = 1), and at an intermediate location ($\underline{x}$ = 0.4). (b) Relative gene dose. This was defined as the number of copies of a gene relative to all genes (total DNA) in genome equivalents (Chandler and Pritchard, 1975) and is calculated from the following expression:

$$\bar{F}/\bar{G} = \frac{C \ln2 \; 2^{C(1 - x)/\tau}}{\tau(2^{C/\tau} - 1)}$$

studies we found that the concentration of DNA could be reduced about twofold without a detectable effect on growth rate. DNA concentration could, of course, be rate limiting in the case of genes located near the chromosome origin, because their concentration is not affected by thymine limitation. Indeed, this could be why genes making major demands on the protein-synthesizing system, such as the ribosomal RNA operons, are clustered in this area of the map. However, in later work we showed that a similar reduction in DNA concentration could be achieved in *dnaA* strains incubated at subinhibitory temperatures, with only a slight effect on growth rate (Orr et al., 1979). Also, Choung et al. (1981) have found an initiation mutant of *E. coli* B/r with half the normal DNA concentration which has no effect on growth rate. In both cases, there was a reduction in the concentration of chromosome origins without a significant change in the rate of chain elongation. The concentration of all genes was reduced equally. These results show that no region of the chromosome has genes present at concentrations that limit growth rate.

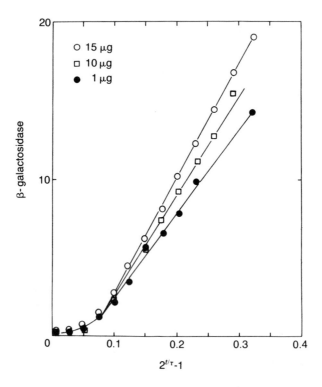

FIGURE 4 Effect of thymine concentration on the differential rate of synthesis of β-galactosidase. A <u>thy⁻</u> E. coli K12 strain was grown in a glycerol-casamino acids medium containing 1.0, 10, and 15 μg/ml of thymine. Inducer was added at time 0. The scale on the abscissa is the calculated increase in mass at time <u>t</u> in a culture with a doubling time τ. The levels of β-galactosidase were corrected for differences in absorbance at time 0 in the three cultures. (From Chandler and Pritchard, 1975.)

A second conclusion from this analysis was that there is no significant competition between promoters for polymerase. If there were, the effect of thymine concentration on gene activity would more nearly conform to the pattern of change shown in Figure 3(b) rather than Figure 3(a). We suggested that the absence of competition was due partly to the fact that changes in gene concentration are balanced by changes in the degree of derepression of many genes, and partly to the fact that the genes that consume the bulk of RNA-synthesizing capacity are located near the origin and are relatively little affected by changes in thymine concentration (Pritchard, 1974).

CHROMOSOME REPLICATION AND CELL DIVISION

Figure 2 illustrates another striking effect of step-time transitions-- they alter average cell size. One interpretation of this effect might be that the number of septa in a culture is related to the number of chromosome termini, as Cooper and Helmstetter (1968) assumed. Since the number of termini in unit mass falls when C increases (Figure 3(a)), the number of septa in unit mass will also fall (i.e., cell size will increase). We used this effect on cell size to try to obtain a more quantitative understanding of the relationship between chromosome replication and division.

It is clear that cell division is coordinated in some way with chromosome replication. At one extreme it has been suggested that termination generates a division signal that is expressed after about 20 minutes. At the other extreme it has been suggested that a division signal occurs at about the time of initiation and is expressed about 80 minutes later. It has proved difficult to investigate these hypotheses, because division is so sensitive to disturbances in cell metabolism (see Slater and Schaechter, 1974) caused by, for example, DNA damage. Therefore, we considered that monitoring the kinetics of cell division in step-up and step-down transitions might provide useful insights into the temporal relationship between termination and cell division. The results of such experiments are striking (Figure 5).

When a step-down is introduced, there will be an immediate reduction in the rate of arrival of replication forks at the terminus,

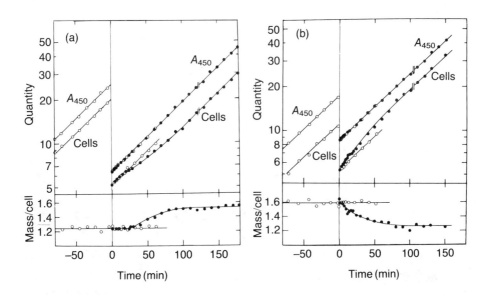

FIGURE 5 Kinetics of the increase in absorbance and cell number in a thy⁻ strain of E. coli K12 following a step-down (a) and a step up (b). A minimal medium was used with proline and alanine as carbon source (τ was 65 min). Thymine concentrations were 6.0 µg/ml and 1.5 µg/ml. Experimental procedure and symbols are as indicated for Figure 2 except that in the lower panel the average cell size (the ratio of absorbance to cell number) is plotted for each point. (From Meacock and Pritchard, 1975.)

but the observed effect on the rate of cell division was delayed for about 20 minutes (Figure 5a). Subject only to the proviso that it is the effect of thymine limitation on the rate of DNA chain elongation that leads to the changed rate of cell division, the response to the step down was that predicted if a late event in replication triggers cell division 20–30 minutes later. It was not consistent with the idea that division is timed from initiation, since thymine limitation does not affect initiation. On the other hand, the corresponding step-up transition (Figure 5(b)) suggests that the situation is more complex. In this case there was an almost immediate response in the rate of cell division to a change in the rate of arrival of replication forks at the terminus. This difference in response to a step-up and a step-down was consistent with other data (Meacock and Pritchard, 1975), which indicated that when the C period is lengthened by thymine limitation, there is a reduction in the length of D. These results taken together favor neither of the simple models referred to above. They would be compatible with the idea that a division signal generated at initiation is expressed after a constant average time interval, but only if chromosome replication has terminated by the time the signal would normally be expressed. When cultures have short C times, it would be the fixed time interval that determines the timing of division. When cultures have long C times, termination would become the limiting factor, thus explaining the difference in behavior of cultures during a step-up and a step-down.

A potential problem in the interpretation of these experiments is that thymine limitation alters the intracellular concentration of thymidine sugars, which may be involved in synthesis of the cell envelope, as well as the triphosphate pools (see Pritchard et al., 1978). Thus, the effect of thymine limitation on cell division might not be due to the change in the rate of DNA chain elongation. It would be instructive to repeat these experiments using a conditional *rep* mutant.

REFERENCES

Chandler, M. G., and Pritchard, R. H. (1975). *Mol. Gen. Genet.*, 138, 127.

Choung, K-K., Estiva, E., and Bremer, H. (1981). *J. Bacteriol.*, 145, 1239.

Cooper, S. and Helmstetter, C. E. (1968). *J. Mol. Biol.*, 31, 519.

Friesen, J. D. and Maaløe, O. (1965). *Biochim. Biophys. Acta*, 95, 436.

Jacob, F., Brenner, S., and Cuzin, F. (1963). *Cold Spring Harbor Symp. Quant. Biol.*, 28, 329.

Lane, D. and Denhardt, D. T. (1974). *J. Bacteriol.*, 120, 805.

Maaløe, O. (1961). *Cold Spring Harbor Symp. Quant. Biol.*, 26, 45.

Maaløe, O. and Kjeldgaard, N. O. (1966). *Control of Macromolecular Synthesis*. Benjamin.

Meacock, P. A. and Pritchard, R. H. (1975). *J. Bacteriol.*, 122, 931.

Oishi, M., Yoshikawa, H., and Sueoka, N. (1964). *Nature*, 204, 1069.

Orr, E., Meacock, P. A., and Pritchard, R. H. (1978). In *DNA Synthesis: Present and Future*. I. Molineux and M. Kohiyama (eds.) Plenum.

Pritchard, R. H. (1974). *Phil. Trans. Roy. Soc. Lond., B.,* 267, 303.

Pritchard, R. H. and Lark, K. G. (1964). *J. Mol. Biol.,* 9, 288.

Pritchard, R. H. and Zaritsky, A. (1970). *Nature,* 226, 5241.

Pritchard, R. H., Meacock, P. A., and Orr, E. (1978). *J. Bacteriol.,* 135, 575.

Slater, M. and Schaechter, M. (1974). *Bacteriol. Rev.,* 38, 199.

Zaritsky, A. and Pritchard, R. H. (1973). *J. Bacteriol.,* 114, 824.

Mechanisms of Genetic Exchange: Natural Transformation of Pseudomonads

John L. Ingraham and Curtis Carlson
University of California, Davis

Several years ago we became interested in the uniquely prokaryotic process known as denitrification whereby (1) the nitrogen content of the earth's atmosphere was formed, (2) large quantities of fixed nitrogen are lost from terrestrial and aquatic environments, and (3) certain bacteria generate ATP by a set of anaerobic respirations for which oxides of nitrogen serve as terminal electron acceptors. For a variety of reasons, the classic denitrifier *Pseudomonas stutzeri* seemed an ideal object of our studies, but since we wanted to use mutant analysis as a primary tool, we first studied this organism with respect to mechanisms by which genetic exchange among various strains might occur. We found that *P. stutzeri* and certain closely related species (Table 1) were capable of natural transformation; i.e., they could take up DNA from their environment and incorporate it into their genome without any artifactual treatment of the culture. Genetic exchange by natural transformation is chromosomally encoded (indeed, it is the only mechanism of genetic exchange among prokaryotes that is so encoded), and the molecular details of the process are quite variable among the various bacterial groups. In this paper we discuss two aspects of natural transformation by *P. stutzeri:* a process we have called *cell-contact transformation,* and the molecular mechanism by which plasmids are transformed into *P. stutzeri* and closely related bacteria.

CELL-CONTACT TRANSFORMATION

Transformation is usually studied by providing a competent culture of bacteria with an appropriate solution of DNA. Since natural transformation is a multigenically encoded bacterial process that presumably evolved to mediate genetic exchange in natural environments, where the half life of extracellular DNA is most probably quite brief, the question arises as to how DNA from the donor cell

TABLE 1   Species of the genus *Pseudomonas* found to exhibit ability of natural transformation and closely related species that do not.

Naturally transformable	Not naturally transformable
stutzeri	*aeruginosa*
alcaligenes	*fluorescens*
pseudoalcaligenes	*putida*
mendocina	*syringae*
	savastanoi

is made available. Does it occur as a consequence of occasional random lysis of cells in the population, or is it a more precisely controlled event coordinated to mesh with the competence of recipient cells in the population? The following experiments with *P. stutzeri* suggest that the latter is probably the case, at least with this organism.

Two strains were cofiltered on a membrane filter and the resulting cell mat was incubated nonselectively on a rich medium. About 16 h later, the mat was suspended, and cells were plated on selective media to score recombinants (Table 2). Resistance to kanamycin (Kan), which was encoded on the conjugative plasmid RK2, was transferred at a high frequency, while the chromosomally encoded markers, *his-1* and *rif-12* were also transferred, but at a much lower frequency. However, if DNase I was added to the cell mat, transfer of Kan was unaffected, while transfer of the chromosomally encoded genes was eliminated. Thus, transfer of the chromosomally encoded markers appears to be unrelated to the conjugative transfer of RK2. Indeed, its transfer fits the usual definition of transformation, because at some point in the process of transfer it is available for digestion by extracellular DNase. Subsequent experiments showed that the transformational transfer of chromosomal markers was completely unrelated to the presence of RK2 in the donor strain: transfer of chromosomal markers occurred with about equal frequency from and

TABLE 2   Effect of DNase treatment on the transfer of plasmid and chromosome-encoded genes

Marker transferred		Frequency of transfer*	
Location	Designation	Without DNase	With DNase
Plasmid RK2	*kan-r*	2.3×10^{-1}*	2.7×10^{-1}
Chromosome	*his1*	2.2×10^{-5}	$<10^{-8}$
Chromosome	*rif12*	2.0×10^{-5}	$<10^{-8}$

*Recombinants per recipient.

to each of the pairs of strains in the cell mat and independently of whether either one of them carried a plasmid. However, cell contact was found to be essential: transfer does not occur in cell suspensions, and it is inhibited in cell mats if the mixture of cells is cofiltered with an inert material such as Celite.

For several reasons the release of DNA by donor cells in the mat does not appear to occur by means of cell lysis. First of all, the process of cell-contact transformation is too efficient for this to be the case. If a subsaturating quantity of DNA (250 ng) is added from solution to a cell mat, approximately 1000-fold fewer transformants are formed than if the same amount of DNA is added in the form of intact cells (Table 3). Moreover, cell-contact transformation occurs only if the donor cells in the population are able to synthesize DNA. In transformational matings by cell contact, in which the recipients are drug-resistant and the donors sensitive, growth-inhibitory concentrations of streptomycin or rifampin had no effect on frequency of transformation, but nalidixic acid inhibited it completely (Table 4).

Thus, when cells of *P. stutzeri* came in contact, certain cells in the population release DNA by a process that is dependent on the ability of the cell to synthesize DNA. Other cells in the population take up this DNA and are genetically transformed by it.

TABLE 3   DNA dependency on the yield of transformants

Method of transformation	Total DNA added (ng)	Frequency of transformation*	Relative frequency
Soluble DNA	250	6.0×10^{-8}	1
Cell contact	260▼	7.0×10^{-5}	1.7×10^3

*Recombinants per recipient.
▼DNA content of donor population.

TABLE 4   Effect of various inhibitors on transformation by cell contact▼

Inhibitor	Concentration (mg/ml)	Process inhibited	Relative frequency of transformation*
Streptomycin	100	Protein synthesis	1.8
	1000		0.97
Rifampin	250	RNA synthesis	2.7
Nalidixic acid	50	DNA synthesis	$<10^{-4}$
	100		$<10^{-4}$

*Frequency of transformation in the presence of inhibitor divided by the frequency in the absence of inhibitor.
▼(Stewart et al., 1983).

TRANSFORMATION OF PLASMIDS

Despite the fairly high frequency at which chromosomal genes are transformed between pairs of strains of *P. stutzeri,* either by a solution of DNA or by DNA contained in intact cells, we were unable to transform a variety of plasmids into this organism by either method. Neither could we transform these cells if they had been previously treated with several of the procedures used to render cultures of *Escherichia coli* and *Salmonella typhimurium* competent.

Studies on another Gram-negative bacterium, *Haemophilus influenzae,* provided a clue as to the reasons for our being unable to introduce plasmids into *P. stutzeri* by transformation. *H. influenzae* bears receptor sites on its outer membrane that, as the first essential step of its natural system of transformation, bind specifically to an 11-bp sequence that occurs every 5 or 6 kb on the chromosome of this organism. If this type of recognition sequence were also required for transformation of *P. stutzeri,* it would probably not be transformable by plasmids, because the probability of their carrying the correct recognition sequence would be low. Accordingly, we prepared a set of plasmids composed of varying lengths of chromosomal DNA cloned into the EcoRI site of the IncW plasmid, pSa151, and tested their ability to be transformed into *P. stutzeri* (Figure 1). When provided either with a solution of DNA or intact donor cells, transformation was dependent on the plasmid containing chromosomal DNA. However, for at least two reasons the relation between size of the chromosomal insert and frequency of transformability of a plasmid was inconsistent with the hypothesis that the role of the insert was merely that of providinga recognition site. All of the insert-containing plasmids that were tested, even the smallest one with a chromosomal insert of only 200 kb, could be transformed into *P. stutzeri,* and the frequency of transformation increased with the size of insert up to the largest plasmid tested (insert size = 18 kb). For such results to be a consequence of dependency on a recognition site, one must assume that recognition site occurs with such frequency that even the very small pieces of DNA have a high probability of containing one and that the frequency of transformation is dependent on the number of recognition sites, even up to very high values.

Further analysis of the relationship between the size of the insert and the frequency of transformation has revealed that the frequency is proportional to the square of the size of the insert (Figure 2), which suggests two roles for the insert in the process of transformation of a plasmid. Such a situation is reminiscent of the model proposed to explain the dependency of transformation of plasmids into the Gram-positive bacteria *B. subtilis* on their containing a region of sequence homology with the endogenote. The model assumes (Figure 3) that DNA, including plasmid DNA, is obligatorily cut as it enters the cell in the process of transformation, and that a plasmid can only become recircularized within the cell by pairing with a homologous region of the endogenote, thereby placing the cut ends in an appropriate juxtaposition to be ligated. Thus, only if the cut occurs within the insert is recircularization possible and the size of the region of homology will be related to the probability of recircularization. Taken together, these two contributions of insert size to frequency of transformation are consistent with the squared relationship observed between the former and the latter.

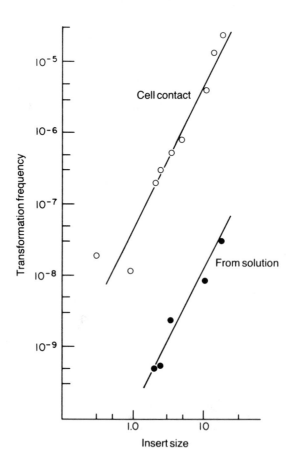

FIGURE 1 Frequency of transformation by DNA in solution and by cell contact of <u>Pseudomonas stutzeri</u> by plasmid pSa151 with various lengths of DNA (insert size in kb) inserted in the EcoRI site. Without an insert, transformation was undetectable by either method (Carlson and Ingraham, unpublished data).

The model leads to three other predictions: (1) A *recA* strain should not serve as a recipient in plasmid transformation. (2) Cutting the plasmid *in vitro* prior to testing its ability to transform would eliminate the transformability of the plasmid, if the cut were made within the vehicle, but only decrease it, if the cut were made within the insert. (3) The homology between plasmid and endogenote that is apparently required need not be restricted to chromosomal DNA; for example, the same transposon on the plasmid and the chromosome should serve well in promoting transformability of a plasmid as would the same-sized insert of chromosomal DNA.

These three predictions were tested and all were found to hold. (1) A relatively uncharacterized *rec* mutant of *P. stutzeri* was unable to serve as a transformation recipient for plasmids including those

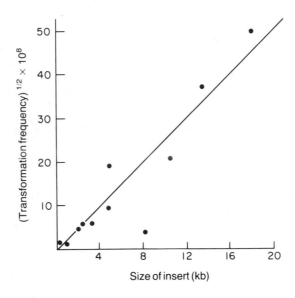

FIGURE 2 Transformation of plasmids by cell contact. Data obtained by cell-contact transformation, shown in Figure 1, plotted as a function of the square root of the frequency of transformation.

Transformation of Hybrid Plasmid DNA: A Model

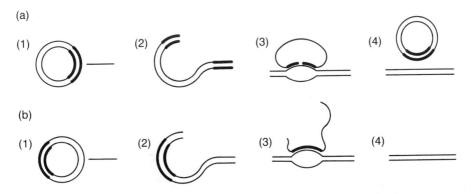

FIGURE 3 Scheme of the molecular details of transformation of plasmids by Bacillus subtilis. Thickened portion of the concentric circles represent the portion of DNA that is homologous with the endogenote. Thin portions represent the vehicle portion. Consequences of the obligatory cut being made in the homologous DNA (a) and in the vehicle (b) are shown at various stages of the process: (1) orientation of the plasmid as it binds to the cell receptor (horizontal line), (2) molecular form of the plasmid as it enters the cell, (3) pairing of the homologous region of a (presumed) single-stranded form of the plasmid with the endogenote, and (4) final molecular form of the plasmid. (After Canosi et al., 1981.)

that contain chromosomal inserts. (2) One of the plasmids with a short chromosomal insert contained a unique restriction endonuclease site (SalI) in the insert and another (SstII) in the vehicle. Treatment of this plasmid with SalI decreased its transformability about 50-fold, but treatment with SstII decreased it below detectable limits. (3) A derivative (pSa151::Tn10) of pSa151 was constructed that contained the transposon Tn10 inserted at an undetermined site within the plasmid. This plasmid could be transformed into a strain of *P. stutzeri* that contained a Tn10 insertion in the chromosome at a frequency equal to that obtained with plasmids containing the same-sized chromosomal insert, but pSa151::Tn10 could not be transformed into strains of *P. stutzeri* that lacked a Tn10 element. The ability of plasmids to be transformed into *P. stutzeri* when the only homology between plasmid and endogenote is the transposon Tn10 strongly suggests that another characteristic of the system of transformation in *P. stutzeri* is that a specific recognition site in the DNA is not required for plasmid transformation into *P. stutzeri*.

REFERENCES

Stewart, G. J., Carlson, C. A., and Ingraham, J. L. 1983. *J. Bacter.*, 156, 30.

Canosi, U., Iglesias, A., and Trautner, T. A. 1981. *Molec. Gen. Genet.*, 434.

PART FOUR

GROWTH AND THE HISTORY OF GROWTH

INTRODUCTION

This book is a testimony to the endeavor of Ole Maaløe during three decades from the 1950s to the 1980s to revive the growth physiology of bacteria, and to establish the molecular biology of bacterial growth. But furthermore, this book and this part of the book is also a testimony to the international collaboration that indeed was a decisive factor in making this development possible. The studies of bacterial growth by the Copenhagen group had the great fortune to be initiated, as emphasized by Jim Friesen, during the very happy period when a large number of young American postdoctoral fellows had the curiosity and the financial backing to travel abroad, and when Europe in a broad sense still seemed to offer an opportunity for these young scientists to expand their background of knowledge.

It is like a tribute to this international perspective that the early history of what has been called the Copenhagen School is recalled in this part of the book by contributions from Berkeley, Boston, and Toronto, as well as by the recollections of history by Ole Maaløe himself. Furthermore, as emphasized by Gunther Stent, the Copenhagen Spirit of physics animated by Niels Bohr returned to Copenhagen biology from the U.S., transformed through the mighty influence of Max Delbrück. His Pasadena Phage Group saw Ole Maaløe as one of its earliest adepts, who adopted the quantitative thinking and experimental accuracy of the phage work to the studies of bacterial cultures. As pointed out by Elio Schaechter, it was this experimental accuracy that in reality was the almost accidental background for the rediscovery of the size and structure variations of bacteria as a consequence of their growth conditions. In fact, this phenomenon was observed three decades earlier, now almost 60 years ago, by the Minnesotan bacteriologist Arthur Trautwein Henrici (1889–1943). In 1928 he summarized his experiences of the growing bacteria by these farsighted remarks:

"Contrary to the orthodox teaching, the cells of bacteria are constantly changing in size and form and structure; but that instead of these changes occurring in a haphazard or meaningless fashion, or instead of being phases in a rather vague and complex life cycle, they occur with great regularity and are governed by simple laws which, after more data have been accumulated and analyzed, may probably be very precisely formulated."

It was only within the framework of the developing molecular biology that it was possible for Ole Maaløe and his collaborators to approach such a formulation of simple laws. Thus, it became possible to look into some of the details of the regulatory circuits for macromolecular synthesis that govern the growth characteristics of the bacterial cell, and to formulate plausible models for the interaction of these circuits. But, as suspected by Jim Friesen, before such integrated and not so very simple laws will get beyond the stage of plausibility, we still might have to wait another three decades.

N. O. Kjeldgaard

Going After the Growth Curve

Moselio Schaechter
Tufts University; Boston, Massachusetts

A student of microbiology who wished to follow the subject of bacterial growth in the early 1950s was hard pressed to find many guideposts. The focus of the relevant chapters in textbooks, then as now, was the growth curve, with its depressingly unintelligible sequence of phases and its hint of life cycles with obligatory stages. Yet, there are fine examples of lucid thinking on the subject of growth from the earliest days of microbiology. One of Pasteur's first three students, Raulin, carried out precise quantitative growth measurements with the mold, *Aspergillus niger*, which demonstrated the requirements of this organism for trace metals. Except for using "sucri candi" as the sole organic compound, his minimum medium was not too different from M-9 or C (chemically defined media now used for *Escherichia coli*). Despite many such mechanistic examples, the fog surrounding growth physiology remained, aided by tantalizing but fanciful notions. Thus, the yield of cultures was thought to be limited by "biological space." Or, the growth curve of bacteria was seen as sufficiently S-shaped to be inevitably determined by a logistic equation. The focus on the sanctity of the growth curves was unmistakable. In a 1949 review on growth, even Van Niel stated that: "Nearly all that is known about the kinetics of growth of microorganisms has been learned from studies of so-called growth curves."

The fog began to be lifted by Jacques Monod, who reduced the growth response of whole cultures to kinetics analogous to those exhibited by enzymes. The rate of growth was shown to be dependent, in Michaelis–Menten fashion, on substrate concentration, and the yield was shown to vary with the amount of substrate available. However, Monod was soon looking elsewhere: "The study of the growth of bacterial cultures does not constitute a specialized subject or branch of research: it is the basic method of microbiology" (Monod, 1949). As a discipline, the physiology of bacterial growth came close to passing from confusion to oblivion in a single leap.

As is often the case, subsequent work was facilitated by a clear definition as well as by novel experimental manipulations. In defining "balanced growth," Campbell (1957) dignified what was previously just a phase in the growth curve by a physiologically meaningful generalization. The difference between "exponential phase" and "balanced growth" is the difference between watching apples fall and thinking of gravity. The importance of growth at a steady state had been realized earlier by many workers, but Campbell's novel term and precise definition helped remove the aura of immutability from the growth curve. It went along with the freedom to manipulate cultures, which was being exercised by quite a few investigators. In the early 1950s the most cogent of these manipulations was continuous culture growth in the chemostat (Monod, 1950; Novick and Szilard, 1950).

How did the nutritional shift experiments done in Copenhagen in the late 1950s contribute to our understanding of what growth curves really mean? It must first be mentioned that this work was made possible, as much as anything, by the simplicity and precision that Ole Maaløe brought to experimental measurements. Viable counts of bacteria were carried out so easily and accurately that the experimental error was consistently smaller than the random sampling error. In other words, dozens of plates each with 300 to 500 colonies could be produced per experiment, with pipetting and dilutions reproducible to within 5 percent. These simple but powerful tools were used with well-aerated cultures at low cell density, befitting the origins of that laboratory in phage biology. Careful measurements allowed one to believe that these cultures were, as nearly as possible, in balanced growth. With such confidence it became possible to exit from the limitations imposed by studying a culture in a single medium and to systematically compare cultures growing in different media or in a chemostat. The new conclusion was that bacteria (or, at least, the *Salmonella typhimurium* used) exist in any of a number of physiological states, characterized by defined cell size and composition. At any given temperature, these characteristics are a simple function of the growth rate (Maaløe and Kjeldgaard, 1966).

Shifts between rich and poor media and vice versa dealt a final blow to the growth curve as a mystical entity. Lag phase could now be understood as the expected outcome of a shift-up experiment, and stationary phase could be seen as a series of shifts-down. Both of these phenomena could be reproduced by manipulation of cultures in defined states of balanced growth. The mystery of the lag had been partially lifted as early as 1938 by Hershey, who showed that, using fully viable cultures in stationary phase, cell mass increased immediately upon inoculation into fresh media, but cell division lagged behind. The impact of this work was not as great as that of the Copenhagen reports 20 years later, one reason being that Hershey did not attempt to correlate his finding with Henrici's classic report (1928) on the changes in size of bacteria throughout the growth cycle. The papers from Copenhagen made the connection between the response measured by changes in growth rate or in mean cell size. Thus, their impact may well be attributed to the fact that they presented a unitary explanation.

Due to Maaloe's insight, the work was extended to the concerns of the day about the relationship of nucleic acids to protein synthesis. The Copenhagen lab was not alone in this. Extensive experiments on the relation of RNA content to the growth rate were reported by

Herbert (1958) and by Neidhardt and Magasanik (1960). Thus, a window was opened to molecular mechanism, and Monod was proved at least partially wrong.

Still, the limitations of growth physiology are real. Crick understood it disgustingly early. When I visited the Cavendish labs in 1958, he told me: "Congratulations! You people started a new field, but it will end with what you did." In a narrow sense this is true, but more broadly speaking, the physiological focus on the growing cell had contributed a needed counterpoint, as well as a complement, to molecular reductionism. That may well be the lasting contribution of the "Copenhagen School."

REFERENCES

Campbell, A. (1957). *Bacteriol. Revs.,* 21, 263.

Henrici, A. T. (1928). *Microbiological Monographs,* London, Bailleres, Tindall and Cox.

Herbert, D. (1958). In: *Recent Progress in Microbiology,* Almquist and Wiksell, Stockholm.

Hershey, A. D. (1938). *Proc. Soc. Exptl. Biol.,* 38, 127.

Maaløe, O. and Kjeldgaard, N. O. (1966). *Control of Macromolecular Synthesis,* Benjamin.

Monod, J. (1949). *Ann. Rev. Microbiol.,* 3, 371.

Monod, J. (1950). *Ann. Inst. Pasteur,* 79, 390.

Neidhardt, F. C. and Masasanik, B. (1960). *Biochim. Bioph. Acta,* 42, 99.

Novick, A. and Szilard, L. (1950). *Proc. Nat. Acad. Sci.,* 36, 708.

Van Niel, C. B. (1949). In *The Chemistry and Physiology of Growth,* A. K. Parpast, (ed.), Princeton Univ. Press.

The Balanced Growth of Balanced Growth

James D. Friesen
University of Toronto, Ontario

When I first stepped into the University Institute of Microbiology in the early 1960s as an insecure young postdoc, first time in a foreign land, trained as a somatic cell biologist after an undergraduate degree in physics, knowing absolutely nothing about bacteria save that some of them made one uncomfortable at times, I was imprinted by two sensations: the look and feel of the enormous teak planks that were so important in the interior design of the laboratory, and the pungent essence of cigar that was the laboratory's olfactory basis. I have never lost these two memories, especially since I have renewed them in frequent return visits to the Copenhagen lab. Lasting as these have been, however, there is another legacy, less individual and far more important, left by the prime-time Maaløe years—the value of balanced growth and the lessons that can be learned about the inner workings of the cell by presenting it with feasts, famines, and minor embarrassments, then carefully observing its response.

I think there were two elements that placed the Copenhagen lab in a central position in bacterial regulation studies and kept it there for two and one-half decades. The first was the strong belief, consistently put into practice, that studies of cellular regulation are essentially physiological, which means that one can interpret experimental data only if they are collected on healthy cells in exponential growth, renewing all of their various components at a uniform rate. The second was that the people in the group always bent every effort to use the latest technology in seeking answers to questions that interested them. Indeed, these two elements were linked.

The Copenhagen Institute became a northern Mecca for disciples of cellular regulation. Sooner or later, it seems, everyone who had an interest in studying how cells governed themselves passed through Copenhagen; many stayed for weeks, months, or years. Postdoctoral fellows from all over the world, but especially from the United States, in the early days when Uncle Sam's largesse extended to supporting

373

American postdocs on foreign soil, took up residence in Copenhagen. Many of these visitors brought with them new techniques and experimental approaches for tackling problems of cellular regulation. There was also traffic in the opposite direction as residents of the Copenhagen lab brought back news and views from sojourns in the United States and England. Altogether, the net result was a group that began with a unique set of precepts regarding the life and times of the cell, then explored and modified these with the latest methods and in the light of current ideas. Underpinning all this, of course, was the tolerant, easy acceptance of strangers by everyone in the lab, the ready willingness for scientific collaboration, not to mention smørrebrød, øl, and the charms of the city and the countryside. If the winter nights were long at 56 degrees North latitude, we could brighten them with Jens Ole Rostock's gløgg at the Christmas party and devastating Paaske Bryg at Easter.

As with any scientific proposition worth its salt, the early formulation of balanced growth both metamorphosed, as new data and ideas appeared, and also provided insights for other areas. One of the earliest and most important contributions of the Maaløe group was the demonstration that simple but careful measurements of the content of the fundamental macromolecules (DNA, RNA, protein) that made up the cell, when analyzed as a function of culture doubling time, yielded a surprising amount of information about cell growth. The two papers from the late fifties, one on the analysis of cells in balanced growth, and the other describing the effect of sudden shifts in growth milieu on macromolecular synthesis, were landmarks in the field and formed the basis for the next decade of research in the Copenhagen lab. These two early papers from the Copenhagen group formed the basis of the conclusion that the two functions of cell growth that are most important for the understanding of cell regulation are the translation apparatus and DNA replication.

The bi-authored monograph published in 1966 by Maaløe and Kjeldgaard set forth the accumulated philosophy of the Copenhagen group. It is quite striking, on re-reading this slim volume, to realize that most of the ideas on balanced growth, the central importance of ribosomes, the invariance of ribosome efficiency with growth, and the proportionality of ribosome number with growth rate were developed by that time. The tone of the book, as was always the central theme of the laboratory, stressed the "compleat cell." Reductionism was welcomed as a necessary means to gain data, but results were always put in the context of the cell in its physiologically prime state of balanced growth. The analysis presented by Maaløe and Kjeldgaard stressed the quantitative approach in two ways. First, measurements of cell number, RNA, DNA, and protein were transfigured into values for the number of molecules per genome, the number of ribosomes per cell, the number of tRNA molecules per ribosomes, the time required for polymerization of a single amino acid to the growing polypeptide chain (a constant at nearly all growth rates), and coordinate synthesis of ribosomal components. It was astonishing that so much insight into cellular molecular happenings could be derived from such simple measurements. But it was the care and accuracy with which these measurements were taken that made the insights possible and believable. It was a credo of the Copenhagen laboratory that assays must be closely reproducible, that curves be defined by many experimental samples, and that consequently a lump in a curve or a change of slope was likely to mean something. This stress on quantitation, as applied to studies of microbial physiology, was the hallmark of the Copenhagen style of microbiological science. It reminds one of the approach taken by another microbiologist a generation earlier when Max Delbrück and his colleagues demonstrated the power of quantitation as applied to simple measurements of the interaction of bacterial viruses with their hosts. It is no coincidence that Maaløe had a high regard for the physicist Delbrück and also appreciated the forays into biological thought made by Delbrück's physicist colleagues, Niels Bohr and Erwin Schrödinger.

Following the publication of the book by Maaløe and Kjeldgaard, the late sixties and early seventies saw an extension and refinement of the measurements and inferences that could be derived from general considerations of cellular macromolecules: global mRNA measurements in rifampicin "run-off" experiments, further nuances in the measurement of ribosomal events in the shift up, and the use of fusidic acid in limiting concentration to study ribosomal metabolism. Throughout this period, the quantitative aspects of rRNA, tRNA, mRNA, ribosomal proteins, and DNA metabolism were codified. The experiments on which they were based formed the foundation of a rigorous and highly regarded graduate course that was the incubator for many young microbiologists who now populate laboratories practically the world over. During this time, Maaløe's personal interest remained centered on the interplay of macromolecules in the cell. Also during this time the technologies of molecular biology were expanding; of particular importance were the detection and quantitation of nucleotide pools and techniques of gene isolation, the latter in turn made possible the study of regulation of particular genes or gene clusters, especially the genes for ribosomal components. The former made possible a series of studies of the stringent response and its peculiar associated nucleotide, guanosine tetraphosphate, and the ribosomal functions necessary for its synthesis.

Guanosine tetraphosphate was (and is) a tantalizing candidate for a regulator. It charmed some of Maaløe's junior colleagues in the Copenhagen laboratory. So did gene cloning. Thus, by the mid-1970s, the laboratory was moving in new directions, inevitably reductionist, because that was clearly the direction that the new technology was increasingly pushing molecular biology. Some of the first cloning of structural genes encoding ribosomal components was carried out in the Copenhagen laboratory; the first suggestion that the ribosomal protein genes might be post-transcriptionally regulated also came from Copenhagen.

Since the late seventies, Maaløe has no longer been actively participating in the day-to-day experimental routine of the laboratory. During the last ten years of his scientific career he has remained fixed to his life-long view—the whole cell in balanced growth. The logical extension of this, Maaløe's *idée fixé* during the 1970s, is his model of passive regulation. This idea, attractive in its simplicity, yet profoundly upsetting to those who find it difficult to believe that a function as central as ribosomal regulation could be that elementary, holds that the synthesis of ribosomal components takes what synthetic capacity is left over after all the cellular genes that are actively regulated have either gulped their share (as in slow growth) or have declined to participate (as in rapid growth). As a consequence there is either relatively little capacity remaining in poor medium for ribosomal synthesis and thus few ribosomes, or a large opportunity for ribosome biosynthesis in rich medium when the cell, released from its obligation to expend energy in the synthesis of materials that are freely available for import from its universe, devotes its spare capacity to the synthesis of ribosomes, to the ultimate good of its ability to compete in society. Although Maaløe champions this theme (see his article in this volume), he never denies the importance of other, more particular regulatory mechanisms as responses to stress or random fluctuations in cell metabolism.

If the 1966 monograph was the seed for the formulation of the cell in balanced growth, the 1983 textbook by Ingraham, Maaløe, and Neidhardt is its full-blown flowering. It is interesting to compare the two books, particularly those sections that bear the Maaløe stamp. What is striking is that the concepts set forth in the earlier book, though made more pointed and fleshed out in the later book with the detail that nearly two decades of research has added, are basically unchanged. It is at once impressive to realize the depth of insight displayed in the 1966 book, despite the relatively primitive state of detailed molecular knowledge at the time, and at the same time disconcerting to know that despite a massive accumulation of facts covering gene structure, regulation, and interaction, we still are a long way from understanding what makes a cell run as a successful negative entropy machine in a universe running down. In the end it could be that biosynthetic regulation in the cell is much too complex and interactive a process to be understood in a single sweeping "unified field theory" of growth. We might be faced with a Chef's Salad of mechanisms—here the effect of guanosine tetraphosphate on transcription initiation, there feedback inhibition of translation, a dash of attenuation control, plus some variation in mRNA decay. In this culinary analogy Maaløe would, I am sure, maintain that passive regulation is the main course to which the Chef's Salad is adjunct. Nevertheless, despite our accumulated knowledge we could be forgiven for thinking that the integrated picture of cellular regulation that Maaløe sought throughout most of his scientific career still eludes us.

The question that Maaløe posed is, of course, an enormous one. Indeed, it is the most important in biology. We should feel no regret that it has not yet been answered; after all science is long and life is short. Instead, we should feel gratitude that the question was asked at all. Problems of biosynthetic regulation are absolutely central to most modern biological research. The demonstration by Maaløe and a handful of his like-minded contemporaries in the late fifties that one can make precise measurements of simple cellular systems and that one can gain remarkable insight into the whole cell, along with studies by others of individual genes and operons, was the recognizable beginning of the pervading theme of modern biology. Maaløe played a leading part in laying the foundation for a generation of molecular and cellular biologists whose interests lay in global cellular regulation mechanisms. A notable contribution indeed!

The Copenhagen Spirit

Gunther Stent
University of California; Berkeley

Not long ago, at a party in Washington, D. C., I happened to run into the distinguished organic chemist, Nelson Leonard, who, forty years earlier, had been my teacher in a sophomore chemistry course at the University of Illinois. I always remembered Leonard as one of the most inspiring and sympatico professors of my undergraduate days, but I had not seen him since 1948, when I forever quit the Champaign-Urbana scene as a newly minted Ph.D. in Physical Chemistry. Leonard, it seems, had not forgotten me either. He greeted me like a long-lost son and said: "You know, Gunther, one question has been bothering me all these years. Namely, how in hell did a young squirt like you manage to know in the 1940s that molecular biology was going to be the field of the future, when none of us knew it?" "Nelson," I answered, "Stop worrying: I didn't know it either." I explained to him that it was just a fantastic stroke of luck that during my student days at Illinois our mutual friend Martha Baylor gave me a copy of Schrödinger's *What Is Life;* that I thus learned about, and was fascinated by, Max Delbrück's "quantum mechanical model" of the gene; that Martha knew that Delbrück was not, as Schrödinger intimated, in Germany, or maybe even dead, fallen on the Russian front, but instead had just moved, alive and well, from Nashville to Pasadena, to the Californian Eldorado of my dreams; and that in 1948, for mysterious, never-explained reasons, I was awarded one of the first Merck Postdoctoral Fellowships, to join Delbrück at Cal Tech. I told Leonard, moreover, that I am by no means the only veteran of those far-away days (to which I have previously referred to as the "Romantic Period" of molecular biology), whose switch from physics or chemistry to bacterial viruses was attributable more to sonambulism than to superior strategic insights. It was only in the spring of 1952, upon the discovery by Alfred Hershey and Martha Chase that the DNA of the bacterial virus enters the host cell, while its protein stays outside, that it finally became clear to me (and, I think, to most of my fellow proto-molecular biologists) what it actually was that we were after: to fathom how the viral DNA molecule manages to replicate itself several hundred-f old and to induce, or preside over, the synthesis of as many viral protein complements within the brief half-hour latent period prior to dissolution of the infected bacterium. Although by then Delbrück had already stopped working with viruses, it was his pioneering work that had allowed this central question of the then nascent molecular biological revolution to be brought into clear focus.

I think it is likely that Ole Maaløe also didn't have a very clear idea that he was about to join a revolutionary biological movement when he first appeared in Delbrück's lab in January 1949, on a study tour of American microbiological laboratories as a Rockefeller Fellow, no more than did Jean Jacques Weigle, then recently retired as Head of the Physics Institute of the University of Geneva, who also turned up in Pasadena at just about the same time. Since both visitors expressed an interest in bacterial viruses, Delbrück delegated his only postdocs, namely Elie Wollman (from the Institut Pasteur) and me, to run them through the hoops of his Cold Spring Harbor Phage Course. They stayed with us for about six weeks and, whatever scientific benefits they may have derived from Wollman's and my tuition, both had a tremendous formative influence on me. For Maaløe and Weigle were the first urbane, Continental *bon vivants* with whom I happened to have come into contact. In stark contrast to the generally frugal, not to say dowdy, scientific community in which I had moved at Illinois, Cold Spring Harbor and Cal Tech, Weigle drove a brand-new Cadillac convertible, Maaloe bought, with his last leftover Rockefeller dollars, the most expensive suit he could find in the finest clothing store in Pasadena, and both were connaisseurs of good food and wine. So I tried to style myself after them, although it took me another 15 years until I finally managed to get my own (and even then, second-hand) Cadillac convertible.

Delbrück suggested that, after my two-year Merck Fellowship was over, I should continue my postdoctoral training in Copenhagen, to learn something about nucleotide chemistry (which he had reluctantly come to accept as possibly relevant for genetic processes) from his friend Herman Kalckar. I enthusiastically agreed because, before leaving Pasadena, Maaløe had proposed that I come to his lab some day, so that we could do some virus experiments together at the State Serum Institute. And the prospect of joining forces with my recently found Danish paragon of elegance and microbiological savoir-faire had tremendous appeal for me. My idea was more or less that I would combine nucleotide studies in Kalckar's lab with virus work in Maaløe's. As it turned out, for a variety of reasons, I hardly spent any time with Kalckar at all, and instead worked full time with Maaløe, whose crew that year included also two future superstars of 20th century biology: James Watson and Niels Jerne. Although I had little or nothing to do with nucleotides, Maaløe and I did study the kinetics of uptake of radioactive ^{32}P into virus DNA in the infected host cell and thus became, willy nilly, premature investigators of the process of DNA replication. On several occasions that year, Niels Bohr came over from his Physics Institute to visit Maaløe's laboratory, in order to find out what we were doing with viruses. Bohr was interested in our work because Delbrück had been his pupil in the 1930s.

FIGURE 1 Lunch at Kerckhoff Laboratory, Cal Tech, Spring, 1949. Left to right: J. J. Weigle, O. Maaløe, E. Wollman, G. Stent, M. Delbrück, and G. Soli.

Before leaving Pasadena for Copenhagen, Delbrück gave me some fatherly, albeit cryptic, advice about the importance of what he called the "Copenhagen Spirit." I did not really understand what he was talking about, except that I realized, much to my disappointment, that Copenhagen Spirit relates to epistemology rather than to the *dolce vita* of the Venice of the North, of which, having just passed my 26th birthday, I was eagerly looking forward to gaining first-hand knowledge under Maaløe's tutelage. All the same, I promised Delbrück that I would strive to follow his advice. My first encounter with the mysterious Copenhagen Spirit occurred soon after my arrival in Denmark in the fall of 1950, when Kalckar's lab made a weekend excursion to Frederiksborg Castle. On the walls of the chapel of that castle are hung the coats of arms of the Knights of the Royal Danish Order of the Elephant, including that of the then recently knighted Bohr. I was astonished to see that Bohr's shield features the Taoist symbol of Ying and Yang, under the motto "CONTRARIA SUNT COMPLEMENTA." Since I couldn't make any sense out of the strange motto--what does it mean that contraries are complements?--nor out of the presence of that to me then still unfamiliar Asian device, I asked one of Kalckar's students--maybe it was Niels Ole Kjeldgaard-

-why Bohr had chosen to serve under these oddly blazoned arms. He replied that Bohr, with his famous sense of subtle humor, had probably just played a little joke on King Christian X. But, as I was to learn eventually, like all good jokes, the one Bohr played on his sovereign embodied a profound truth: that truth is nothing less than the evidently self-contradictory proposition, forbidden by the rules of formal logic passed on to us by the Greeks, but permitted under the Ying and Yang of the Chinese, and that a statement and its negation (*contraria*) can both be true (*sunt complementa,* as are Ying and Yang). Thus, the truth symbolized in Bohr's shield leads to the recognition of a peculiar feature of human reason, namely that reason is paradoxical.

In the summer of 1932, one year after Delbrück had spent six months as a postdoctoral fellow in Bohr's laboratory, he returned to Copenhagen to hear Bohr deliver a lecture entitled *Light and Life*. In that lecture, Bohr outlined the philosophical implications for the life sciences of the fundamental changes that the quantum theory had brought to the conception of natural law. One of the most profound of these changes was that the quantum theory forces us to renounce the possibility of a complete, causal account of phenomena and to be content with probabilistic, rather than deterministic, natural laws. This revised view of the foundations of natural law, which extends to the very idea of the nature of scientific explanation, Bohr thought not only to be essential for the full appreciation of the new situation brought to physics by the quantum theory, but also to have created an entirely new background for the problems of biology, in their relation to physics. That new background was, in fact, the Copenhagen Spirit, whose central lesson was that we face a fundamental limitation in defining the objective existence of phenomena independently of the (subjective) means we employ for their observation. Five years later, in the fall of 1937, Bohr presented a lecture in Bologna entitled "Biology and Atomic Physics," in which he likened this aspect of his Copenhagen Spirit to the world view taught by Buddha and Lao Tzu (the founder of the Taoist School of Chinese philosophy): long before him, but just as he, these ancient Far Eastern sages addressed the epistemological problems arising from our being both observers and actors in the great drama of existence. In *Light and Life* Bohr conjectured that we might still have to discover some as yet missing fundamental traits in the analysis of natural phenomena before we could reach an understanding of life in physical terms. Fascinated by this conjecture, the young Delbrück's principal scientific interest turned from physics to biology.

Most contemporary philosophers of science are quite familiar with Bohr's Copenhagen Spirit, and they are fully aware of the role it has played in the development of modern physics. Indeed, a generation ago many of them wrestled with its epistemological implications for their Ph.D. theses. But few, if any of them, have taken the Copenhagen Spirit seriously as a general world view, whose implications transcend physical sciences and can inform our ideas in nearly all domains of human interest. It was with Delbrück that Bohr eventually found his most influential philosophical disciple outside the domain of physics. Through Delbrück, the Copenhagen Spirit became the intellectual infrastructure of molecular biology and of its hegemony over 20th-century life sciences. It provided for molecular biologists the philosophical guidance for navigating between the Scylla of crude

FIGURE 2 Copenhagen, August, 1951. Left to right, first row: M. Delbrück, Florence Goldwasser, Mogens Westergaard. Second row: G. Rasch, Gunther Stent, Aase Maaløe, Herman Kalckar. Third row: James Watson (face covered by camera flash attachment), Ada Jerne, Ole Maaløe, Carsten Bresch, (?), Inga Loftsdottir (later Stent), Barbara Wright (later Kalckar), Andre Lwoff, E. Goldwasser, (?). Standing: (?), (?), Niels Jerne and son Donald, (?).

biochemical reductionism, inspired by 19th century physics, and the Charybdis of obscurantist vitalism, inspired by 19th century romanticism.

Although the Copenhagen Spirit found life-long reflection in Delbrück's personality and scientific attitudes, by the time he was in his sixties he had published only one explicit statement of his philosophical views, namely the 1949 essay "A Physicist Looks at Biology." Most of the pioneer molecular biologists of Delbrück's Phage Group, then laying the groundwork for the eventual rise of their as-yet-unnamed discipline, found that essay hard to read. They couldn't see what philosophical point he was trying to make and tended to consider his fascination with paradoxes (*contraria sunt complementa*) as one of those foibles against which even the greatest minds are not immune. It was only in 1975, when nearing retirement, that Delbrück presented a more fully developed account of his interpretation of the Copenhagen Spirit, in a course at Cal Tech, which he epitomized as an "investigation into human cognitive capabilities as expressed in various sciences." Delbrück intended to turn the transcript of these lectures into a little book, to be entitled *Mind from Matter?*, but his fatal illness prevented him from completing that project. A group of Delbrück's students and friends have recently edited the transcript of *Mind from Matter?* for posthumous publication in the fall of 1985.

As for myself, I came to appreciate the philosophical implications of the Copenhagen Spirit, not in Copenhagen, but only ten years later in Kyoto, where I spent part of my first sabbatical leave. There, working (rather than sightseeing) in a modern, yet utterly strange, social setting still governed as much by the world view taught by Buddha and Lao Tzu as ours is by the teachings of Moses and Plato, it was easy to see that the set of basic concepts about the world intuited by human reason *is* inherently paradoxical. And since not only science but also ethics is grounded in this paradoxical set, both of these rational approaches to structuring what would otherwise be a chaotic human existence are internally inconsistent and mutually incompatible. These internal inconsistencies are not so grave that they prevent us from drawing a superficially coherent picture of reality for a rational conduct of everyday affairs. They become troublesome only when scientists and philosophers pursue the analysis of that picture to the bottom of the night. Then, once the analysis has gone too far, contradictions (*contraria*) come into view whose resolution only alienates us from the reality that we construct in the service of a sane human life.

By means of the scientific approach to the world, we seek to understand a reality of material objects governed by the laws of Nature, whereas by means of the ethical approach, we seek to provide norms for the interpersonal relations of a reality of human subjects. In various cultures this global ideology with its twin approaches of science and ethics has found a variety of concrete realizations, some religious, some secular. In many, or even most, of these cultures, the distinction between the scientific and the ethical approach is either not recognized or is expressly denied. The failure to distinguish between science and ethics has the virtue that it permits a global ideology relatively free of conflicts between these two domains. It has the drawback, however, of preventing the development of science as an autonomous intellectual activity and thus presents an insurmountable obstacle to getting very far towards understanding, and therefore gaining mastery over, the reality of material objects. Western culture has emphasized the distinction between science and ethics since the time of the Greeks. It is this distinction that made possible the eventual flowering of Western science.

But, as is evident by the current debates about conflicts between science and ethics, which seem to be interminable and quite beyond the reach of clear rational solutions, science is not really separable from ethics. As was pointed out by Immanuel Kant, many of the conflicts and contradictions that are encountered at the intersection of scientific and ethical approaches to the world have their root in our contradictory attribution of freedom of will to the person and of causal necessity to Nature. According to Kant, this contradiction arises because the relationship between the person and his action is fundamentally different according to whether we think of the person as a moral subject or as a natural object. Thus, reason has saddled us with a dualistic, or in the parlance of the Copenhagen Spirit, complementary global ideology that obliges us to think of ourselves as existing simultaneously in two worlds: on the one hand, we are material objects forming part of the causally determined events of the natural world; on the other hand, we are intelligent subjects belonging to a world of thought independent of the laws of Nature. But, however logically compelling that complementary ideology may be for philosophers, it is unfortunately not a serviceable outlook for a sane, everyday life.

Can the paradoxes arising from the concepts imminent in human reason be resolved? As Bohr had realized, for this purpose we may look to Far Eastern philosophy. Evidently, the Chinese managed to develop a global world view from which the troublesome separation of subject and object is largely absent. As conceived by the Chinese, the world is an organismic whole of which each person forms an inseparable part. Hence, being on their inside rather than on their outside, each person is thought to have as his birthright the potential power and insight to penetrate things-in-themselves. This power to gain direct knowledge of reality derives from introspection rather than logical reasoning or inference. Accordingly, knowledge of the world is not viewed as a cognitive grasp of a given structure of objective truths; rather it is an understanding of one's own mental states and an appreciation of one's own inner feelings. Thus, to gain sagehood is to know oneself; and with sagehood comes the knowledge of The Way, or *Tao,* that allows the person to harmonize himself constantly with an ever-enlarging network of relationships.

The belief in the superiority of introspection over logical analysis of objective data for the acquisition of knowledge caused Chinese science to take a course very different from that of Western science. Because Taoism regards the workings of Nature as inscrutable for the theoretical intellect and lacks the concept of natural law, Chinese science developed along mainly empirical lines. As it turned out, the scepticism of the Taoist doctrine had been somewhat exaggerated: the workings of Nature are not all that inscrutable for the intellect. Provided that the questions one asks of Nature are not too deep, satisfactory answers can often be found. Difficulties arise only, when, as Bohr had noted, the questions become too deep and their answers are no longer fully consonant with our intuitive picture of reality. So Taoism, though wrong in the short run, turned out to be right in the long run.

And just as Far Eastern philosophy has given rise to a fundamentally different form of science, so has it also engendered a view of the moral life that is very different from that of the West. Just as the Chinese tradition does not regard scientific knowledge as a body of objective truths, so does it not regard virtue as the capacity to make objectively correct moral judgments. Here moral behavior is not seen primarily as involving choice and responsibility at all, since The Way to social harmony is not thought of as having any crossroads. One may start and stop along The Way, or even deviate from it, but there are not alternative directions open to choice. Therefore, the central ethical issue is not the responsibility of persons for deeds done by their own free will, but the factual questions of whether a person is properly taught The Way and whether he or she has the desire to learn it diligently.

Bohr thus already appreciated the relevance of Far Eastern philosophy for the new world view brought by modern physics long before this notion appeared in Western pop-philosophy of the 1970s. (It has been hypothesized that Bohr first learned about the Ying and Yang from Wolfgang Pauli, who in turn had been told about it by his Zurich psychoanalyst, Carl Jung, who in turn had collaborated with Richard Wilhelm, the German translator of the *I Ching*.) But despite its obvious relevance for our postmodern age, the Chinese attempt to create a world view free of conflict seems to come at a very high price: the demolition of the foundations of Western ethics and Western science. As a dyed-in-the-wool Westerner, I find this price too high, and I imagine that Bohr, Copenhagen Spirit and Ying and Yang in shield notwithstanding, would not have been willing to pay it either.